8-7-72

Drying of
Milk and Milk Products

Second Edition

other AVI books

Drying of Milk and Milk Products

Second Edition

by CARL W. HALL

Dean, College of Engineering
Washington State University
Pullman, Washington

and T. I. HEDRICK

Formerly Manager, M.S.U. Dairy Plant
and
Professor,
Department of Food Science
Michigan State University
East Lansing, Michigan

WESTPORT, CONNECTICUT

THE AVI PUBLISHING COMPANY, INC.

1971

Printed in the United States of America
BY MACK PRINTING COMPANY, EASTON, PENNSYLVANIA

Acknowledgments

The authors are grateful to the many people in all phases of the dry milk industry who willingly assisted during the preparation of this manuscript. Specifically sincere appreciation is presented to:

DR. FRED W. BAKKER, Assistant Professor, Agricultural Engineering, Michigan State University, East Lansing, Michigan.

MR. BARCLAY BEAHM, Instrumentation Engineer, Taylor Instrument Company, Rochester, New York.

DR. J. R. BRUNNER, Professor, Department of Food Science, Michigan State University, East Lansing, Michigan.

DR. M. H. CHETRICK, Professor and Chairman, Chemical Engineering, Michigan State University, East Lansing, Michigan.

MR. FLOYD FENTON, Chief, Standard Branch, Dairy Division, Consumer and Marketing Service, U.S. Department of Agriculture, Washington, D.C.

DR. J. C. FLAKE, Director of Sanitary Standards of Evaporated Milk Association, Chicago, Illinois.

DR. D. R. HELDMAN, Assistant Professor, Agricultural Engineering and Food Science, Michigan State University, East Lansing, Michigan.

MR. W. J. KETCHAM, In Charge, Laboratory and Quality Control, Mid-West Producers' Creameries, Inc., South Bend, Indiana.

MR. A. A. ROGERS, President, C. E. Rogers Company, Detroit, Michigan.

MR. R. G. SEMERAD, Assistant Regional Supervisor, Grading and Inspection Branch, Dairy Division, Consumer and Marketing Service, U.S. Department of Agriculture, Chicago, Illinois.

MR. T. S. SIMMONS, Engineer, C. E. Rogers Company, Detroit, Michigan.

MR. ROBERT SODERLUND, Manager of Engineering, DeLaval Separator Company, Poughkeepsie, New York.

DR. C. M. STINE, Associate Professor, Department of Food Science, Michigan State University, East Lansing, Michigan.

MR. J. W. TATGE, Sales Manager, Dairy Equipment, Blaw-Knox Company, Buffalo, New York.

Mr. A. M. Walker, President, Marriott Walker Corporation, Birmingham, Michigan.

Mr. J. T. Walsh, Executive Director, American Dry Milk Institute, Chicago, Illinois.

Dr. B. W. Webb, Chief, Dairy Products Laboratory, Eastern Utilization and Development Division, Agricultural Research Service, U.S. Department of Agriculture, Washington, D.C.

Carl W. Hall
T. I. Hedrick

East Lansing, Michigan
July 1966

Foreword

The food shelves of our modern stores give evidence of the extensive changes which have taken place in the quality technology of dried food products. The files of the processing plant and the equipment supplier give equal evidence of the demand for increased efficiency in the utilization of human and natural resources. The substantial strides taken toward realization of these requirements in the past decade are the more remarkable considering that the nature of these goals are unfortunately often in mutual conflict.

Furthermore, as the national conscience has been directed toward nutritional assistance to the less fortunate areas of the world, the advantages of food preservation through drying have become more obvious, and the need to strive for further production economies more pressing.

The advances to date have been the result of the informal partnership between laboratory, equipment manufacturer, and processing plant; that is, between the academic and the industrial provinces. Moreover, certainly to those of us connected with the equipment design phase of the industry, it is particularly apparent that this liaison must be made even more intimate if speediest response to technological requirements and fullest use of technological achievements are to be realized.

Thus, we are gratified to see this new book as the combined effort of a dairy technologist and an engineer, and the combination of both academic and industrial experience. Written to include both basic principles and actual operating data, it can serve as a text for the student, a handbook for the operator, and a counselor for the equipment manufacturer.

Although directed primarily toward the drying of milk and milk products, the basic principles outlined will be of benefit to all who are concerned with the drying of other food products.

We are pleased to have had this small part in its presentation.

A. A. ROGERS, *President*
C. E. Rogers Company

Preface to the Second Edition

This book was prepared to present the pertinent facts on the basic principles and the numerous ramifications of the dehydration of milk and related products. Concise chapters are included on evaporated milk and sweetened condensed milk. The treatise was developed to serve as a complete guide to the technologist, engineer, student, and all other personnel interested in the basic principles, practical applications, and related parameters of the milk concentration and drying industry.

With the increasing importance of the dry milk industry, numerous technological and scientific changes are occurring. The revisions included in this Edition add principally the production and utilization data for several recent years. Other modifications were effected to reflect a more accurate delineation of current knowledge. However, in order to minimize publishing costs the changes were limited to the essentials.

The authors emphasize acknowledgement and appreciation to the American Dry Milk Institute and dairy related Government agencies, as well as to many authors for most of the information in this Edition.

C. W. HALL

T. I. HEDRICK

January 1971

Contents

CHAPTER 1

History and Growth

INTRODUCTION

Dry milk production has become an increasingly important segment of the dairy industry. In 1969, nonfat dry milk manufacture used roughly 14% of the 116,200,000,000 lb. of milk produced on the United States farms. Dry whole milk required only 0.4% of the fluid milk. The indications suggest continued growth in the next decade. These include such features as better keeping quality, less storage space, and lower shipping costs which result in attractive economies.

Dry milks provide a means of handling the excess milk produced for other dairy products, especially market milk. Nonfat dry milk serves the same purpose for milk solids-not-fat that, traditionally, butter has done for milk fat. That is, nonfat dry milk provides a means of handling the surpluses from the processing of other dairy products, especially market milk.

Legal standards of identity established by federal law require nonfat dry milk and dry whole milk to contain not more than five per cent moisture. Other dry dairy products do not have specific federal standards for moisture.

The ultimate aim of the industry is to obtain dry products which if recombined with water give little or no evidence of detrimental change compared to the original liquid product. Reconstituted nonfat dry milk is nearly comparable in flavor to the original skimmilk, but much improvement needs to be made with dry whole milks and dry cream. Other aims are to have good keeping quality in all respects and a low manufacturing cost.

Various names have been applied to the same dry milk product. For example, nonfat dry milk also has been called skimmilk powder, dried skimmilk, nonfat dry milk solids, and dehydrated skimmilk. Dry whole milk, dry cream, dry buttermilk, and others frequently are called dried or powdered whole milk, etc., according to their identities. In this text "dry milks" is a generic term and thus includes more than one specific dry milk product.

Trade associations have been formed in various countries to assist their members to properly manufacture condensed milk and dry milk, to provide information of an educational nature to the manufacturers and their customers, and to obtain industry statistics.

1

DRY DAIRY PRODUCTS

Most common dairy products have been dried. The list includes:

Nonfat dry milk

Dry whole milk
Dry buttermilk
Dry whey

Dry ice cream mix

Dry creams
  (sweet or sour)
Dry high acid milks

Dry skimmilk-vegetable
  fat products

Malted milk powder

Cheese powders
  (Cheddar, Blue, etc.)
Sweetened chocolate flavored
  nonfat dry milk

Coffee creaming products
Sodium or calcium caseinates

Miscellaneous products that have been processed experimentally or commercially in small amounts are: dry creamed cottage cheese, dry butter, and dry yogurt. Modified equipment and special methods may be required for their successful manufacture.

HISTORY OF CONCENTRATED AND DRIED MILK

Early Concentration Attempts

Dry milk has been known in some parts of the world for many centuries. Marco Polo in the 13th century reported soldiers of Kublai Khan carried a dried milk on excursions. Before it was used, water was added to a portion of the dried material. Mixing was accomplished by movement of the horse during riding on the trip. The recombined product was then consumed at mealtime. The belief is that part of the fat was removed from the milk before drying and dehydration was accomplished by solar heating (Eckles *et al.* 1951).

The Japanese had a concentrated milk product in the 7th century, but there is no reference to a dry product (Miyawaki 1928). In the United States the development of evaporated or concentrated milk and dry milk as known today began in the early 1800's. But progress was slow during that century. In 1809, Nicholas Appert, a Frenchman, announced his discovery that milk could be reduced to one-third the original volume by evaporation of the moisture from an open vessel. Cooking followed by sealing the product in a container greatly enhanced the keeping qualities. He also developed a dried milk in tablet form by air drying of milk solids concentrated to a "dough" consistency (Olson 1950).

Gail Borden (Fig. 1) was one of the leading pioneers in originating a process of milk condensing (Goodale 1856). By applying a partial vacuum to remove moisture from milk he obtained a product that was superior in flavor to one obtained by boiling in open vessels. Although ini-

Courtesy of the Borden Company

FIG. 1. GAIL BORDEN, INVENTOR OF CONDENSED MILK

tially he encountered difficulties in marketing condensed milk, persistence eventually resulted in a commercially successful business.

Meyenberg, (1884, 1887) who came to the United States from Switzerland, introduced a unique concept for the keeping quality of evaporated milk. He obtained patents in 1884 and 1887 for sterilizing evaporated milk by heating to 240°F. under pressure with steam while the sealed cans were in continuous motion. This innovation provided the basis for a new industry. It grew gradually until 1946, the year of the highest per capita consumption in the United States. Since then a steady decline has occurred.

Attempts to Process a Dry Product

During the last half of the 19th century attempts to produce a dried milk involved the addition of other dry products to concentrated milk. Sugar, cereal products, and sodas, singly or in combinations, were added. In 1850, Birdseye concentrated milk with added sugar until a solid was obtained. A British patent was granted in 1855 to Grimwade who developed a modified dry product from highly concentrated milk to which was added sodium (or potassium) carbonate and sugar (Hunziker 1949). This semi-solid material was extruded into thin streams and dried in trays. Malted milk powder was made from whole milk, extract from malted barley and wheat flour. This powder has been marketed since 1887 although announced four years earlier. Others who pioneered in methods of moisture removal from milk were Heine, Newton, Horsford, Dalson, Gallois, and Deauve (Miyawaki 1928).

Patents and reports which emphasized processes for dry milk manufacture without the addition of other products began about 1898. In 1901, Campbell of the United States and Wimmer of Denmark dried concentrated milk on trays (Hunziker 1949). Later they used concentrated milk and a steam heated cabinent. In 1902, Hall obtained a patent on a new system of manufacturing dry condensed milk.

Development of Drum and Spray Drying Equipment

Just (1902) was among the first inventors to receive patent rights on a drum drier with two rolls. Hatmaker of England improved Just's model. Numerous other types of drum driers were invented, but most of these were used only to a limited extent. These types included Gathmann's single cone drum with a grooved surface, and Mignot-

Courtesy of Blaw-Knox Company

Fig. 2. Drum Drying Skimmilk About 1930

Plumey's drum with a smaller unheated drum for film application to the larger drum (Hunziker 1949). Vacuum drum driers were designed by Ekenbery, Sweden, 1889; Passburg (1903), Germany; and Govers (1909). Figure 2 shows a double drum drier processing nonfat dry milk for animal feed.

Among the early inventors of spray drying equipment was Percy, who in 1872 combined atomization of a fluid and heated air. Stauf, a German, received a U.S. patent on an improved design based on this principle in 1901. Four years later MacLachlan (1905) processed dried products by spray atomizing milk, skimmilk, eggs, or blood in heated air. Improvements and modifications of spray drying were developed and patented by Merrell (L. C.), Merrell (I. S.) and Gere, 1907; Gray and Jensen, 1913; and Rogers, 1917 (Beardslee 1948). C. E. Gray (Fig. 3) was one of the outstanding contributors to the progress of the dry milk industry through inventions, business foresight, and industry services.

Courtesy of Foremost Dairies, Inc.

FIG. 3. C. E. GRAY, A RESPECTED LEADER IN THE DRY MILK INDUSTRY

Courtesy of C. E. Rogers Company

FIG. 4.   C. E. ROGERS, PIONEER MANUFACTURER OF SPRAY
DRIERS

Another was C. E. Rogers (Fig. 4) who did much for the industry growth
with his engineering improvements and fabrication of spray drying equipment.

Improvement of Reconstitutability

Agglomeration.—As a result of pioneering research to improve the
reconstitutability of nonfat dry milk in water, Peebles (Fig. 5) was issued
patents in 1936, 1955, and 1958. In the Peebles method, regular spray
dried milks (usually nonfat dry milk) are reprocessed by rewetting the
surface of the particles in turbulent air which causes the wetted particles
to collide forming clusters. In the next stage, the moisture content is
reduced with hot air. Products treated by this system became known as
an "instantized" product. Instantized nonfat dry milk from the Peebles
process appeared for general distribution on the retail markets in 1954.
Subsequently, other systems and methods (see Chapter 7) were developed.

Foam Spray System.—The U.S. Department of Agriculture's foam spray
drying technique is an important contribution to the industry. Besides

Courtesy of Foremost Dairies, Inc.

FIG. 5. D. D. PEEBLES, INVENTOR OF INSTANTIZING AND MANY
OTHER PROCESSES IN THE DRY MILK INDUSTRY

providing the advantage of one-step processing, the method is more satisfactory for products sensitive to heat damage during dehydration. Foam spray dried dairy products have very good dispersibility, but poor sinkability.

AMERICAN DRY MILK INSTITUTE

The first organized effort to establish a national trade organization for the dry milk industry was taken at the National Dairy Show in Milwaukee, Wisconsin, October 1924 (Jones 1948). Approximately 30 leaders, representing 22 firms, initiated the action by appointing a committee to develop plans. The committee members were N. J. Dessert, Austin Griffiths, Walter Page, R. G. Soule, and B. D. White.

On March 24, 1925, the group made its report and as a result the American Dry Milk Institute (ADMI) was created with 19 charter member companies. More permanent incorporation was effected in June 1925. C. E. Gray was the first chairman of the Executive Committee and

H. E. Van Norman the first executive officer. Both of these leaders contributed much to the growth and effectiveness of ADMI.

Roud McCann succeeded Van Norman. McCann developed ADMI into one of the outstanding trade organizations in the dairy industry.

The ADMI provides a number of worthwhile services. It operates a laboratory for quality and composition control analysis and research. Field service is available to members on quality, sanitation, and processing problems. Special aid to bakery, meat, and other industries is provided. The collection and yearly distribution of current information, especially statistics on production and utilization of dry milk products, has been a beneficial service. Its office and laboratory are located at 130 North Franklin Street, Chicago, Illinois.

CANADIAN MILK POWDERS MANUFACTURERS ASSOCIATION

The Canadian Milk Powders Manufacturers Association was organized March 2, 1939, by 24 representatives of the dry milk industry (Webster 1965). Principal leaders were: W. R. Aird, Lea Marshall, George W. Rogers, A. S. Thurston, and S. B. Trainer. The purpose of the organization is to promote the marketing of dry skimmilk. The Association also functions as a liaison with its federal government in the development of standards and other regulations for the use of dry milk in foods and for the problems associated with export markets. H. G. Webster is the secretary of the Association at Woodstock, Ontario.

PRODUCTION STATISTICS

Similar to the demand for the manufacture of evaporated milk caused by World War I, the second World War encouraged rapid expansion of production of several dry milk products (Fig. 6). Concomitantly a large shift occurred from animal feed to human consumption of nonfat dry milk and dry buttermilk.

Since the 1940's, the nonfat dry milk industry has become the main outlet along with butter for surplus supplies of fluid milk. Decreasing per capita consumption of several milk products along with increasing farm production has caused huge surpluses. These were converted to nonfat dry milk.

Nonfat Dry Milk and Dry Whole Milk

In 1952, the production of nonfat dry milk was less than 1,000,000,000 lb., but it surpassed this milestone in 1953. Eight years later, 2,020,-000,000 lb. were manufactured for human consumption (including 99,000,000 lb. of drum dried) plus an additional 28,000,000 lb. for animal

Courtesy of Lake to Lake Dairy; and Niro Atomizers, Inc.

FIG. 6.   DRY MILK PLANTS IN WISCONSIN (A) AND DENMARK (B)

feed (Table 1). Dry whole milk production in the United States was roughly 217,000,000 lb. in 1945. After World War II, it decreased and has not changed substantially since that time although yearly fluctuations are evident. The production in 1969 was roughly 66,080,000 lb.

Other Dry Dairy Products

During the last two decades dry buttermilk has increased approximately 36% and dry whey has more than doubled. The recent development of a commercially feasible dialysis method for the removal of salt ions should stimulate greater utilization of whey in concentrated and dry forms. In contrast, malted milk powder is declining slowly in volume. Dry cream manufacture has been low and very erratic if available data are correct. Dry ice cream mix production declined after World War II and has remained fairly steady since then. The special dry milk products

TABLE 1

U.S. PRODUCTION OF DRY MILK PRODUCTS
(add 000 lb.)

Year	Nonfat Dry Milk		Dry Whole Milk	Dry Cream	Dry Butter-milk	Dry Whey	Dried Malted Milk	Dry Casein: Skim- or Butter-milk Product
1916	16,463		2,123		342		11,654	
1917	22,624		3,139		2,557		13,852	
1918	26,202		4,006	621	4,951		15,623	11,239
1919	34,945		9,042	607	5,279		17,436	14,407
1920	41,893		10,334	309	5,704		19,715	11,526
1921	38,546		4,242	130	7,708		15,652	8,076
1922	40,617		5,599	118	9,007		13,659	6,927
1923	62,251		6,560	328	13,032		15,331	14,548
1924	69,219		7,887	1,018	18,058		15,889	20,759
1925	73,317		8,931	339	20,246		18,050	16,660
1926	91,718		10,768	331	31,378		20,673	16,953
1927	118,123		11,464	338	38,435		22,116	18,033
1928	147,990		9,605	673	45,502		21,128	22,151
1929	207,579		13,202	294	54,215		22,850	30,537
1930	260,675		15,440	400	64,601		22,691	41,965
1932	270,194		11,983	80	48,712		13,215	24,428
1934	294,935		15,869	65	53,636		13,569	37,331
	Human Use	Animal Use						
1938	289,121	160,170	21,496	40	63,910	47,384	15,394	48,549
1940	321,843	159,962	29,409	54	67,931	90,996	20,021	46,616
1941	366,455	110,042	45,627	43	75,614	111,316	23,242	47,346
1942	565,414	61,148	62,167	54	69,637	124,479	34,679	42,268
1943	509,620	24,279	137,766	216	60,995	110,158	49,435	18,368
1944	582,912	16,407	177,754	193	56,683	141,553	40,549	15,264
1945	642,546	17,508	217,276	203	49,578	135,920	42,751	12,333
1946	653,465	13,704	188,406	567	38,627	147,953	45,029	18,319
1947	677,941	22,149	164,888	320	45,437	157,583	37,354	35,831
1948	681,532	13,145	170,087	312	41,839	125,185	31,361	14,372
1949	934,934	21,244	125,541	178	49,359	159,358	25,369	18,348
1950	881,492	17,404	124,986	459	48,837	155,579	30,710	18,531
1951	702,476	14,299	131,017	1,070	45,467	139,946	33,391	21,620
1952	863,220	25,306	102,318	874	47,067	164,105	29,811	7,482
1953	1,213,774	19,967	101,179	717	57,424	174,656	32,482	5,532
1954	1,334,043	20,396	92,700	839	56,261	177,617	31,415	5,175
1955	1,365,772	20,145	108,317	782	58,333	211,026	33,536	3,147
1956	1,489,894	19,262	111,315	756	64,269	211,984	31,223	2,533
1957	1,623,880	21,203	103,174	539	70,358	232,380	34,606	1,700
1958	1,709,664	21,182	87,702	628	77,160	233,261	32,768	600
1959	1,723,212	23,295	90,383	NA*	81,475	247,329	26,599	100
1960	1,818,605	25,870	97,998	NA*	86,441	276,860	24,542	900
1961	2,019,848	27,799	81,695	241	89,036	271,485	23,986	600
1962	2,230,269	30,474	86,117	659	86,375	284,845	23,111	1,200
1963	2,096,494	23,218	91,015	1,018	86,182	311,779	22,495	1,800
1964	2,176,800	21,967	87,622	569	86,100	361,511	22,369	2,100
1965	1,999,000	23,600	84,750	982	87,400	404,301	22,184	3,000
1966	1,610,000	26,000	94,000	528	76,200	470,931	22,904	1,800
1967	1,678,000	30,329	74,348	NA	72,600	492,815	15,197	1,100
1968	1,594,000	26,997	79,821	NA	70,400	495,173	20,354	NA
1969	1,432,000	24,423	66,080	NA	66,089	502,533	18,636	NA

Sources: American Dry Milk Institute 1954 to 1970. U.S. Dept. Agr. 1923, 1927, 1932, 1935, 1940, 1946, 1952, 1957, 1961, 1964, 1968, 1970.

have increased in numbers during the last 25 years. The quantities of sweetened chocolate flavored nonfat dry milk, and cheese powders have been gaining since their development in recent years.

Leading States and Countries

In the United States most of the dry milk is produced by the midwestern states of Minnesota, Wisconsin, Iowa, Michigan, and Ohio (Table 2). But, New York 4th, California 5th, and Idaho (8th) ranked in the first eight for total 1968 production.

Table 3 represents the production data on dry milk (nonfat dry milk and dry whole milk) in the leading countries; the United States heads the list by a wide margin. Other major producers are Canada, France

TABLE 2

PRODUCTION OF NONFAT DRY MILK AND TOTAL DRY MILKS IN LEADING STATES SINCE 1950

State	1950 Nonfat Dry Milk	Total	1952 Nonfat Dry Milk	Total	1954 Nonfat Dry Milk	Total	1956 Nonfat Dry Milk	Total
Wisconsin	257,752	305,324	266,314	323,351	437,762	483,001	479,109	530,285
Minnesota	160,474	197,619	187,679	216,118	268,273	287,730	383,461	414,886
New York	99,736	128,108	104,180	122,859	152,973	178,410	148,203	179,061
Michigan	51,451	67,376	41,309	56,240	67,331	76,479	50,411	66,138
California	44,920	54,840	25,523	32,673	73,105	80,759	48,692	53,544
Ohio	36,679	53,080	33,945	46,824	63,181	81,189	56,408	70,567
Idaho	32,603	33,254	31,908	32,211	52,764	53,068	54,164	54,523
Missouri	32,485	33,542	31,069	32,590	38,202	39,537	39,449	41,776
Indiana	32,185	36,728	33,131	35,654	35,866	37,328	22,341	25,829
Vermont	20,852	20,852	14,521	14,521	22,142	22,593	21,801	22,151

State	1957 Nonfat Dry Milk	Total	1958 Nonfat Dry Milk	Total	1959 Nonfat Dry Milk	Total	1960 Nonfat Dry Milk	Total
Minnesota	446,111	475,331	477,809	505,762	493,367	530,339	500,794	548,313
Wisconsin	518,707	575,593	475,338	526,878	447,131	474,704	424,938	468,362
Iowa	106,202	115,103	127,960	138,091	148,422	159,787	174,841	188,017
New York	121,075	146,835	142,838	163,905	130,421	132,056	157,323	186,099
Michigan	57,527	79,646	72,870	94,861	71,541	74,504	74,545	92,289
California	63,286	67,052	53,407	57,082	66,854	69,033	71,681	76,154
Idaho	55,344	55,707	58,510	59,035	58,994	59,845	60,563	62,049
Pennsylvania	27,643	28,148	34,185	34,482	35,374	35,878	47,284	47,904
Ohio	50,192	64,206	40,217	42,093	36,557	38,546	43,069	53,539
South Dakota	...	...	18,285	20,595	27,634	29,973	32,364	35,043

State	1962 Nonfat Dry Milk	Total	1964 Nonfat Dry Milk	Total	1966 Nonfat Dry Milk	Total	1968 Nonfat Dry Milk	Total
Minnesota	577,803	624,353	632,408	719,456	529,854	594,206	560,445	615,958
Wisconsin	510,206	546,956	468,158	694,225	272,470	305,151	253,538	294,118
Iowa	203,434	214,948	248,073	276,554	163,639	202,597	164,889	175,985
New York	209,031	226,768	193,086	208,579	123,729	145,517	91,641	109,863
Michigan	97,648	120,033	96,918	126,773	44,996	63,353	39,623	43,694
California	81,454	84,784	72,320	74,398	46,687	47,829	72,789	74,159
Idaho	63,894	64,457	50,733	51,095	45,312	45,686	46,158	46,532
Ohio	63,390	66,273	48,734	55,209	26,088	29,557	31,817	33,088
Pennsylvania[1]	57,742	59,442	47,920	57,444	32,812	37,284	32,905	33,147
South Dakota	33,953	36,354	46,351	49,240	44,570	48,274	51,154	55,298

Sources: American Dry Milk Institute 1952 to 1970. U.S. Dept. of Agriculture 1965–1969.
[1] Nebraska ranked 7th with 48,328,000 lb. of total dry milk in 1968. Pennsylvania was tenth.

TABLE 3

YEARLY PRODUCTION OF DRY MILK IN LEADING COUNTRIES[1]
(add 000 lb)

Year	Australia	Belgium	Canada	France	Nether-lands	Sweden	United Kingdom	United States	West Germany	New Zealand[2]
1947	44,256	5,509	69,911	2,621	34,791	23,382	55,372	838,250	...	...
1948	49,752	...	81,921	4,173	53,945	24,733	77,505	831,495	...	...
1949	67,109	8,059	75,942	...	62,472	33,024	60,479	1,059,203	...	...
1950	62,631	9,397	68,536	15,432	91,134	19,841	83,507	844,180	...	...
1951	57,016	17,279	70,179	...	75,432	17,789	50,086	841,124	...	...
1952	70,481	21,932	99,150	...	109,625	27,978	57,568	965,111	...	...
1953	83,999	34,622	99,928	...	129,707	25,496	95,424	1,317,347	...	...
1954	84,225	34,513	100,259	35,274	130,673	25,862	110,432	1,379,025	...	...
1955	96,566	34,869	107,976	22,000	117,757	23,215	99,904	1,499,464	83,224	...
1956	110,161	41,472	98,829	22,000	136,906	31,687	154,784	1,625,522	102,802	...
1957	102,030	55,465	142,885	33,100	149,692	39,147	154,112	1,755,007	127,730	...
1958	103,237	66,453	205,338	79,400	181,879	39,769	126,560	1,819,152	121,041	...
1959	135,239	62,454	197,101	110,984	173,405	37,685	108,192	1,834,765	163,702	104,743
1960	127,996	90,345	217,798	210,637	233,531	55,836	187,936	1,942,473	202,724	106,514
1961	132,733	148,127	238,651	239,616	229,089	62,066	204,736	2,122,160	224,794	98,194
1962	126,530	120,217	215,602	313,053	255,465	75,197	220,640	2,335,161	258,280	97,687
1963	121,477	127,225	197,993	441,361	236,922	77,485	168,000	2,210,695	320,145	123,913
1964	141,013	139,185	225,377	525,797	236,779	74,295	125,440	2,286,409	375,362	160,406
1965	135,018	189,408	244,104	727,298	236,106	94,136	207,648	2,107,350	492,826	183,098
1966	190,014	231,168	271,240	901,461	279,632	97,884	178,976	1,698,800	600,597	255,269
1967[2]	198,637	207,852	316,378	1,163,588	142,944	68,583	162,400	1,678,000	743,832	324,486
1968[2,3]	133,327	264,995	346,578	1,496,923	229,278	83,775	211,904	1,604,000	875,447	298,836

Source: U.S. Dept. Agr. 1948 to 1969.
[1] Includes nonfat dry milk and dry whole milk.
[2] Nonfat dry milk only
[3] Preliminary

Netherlands, United Kingdom, West Germany, Belgium, Australia, Sweden and New Zealand. Information on dry milk production in the Soviet Union has been scarce, but it was considered to be among the first ten countries.

TRENDS IN DRY MILK INDUSTRY

Plant Growth

In the incipient stages of the dry milk industry, production of dry milk was only one of several dairy product processing operations in a plant. This is currently much less true. The trend for the last 20 years has been toward larger and larger plants designed specifically to manufacture dry milk and possibly butter. This change to specializing in the manufacture of one product is an effort to obtain greater efficiency in processing and marketing. The huge capital investment in drying facilities necessitates a large volume operation. Nevertheless, nonfat dry milk and butter probably will continue to handle surpluses and scarcities for other dairy products.

The dry milk industry has an excellent record of ready acceptance of mechanization in the processing and handling of its products. This progressiveness is likely to continue. It will affect more economies in drying and packaging costs. Vertical and horizontal mergers will continue although subjected to more scrutiny by government antitrust officials.

Farm Milk Production

The grade of raw milk for dry milks is changing. A survey by Hedrick and Hall (1964) revealed that 26.4% of the drying plants handled 100% Grade A raw product. This does not necessarily mean that approximately a fourth of the dry milk production came from Grade A milk. The implication is that surplus market milk (Grade A) constitutes an important source of nonfat dry milk. Only 25.5% of the drying plants processed manufacturing grade milk exclusively.

Accurate predictions of future U.S. production trends of nonfat dry milk are difficult because of the uncertainty of political activities associated with government price support programs and Federal milk orders. Federal Government purchases of surplus (Table 2) have been in effect since the termination of enforced setasides during and shortly after World War II. Such government programs have been an important factor in causing dairy farmers to shift from cream to milk delivey to the dairy plants. The completion of the shift and greater production per cow should result in a general trend of increasing total milk production. However, cycles of reduced dry milk manufacture will occur.

Dry Milk Sales

With domestic and nonsupported export sales making up only roughly half the total production, the industry has serious problems that undermine its stability. Considerable emphasis should be exerted on developing steady markets with increased consumption of these nutritious products. But the possibilities of enlarging industrial or retail demands of the dry milk products with milk fats are not encouraging in the United States. Except for a national or world emergency, demand for dry products containing milk fat is not likely to increase during the next few years. Much more research and development are needed to overcome the inherent problems of stability, reconstitutability, and sensory response equal to the original fluid products for the stimulation of utilization. The necessity for creation of new markets for these products offers a great challenge to the industry.

Fluctuation in demand must be taken into account. Some, but not all, of the industrial outlets for nonfat dry milk may be expected to increase in the next decade. For example, the baking industry is highly sensitive to even a small cost increase in dry milk. Consequently, bakers may purchase a decreasing proportion of the total production. Utilization in dairy plants should continue to enlarge.

The degree of success of the "War on Poverty" will be reflected in the usage of dairy products directly and indirectly. There is little evidence to support the possibility of a drastic improvement in the demand by the meat industry. Its present usage is approximately three per cent of the total production (1968).

One of the brightest changes has been the success of instant type nonfat dry milk. It was immediately accepted by the housewife. Within the brief period of four years it replaced regular nonfat dry milk in retail packages. Agglomeration was applied to other dry dairy products (less successfully if they contained milk fat) and to many other food products such as coffee, cocoa, dry vegetable juices, dry yeast, tea, dry potatoes, dry baby foods, dry eggnog, dry cereal products, dry 900-calorie products, and flour.

Per capita retail sales of nonfat dry milk and other low-fat dry milk products are expected to increase gradually during the next few years. Competition from other food products will become more intense. The marketing of substitutes and imitations in the dry form for direct replacement of dry dairy products will increase. The dry coffee coloring products made fom highly refined vegetable fat, corn syrup solids, sodium caseinate, and stabilizer are examples of this trend. Regular products such as nondairy beverages and juices are being improved in dehydrated form to attract greater attention from the consumer. Consequently, the dry milk industry in the years ahead will have to solve the marketing problems with the same degree of success as in processing to date if the industry is to progress without government assistance.

REFERENCES

ALLEN, L. A. 1932. The properties of milk in relation to the condensing and drying of whole milk, separated milk, and whey. Hannah Dairy Research Institute Bull. 3. Aird & Coghill, Ltd., Glasgow.

AMERICAN DRY MILK INSTITUTE 1952 to 1970. 1951 to 1970 Census of dry milk distribution and production trends. Bull. 1000. Chicago.

BAUMGARTNER, F. W. 1920. The Condensed Milk and Milk Powder Industries. Jackson Press, Kingston, Ontario.

BEARDSLEE, C. E. 1948. Dry Milks: The Story of an Industry. American Dry Milk Institute, Chicago.

BIRDSEYE, C. D. 1850. Process of preparing cream. U.S. Pat. 7,644.

CAMPBELL, J. H. 1901. Desiccated milk and method of making same. U.S. Pat. 668,159 and 668,161.

COOK, H. L., and DAY, G. H. 1947. The Dry Milk Industry. American Dry Milk Institute, Chicago.

ECKLES, C. H., COMBS, W. B., and MACY, H. 1951. Chapter 13, The manufacture of dairy products-condensed milk, dry milk, milk by-products. Milk and Milk Products. McGraw-Hill Book Co., New York.

GOODALE, S. L. 1856. A brief sketch of Gail Borden. Maine State Board of Agriculture, Augusta, Maine.

GOVERS, F. X. 1909. Desiccating milk. U.S. Pat. 939,495.

HALL, W. A. 1902. Producing dry condensed milk. U.S. Pat. 694,100.

HAMMER, B. W., and BABEL, F. J. 1957. Chapter 13, Bacteriology of milk powder. Dairy Bacteriology. John Wiley & Sons, New York.

HEDRICK, T. I., and HALL, C. W. 1964. The dry milk industry: status and practices. Manuf. Milk Prod. J. 55, No. 12, 5–6.

HERRINGTON, B. L. 1948. Chapter 19, The dry milk industry. Milk and Milk Processing. McGraw-Hill Book Co., New York.

HUNZIKER, O. F. 1949. Condensed Milk and Milk Powder. Published by the author, LaGrange, Ill.

JOHNSON, A. H. 1960. The industrial way. Food Technol. 14, No. 11, 541–546.

JONES, R. E. 1948. Industry Builder. Pacific Books, Palo Alto, Calif.

JUST, J. A. 1902. Preserving milk in dry form. U.S. Pat. 712,545.

LAMPERT, L. M. 1965. Chapter 18, Dry milk products. Modern Dairy Products. Chemical Publishing Co., New York.

MACLACHLAN, J. C. 1905. Desiccating process. U.S. Pat. 806,747.

MEYENBERG, J. 1884. Apparatus for preserving milk. U.S. Pat. 308,421 and 308,422.

MEYENBERG, J. 1887. Preserving milk. U.S. Pat. 358,213.

MIYAWAKI, A. 1928. Condensed milk. John Wiley & Sons, New York.

OLSON, T. M. 1950. Chapter 34, Miscellaneous dairy products. Elements of Dairying. The Macmillan Co., New York.

PASSBURG, E. 1903. Filling or emptying apparatus for vacuum driers. U.S. Pat. 748,414.

PEEBLES, D. D. 1936. Method and apparatus for drying liquid containing materials. U.S. Pat. 2,054,441.

PEEBLES, D. D. 1958. Dried milk product and method of making same. U.S. Pat. 2,835,586.

PEEBLES, D. D., and CLARY, D. D., Jr. 1955. Milk treatment process. U.S. Pat. 2,710,808.

PERCY, S. R. 1872. Process of drying and concentrating liquid substances by atomizing. U.S. Pat. 125,406.

PORCHER, C. 1929. Dry Milk. Olsen Publishing Co., Milwaukee, Wis.

SCOTT, A. W. 1933. The engineering aspects of the condensing and drying of milk. Hannah Dairy Research Institute Bull. 4. Aird & Coghill, Ltd., Glasgow.

STAUF, R. 1901. Method of desiccating milk. U.S. Pat. 666,711.

U.S. DEPARTMENT OF AGRICULTURE. 1923, 1927, 1932, 1935. The Yearbook of Agriculture. Washington, D.C.

U.S. DEPARTMENT OF AGRICULTURE. 1940, 1946, 1952, 1957, 1961, 1964. Agricultural Statistics. Washington, D.C.

U.S. DEPARTMENT OF AGRICULTURE. 1948–1969. Foreign Crops and Markets. Office of Foreign Agricultural Relations, Washington, D.C.

U.S. DEPARTMENT OF AGRICULTURE. 1965–1970. Production of manufactured dairy products. Crop Reporting Board DA2-1 (65). Washington, D.C.

U.S. DEPARTMENT OF AGRICULTURE. 1965–1970. World agricultural production and trade statistical report. Foreign Agricultural Service, Washington, D.C.

VAN ARSDEL, W. B., and COPLEY, M. J. 1963. Chapter 20, Dry Milk products. Food Dehydration. Avi Publishing Co., Westport, Conn.

WASHBURN, R. M. 1929. The first spray dried milk. 18th Annual Report of Inter. Assoc. of Dairy and Milk Insp. 18, 180–186. W. F. Roberts Co., Washington, D.C.

WEBSTER, H. G. 1965. Private correspondence.

WHITTIER, E. O., and WEBB, B. H. 1950. Chapter 5, Dried products. By-products From Milk. Reinhold Publishing Corp. New York.

CHAPTER 2

Theory of Evaporation and Evaporators

INTRODUCTION

The overall objective in evaporating and drying is to remove moisture from a liquid in which there is some solid until only a solid remains containing a small quantity of moisture.

The first step in a commercial operation is to remove the bulk of the water (50 to 80%) in an evaporator; the second step is to remove the remainder of the surface-adsorbed water in the drier. At some moisture content (20 to 40%) the behavior of a material becomes more characteristic of a solid than of a liquid (Fig. 7). The water changes from the continuous to discontinuous phase at about 15% moisture content.

EVAPORATION

Evaporation occurs when molecules obtain enough energy to escape as a vapor from a solution of a solid or nonvolatile liquid. The rate of escape of the surface molecules depends primarily upon the temperature

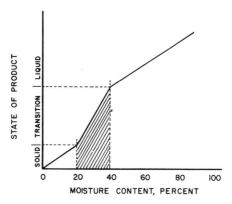

FIG. 7. RELATIONSHIP OF STATE OF NONFAT DRY MILK AND MOISTURE CONTENT

of the liquid, the temperature of the surroundings, the pressure above the liquid, surface area, and type of evaporator (rate of heat transfer to product). Heat is taken from the surroundings as water molecules escape from the surface. The liquid will continue to evaporate until it is gone from an open container. In a closed container with air space above the

17

liquid, evaporation will continue until the air is saturated with water molecules. Removal of water from a liquid product by evaporation is enhanced by adding heat and by removing the saturated air or wet air above the liquid. This is done by forcing air over the liquid or by decreasing the pressure by placing a vacuum on the surface. To avoid excessive entrainment of droplets in the exhaust vapor for an open container, the rate of evaporation should not exceed 30 ± 10 lb. per hr. sq. ft. based on surface area of evaporator.

Boiling

Boiling is evaporation which occurs throughout a liquid as contrasted to a surface phenomenon. Heat is normally added to the bottom of a container in which a liquid is placed. The bubbles form next to the heat

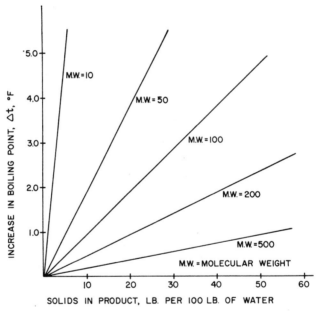

Fig. 8.   Approximate Increase in Boiling Point Due to Dissolved Solids in Water

source, rise to the top of the liquid, and, as heating continues, the entire mass is heated and the vapor bubbles continue rising to the top of the liquid surface. When this occurs, boiling results. Agitation occurs as a result of the formation at the heated surface and movement of bubbles through the liquid. The extent of agitation depends on the rapidity of heating. Even though the speed of heating is increased, the temperature of boiling does not increase, but the rate of evaporation is increased. During boiling, the pressure of the vapor in the liquid must exceed the air pressure by an infinitesimal amount. At normal atmospheric conditions of sea level, the atmospheric pressure is 760 mm. Hg or 14.7 p.s.i.a. (0 p.s.i.g.) which is the same pressure for boiling at 212°F. If the pressure above the liquid is decreased, as through a vacuum, the pressure of vapor in the liquid at boiling is less, and the boiling temperature is lower. However, the quantity of heat required to evaporate a given amount of liquid at a lower pressure is greater. A vacuum is utilized to remove water from liquid/solids at lower temperatures to reduce damage to heat sensitive products which might decompose at higher temperatures.

The boiling point of a solution is greater than the boiling point of water at the same pressure. The elevation of the boiling point due to dissolved solutes is related to the molecular weight of the solute. The addition of 1 gm. molecular weight of substance to 1000 gm. of water increases the boiling point of water 0.9°F. at 760 mm. pressure (Fig. 8). The vapor produced above a solution which boils at 215°F. and 14.7 p.s.i. will, however, be at 212°F. (Table 4).

Latent Heat

The quantity of heat required to change a pound (or a gram) of liquid water to one pound (or gram) of gas vapor is known as the latent heat of evaporation or vaporization. The latent heat value varies for different liquids and depends upon the temperature of evaporation, which is controlled by the pressure (Table 5). As the temperature (or pressure) of evaporation is increased, the quantity of latent heat is decreased. The sensible heat required to reach a certain higher temperature is increased and the total heat (sensible plus latent heat), is greater for a higher temperature of evaporation, except as the critical properties are approached.

TABLE 4

MOLECULAR WEIGHT OF SOME SUBSTANCES

Substance	Molecular Weight
Sucrose ($C_{12}H_{22}O_{11}$) cane sugar	342
Glucose ($C_6H_{12}O_6$)	180
Dextrose ($C_6H_{12}O_6$)	180
Milk fat	696 to 716
Lactose ($C_{12}H_{22}O_{11} \cdot H_2O$) milk sugar	360
Casein (Beta)	24,100
β-Lactoglobulin	35,500
α-Lactoglobulin	16,000

Sources: Richmond (1920) and Jenness and Patton (1959).

TABLE 5

SATURATION TEMPERATURE, PRESSURE, AND LATENT HEAT OF
VAPORIZATION OF WATER

°C.	Cal. per gm.	Pressure, mm. Hg	°F.	B.t.u. per lb.	Pressure p.s.i.a.
0	595.9	4.6	32	1075.8	0.09
10	590.4	9.2	50	1065.6	0.18
20	584.9	17.5	70	1054.3	0.36
30	579.5	31.8	90	1042.9	0.70
40	574.0	55.3	110	1031.6	1.28
50	568.5	92.5	130	1020.0	2.22
60	563.2	149.4	150	1008.2	3.72
70	557.5	233.7	170	996.3	6.0
80	551.7	355.1	190	984.1	9.4
90	545.8	525.8	212	970.3	14.7
100	539.5	760.0	230	958.8	20.8
110	532.9	1074.6	250	945.5	29.8
120	525.7	1489.1	270	931.8	41.9
130	518.5	2026.2	290	917.5	57.6
140	511.1	2710.9	310	902.6	77.7
150	503.5	3570.5	330	887.0	103.1
160	495.6	4636.0	350	870.7	134.6
170	487.2	5940.9	370	853.5	173.4
180	478.6	7520.2	390	835.4	220.4

Sources: Hodgman (1962) and Farrall (1953).

Viscosity

The viscosity of milk decreases slightly during pasteurization. With
the more severe heat treatment, there is an increase in viscosity during
evaporation, with a sevenfold increase in viscosity during sterilization of
evaporated milk. The increase in viscosity during sterilization of evaporated milk appears to be a function of coagulation of protein (Hunziker
1949). The viscosity of evaporated milk decreases with age or thickening
may occur.

Clausius-Clapeyron Equation

The Clapeyron equation expresses the relationship of vapor pressure
to the absolute temperature of liquids (during evaporation) and solids
(during drying):

$$dp/dT = L/Tv$$

where:

p = vapor pressure, lb. per sq. ft.
L = latent heat of vaporization, ft. lb. per lb. (1 B.t.u. = 778 ft. lb.)
T = absolute temperature, °R (°F. + 460)
v = difference between the volume of 1 lb. of vapor and 1 lb. of water at absolute
     temperature, T, cu. ft. per lb.

The volume of the liquid is quite small as compared to the volume of
the vapor, and v can be assumed to represent the volume of 1 lb. of
vapor. The Clausius-Clapeyron equation is obtained:

$$\frac{dp}{dT} = \frac{L}{T(RT/p)} \qquad \frac{dp}{p} = \frac{L}{R} \cdot \frac{dT}{T^2} \qquad \frac{d\ln p}{dT} = \frac{L}{RT^2}$$

where R is the gas constant, ft. lb. per lb. °R.

$$pv = RT$$

where $R = 53.35$ for air; 35.12, carbon dioxide; 55.12, nitrogen; and 85.8, steam, ft. lb./lb. °R.

This equation is useful for determining the change in boiling point due to changes in pressure.

The Othmer method (1940) of determining the latent heat-vapor pressure relationships is based on the Clausius-Clapeyron equation. The Othmer method provides a means of comparing the vapor pressure-latent heat values of an unknown material (such as milk or milk powder) by use of the relationships of a known material, preferably one similar to that of the unknown matter. These relationships are conveniently plotted on log-log paper, usually with the vapor pressure on the ordinate and the saturation vapor pressure on the abscissa:

$$\log p = L/L' \log p' + C$$

(See Chapter 4 for log-log plots and dry milk equilibrium moisture content values.)

EVAPORATORS

Evaporator Usage

Milk and milk products may be treated in the evaporator for removal of moisture to obtain an end product such as concentrated, condensed, evaporated milk, or other milk products. Water is usually removed from liquid milk products in the evaporator before the drying operation. The cost of removing moisture in the evaporator is normally less than in the subsequent conventional drying operation. Milk products are normally condensed from an initial solids content of 9 to 13% down to a final concentration of 40 to 45% total solids, before the product is pumped to the drier.

Evaporation systems may be single-effect or multiple-effect with 2, 3, or 4 or more evaporator bodies or vacuum units. In the dairy industry, the single-effect evaporator is often called a vacuum pan. Four units in a multiple-effect evaporator are the maximum commonly used. In the multiple-effect evaporator, the units operate at decreasing pressure in the direction the product moves through the units.

The usual practice for milk and other food products is to operate the evaporators at a vacuum so that the temperature of evaporation and boiling is lower than it would be at atmospheric pressure. With lower temperatures of evaporation, there is less heat damage to some products.

Evaporator Development and Classification

Evaporator development began in the 1850's and has resulted in many shapes, sizes, and types of units. The major objective is to transfer heat from a fuel or heat source to the product to evaporate water or other volatile liquids from the product. One or more of the following methods may be used for classification of evaporator bodies: (1) source of heat—steam, direct-fired, solar, and other medium; (2) position of tubes for heating—horizontal, vertical, inclined; (3) method of circulation of product—forced, natural; (4) length of tubes—long, short, medium; (5) direction of flow of film of product—upward, downward (rising film, falling film); (6) number of passes of product—1, 2, or more; (7) shape of tube assembly for heat exchanger—coil, basket, straight; (8) location of steam—inside tube, outside tube, or both; (9) location of tubes—internal, external.

Horizontal Tube Evaporator

A simple unit, not used to a great extent on new installations, is the horizontal tube evaporator (Fig. 9). Horizontal tubes from $^3/_4$ to $1^1/_4$ in. diameter extend across the bottom of a cylindrical chamber from 3 to 10 ft. in diameter and 8 to 15 ft. high. Steam enters a chest on one

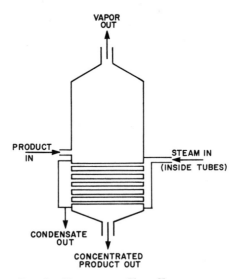

Fig. 9.   Horizontal Tube Evaporator

end of the tubes, moves through the tubes, and the condensate is removed from the chest at the opposite end. The vapor is removed from the top of the cylindrical chamber.

Vertical Short-Tube Evaporator

Tubes carrying the steam internally are placed vertically in the bottom of the cylindrical evaporator chamber (Fig. 10). It is easier to clean the tubes in a vertical unit than in a horizontal tube evaporator. This type of unit is known as the Roberts evaporator in Europe and as the Calandria evaporator in the United States.

In a basket type evaporator the tubes may be placed in the shape of a ring. This unit provides an open space in the center so that the liquid may circulate more freely through the coils, with the liquid moving up through the coils as it is heated and the colder product moving down through a cylindrical volume in the center.

Forced Circulation Evaporator

In natural convection evaporators, the velocity of the fluid is usually less than 3 or 4 ft. per sec. It is difficult to heat viscous materials with a natural circulation unit. Therefore, the use of forced circulation to obtain a velocity of liquid up to 15 or 16 ft. per sec. at the entrance of the tubes is desired for more rapid heat transfer. Forced circulation or agitation can be applied to either horizontal or vertical tube units. The liquid head above the heat exchanger is usually great enough to prevent boiling in the tubes. A centrifugal pump is normally used for circulation of milk products, but a positive pump is used for highly viscous fluids.

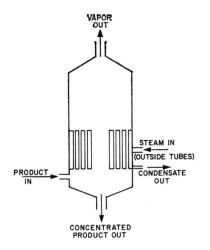

FIG. 10. VERTICAL TUBE (SHORT) EVAPORATOR

Vertical Long-Tube Evaporator

The long-tube vertical (LTV) evaporator uses natural circulation with the flow of product either upward or downward. With upward flow, the unit is known as the climbing or rising liquid film evaporator (Fig. 11A and 11B) and with the downward flow it is known as the falling film evaporator (Fig. 11C). The Kestner is a specific type of LTV evaporator with rising film circulation and with the vapor head concentric with the tube chest. A deflector plate or umbrella on top of the tube bundle is required to deflect the liquid and to reduce entrainment.

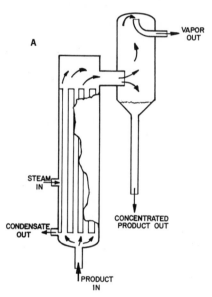

Fig. 11A.   Long Tube Vertical Rising Film (Climbing Film)

Tubes of $1^1/_4$ to 2 in. diameter and 12 to 20 ft. long are used to move liquid on the inside. These are placed in a steam chest so that steam heats from the outside of the tube. The LTV evaporator is normally used with the heating element separate from the liquid-vapor separator. The product enters the bottom of the evaporator body and as it is heated by steam condensing on the opposite side of the tube, the product moves rapidly to the top of the tube and then into a separation chamber. Vapor is removed and the concentrated product removed or recirculated through the evaporation chamber again, depending on the concentration desired.

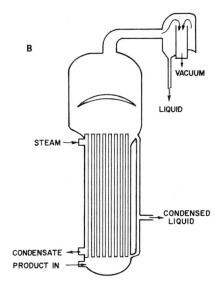

FIG. 11B.   LTV RISING FILM TYPE EVAPORATOR (KESTNER TYPE)

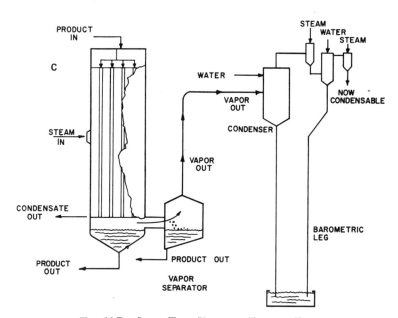

FIG. 11C.   LONG TUBE VERTICAL FALLING FILM

Falling Film Evaporator

The falling film evaporator is used to reduce the amount of heat treatment or exposure to heat of the product. The tubes are from $1^1/_2$ in. up to 2 in. in diameter and up to 30 ft. long in the falling film evaporator. The product is sprayed or otherwise distributed over the inside of the tubes which are heated with steam. Moisture removed moves downward to the vapor separator as does the concentrated product. The product may be recirculated for another concentration or removed from the system. The Reynolds number of the falling film should exceed 2000 for good heat transfer (Lindsey 1953). The falling film evaporator is gaining popularity in the United States and European countries.

Plate Evaporator

A recent development in the evaporation of water from milk and milk products is the use of a plate evaporator (Fig. 12). It operates on the same principle as a vacuum pan. The unit consists of a plate heat exchanger with low pressure steam between every other plate and the product is in the alternate positions. The steam is fed into top openings. The product moves into a space from the bottom of the unit and moves up between the plates, operating on much the same principle as a rising film tube evaporator. Two units may be connected and operated as a two-stage unit.

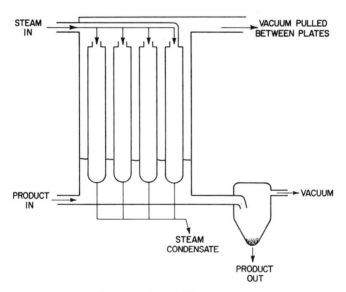

FIG. 12.   PLATE EVAPORATOR

Multiple-Effect Evaporator

The vapor produced through evaporation in the vacuum pan contains considerable latent heat. The vapor from the vacuum pan may be used in much the same way to heat another vacuum pan as steam is used to heat the unit. Thus, two or more effects can be utilized in the evaporator to provide a method of utilizing the useful heat in the vapor and to improve economy of the operation. A common multiple-effect evaporator is the triple-effect system. Each effect can consist of any of the types of several units discussed previously. Fig. 13 illustrates the principle involved and the flow of product and steam, but any of the units can be replaced with a film or other type evaporator. In a multiple-effect evaporation system, the vapor removed from the first effect at a high temperature moves to the heating coils or tubes of the second effects, which is at a lower temperature. Likewise, the vapor removed from the second effect is directed to the heating coils or tubes of the third effect, at a still lower temperature. The temperature must decrease in the direction of flow of the vapor so that the heat will flow from the vapor to the product. Thus, the temperature of evaporation of the product must decrease from the first to the third or final effect. The decrease in the evaporation or boiling point is obtained by maintaining a higher vacuum or providing a lower pressure at the third or final effect as compared to the first effect. A temperature drop of at least 12° to 15°F. is needed to justify the next effect.

The multiple-effect evaporation system is often erroneously given credit for having a larger capacity than a single-effect unit. A triple-effect system operating with the difference in temperature between the first and third effect of 90°F. will have approximately the same capacity as the single-effect unit with the same temperature difference.

The major advantage of the multiple-effect evaporator is that it requires less steam per pound of water evaporated. The major disadvantage is that more investment is required. The cost of the additional equipment and its operation and maintenance must be justified on the basis of the saving in steam for heating. The optimum number of effects is arrived at by an economic analysis. A single-effect unit will require about 1.2 lb. of steam to evaporate 1 lb. of water in the evaporator body; the double-effect, 0.6 lb. of steam per pound of water evaporated; for a triple-effect, 0.4 lb. of steam per pound of water evaporated; and for a four-effect, 0.3 lb. of steam per pound of water evaporated. These figures should be considered as a rule-of-thumb (Table 6). The exact amount of steam required depends upon the design, flow of the system, and many other factors.

TABLE 6

APPROXIMATE QUANTITY OF STEAM REQUIRED TO VAPORIZE 1 LB.
OF WATER IN AN EVAPORATOR, LB.

	Range	Average
Single effect	1.33–1.00	1.17
Double effect	0.63–0.50	0.57
Triple effect	0.40–0.34	0.37
Quadruple effect	0.30–0.26	0.28
Quintuple effect	0.24–0.22	0.23
Sextuple effect	0.20–0.18	0.19
Septuple effect	0.18–0.16	0.17

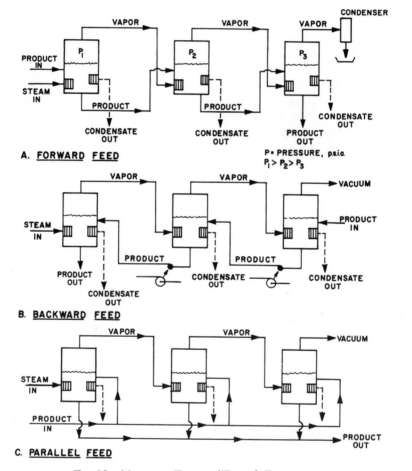

FIG. 13. MULTIPLE EFFECT (THREE) EVAPORATOR

There are three basic methods of feeding (supplying product) in or to a multiple-effect evaporator. The forward feed, backward feed, and parallel feed (Fig. 13) or combinations thereof, may be used for connecting the various effects and for directing product, steam, and vapor flow. In the forward flow, the product and vapors move in the same direction through all three effects. Milk evaporators normally use forward flow systems. In the backward flow, the vapor moves from the first to the third effect and the product enters the third effect and leaves at the first effect. In the parallel feed, the vapor moves from the first to third effect, but the feed is sent directly to each effect with no transfer of material from one effect to another. Thus, only the vapor moves through all three effects. The forward feed is most common, primarily because of its simplicity in that fewer pumps are required. The backward flow is generally more expensive and used in connection with highly viscous products which enter cold.

Considerable labor is required to maintain and clean multiple-effect evaporators for milk and food products. Multiple-effect evaporators have been available for many years. Only recently have evaporators been designed for handling large quantities of milk over practically a 24-hr day operation, so that the multiple-effect can be justified. Clean-in-place systems have decreased labor requirements for cleaning.

Preheaters

Preheating normally precedes evaporation. Tubular heat exchangers are normally used for preheaters. Either double tube or shell and tube heat exchangers are common. The overall heat transfer coefficient, U, ranges from 100 to 350 B.t.u. per hour. sq. ft. °F., but varies greatly depending upon the scale on the steam side and the velocity of flow. The velocity of flow is normally about 5 ft. per sec., but may go up to 10 to 15 ft. per sec. with an additional cost required for equipment.

Accessories for the operation of evaporator consist of the preheater, vapor condenser, pumps, and controls.

Condenser

The purpose of the condenser is to remove the vapor by changing the water vapor removed from the product back to a liquid. Simultaneously, the vacuum is maintained because the volume occupied by the liquid is considerably less than the original vapor. To change vapor to a liquid requires that the latent heat must be removed from the vapor. The heat is abstracted through an indirect or direct heat exchanger.

The surface condenser is an indirect heat exchanger in which cold water on one side causes steam vapor coming from the product to condense on the other side. A common indirect heat exchanger used as a surface condenser is the tubular unit. If the vacuum is not too great, the plate heat exchanger may be used as a condenser surface. A jet condenser is a direct heat exchanger, in which cooling water is sprayed into the volume where vapor is to be condensed (Fig. 14). A jet condenser will use cooling water amounting to from 20 to 50 times the weight of steam. Thus, the vapor being removed from the product and the cooling water are mixed. A jet condenser is normally used in a milk drying operation in preference to a surface condenser, the surface condenser being more expensive. The tubular type may be used with high vacuums often not required in milk drying operations.

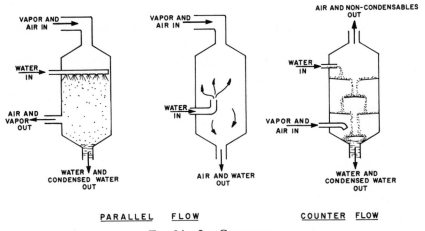

FIG. 14. JET CONDENSER

Condensers may be classified as dry or wet. In a dry condenser, the cooling water is removed by one pump and the noncondensables, including air, removed by another pump. In a wet condenser, the cooling water, condensed vapor, and noncondensables are all removed together. Jet condensers can be further classified as parallel flow or countercurrent flow. The parallel flow condenser is normally operated as a wet condenser and the countercurrent flow as a dry condenser. In the countercurrent flow unit the noncondensables are removed at the temperature of the incoming water, whereas in the parallel flow, the air and water are removed at the same temperature.

The quantity of water required is less for a counterflow type of condenser. Another advantage is that the air and vapor need not enter at the top of the unit as is done with parallel flow. The quantity of air removed from the evaporator system is about 15 to 20% of the volume of the cooling water. Leaks in the system can cause the quantity of air to be considerably higher and result in expensive operation.

Air enters the system through the entering product, steam, leaks in the system, and condenser water.

Vacuum Producing Equipment

Positive pumps of the reciprocating type and steam jet ejectors are commonly used to produce a vacuum. The pump is normally used for producing 24 in. (610 mm.) Hg vacuum or less. The steam jet ejector may be used for one stage to produce 25 in. (636 mm.) Hg vacuum or two stages to produce 28.8 in. (730 mm.) Hg vacuum or three stages to produce 29.8 in. (757 mm.) Hg vacuum, using steam at 100 p.s.i. Condensers are placed between the stages to remove heat. The condenser reduces the amount of vapor to be handled by the stage following it in the multiple-stage steam jet ejector system.

The water may be removed from the condenser with a pump or by the use of gravity. A reciprocating pump of the positive displacement type or centrifugal pump is normally used. A so-called barometric leg may be placed at the discharge or bottom of the condenser of 34 ft. or more in length to remove the water by gravity from the system.

Entrainment Separators

There is a tendency for liquid droplets to move with the exhaust vapor, which is known as entrainment of liquid in the vapor. Excessive entrainment occurs as a result of a high evaporation rate. It is desirable to remove the liquid from the vapor ahead of the vacuum apparatus to avoid loss of product and contamination of the condensed vapors. Entrainment separators are used. The separators consist of a device against which the liquid entrained vapor is impinged and the liquid separated as it hits the surface due to its inertia. Entrainment separators consist of deflectors or centrifugal devices. The separators may be incorporated in the top of the evaporator or may be constructed externally at the discharge from the evaporator. External entrainment separators are connected to return the removed liquid to the evaporator system.

HEAT TRANSFER IN EVAPORATORS

The major consideration in the design and operation of an evaporator is to obtain the maximum utilization of heat energy rapidly and without damage to the product. The basic heat transfer equation is:

$$Q = UA\Delta t$$

where:

Q = rate of heat transfer, B.t.u. per hr.
U = overall heat transfer coefficient, B.t.u. per hr. sq. ft. °F.
A = surface area through which heat is being transferred, sq. ft.
Δt = difference in temperature between the product and steam, °F.

Increasing any of the values on the right hand side of the equation provides increased rate of heat transfer.

Overall Heat Transfer Coefficient

The U-value, or overall heat transfer coefficient, is often available from the manufacturers for particular evaporators. Knowing the U-value, the rate of heat transfer desired and the difference in temperature in the media, the surface area of heat transfer can be determined; whether coils, tubes, or flat surfaces are used. The U-value increases as the difference in temperature increases and with an increase in boiling point, considering uniform concentration. The U-value will decrease as evaporation occurs and as the product becomes more viscous. Average values for the overall heat transfer coefficient for milk are as follows:

Horizontal tube evaporator—approximately 200 B.t.u. per hr. sq. ft. °F.
Vertical and basket type unit—300 to 500 B.t.u. per hr. sq. ft. °F.
Long tube vertical rising or falling film—200 to 600 B.t.u. per hr. sq. ft. °F.
Coil evaporators—400 to 500 B.t.u. per hr. sq. ft. °F.

These values are not to be used for design, because the equipment and the conditions for operation must be evaluated for each system.

Calculation of U-Value

The U-value can be calculated to estimate the overall heat transfer coefficient if the constants in the equation below are known or can be estimated closely.

$$U = \frac{1}{\dfrac{1}{h_1} + \dfrac{x_1}{k_1} + \dfrac{x_2}{k_2} + \dfrac{1}{h_2}}$$

where:

h_1 = film coefficient for steam, usually taken as 1000 or 2000 B.t.u. per hr. sq. ft. °F.
x_1 and x_2 = thickness of scale and tube material, respectively, ft.
k_1 and k_2 = conductivity of the scale and tube material, respectively, B.t.u. per hr. sq. ft. °F. per ft. of thickness
h_2 = film coefficient of the product

The resistance to heat transfer from the steam to the product consists of the film on the steam side, the scale on the steam side, the metal of the tube, and the film on the product side of the heat exchanger surface.

More involved relationships utilizing Nusselts (Nu), Prandtl (Pr), and Reynolds (Re) numbers are available and sometimes used, but are beyond the scope of this writing. The major factor which must be controlled in operation and which affects the overall U-value is the accumulation of scale as represented by x_1. An increase in the thickness of the scale decreases the U-value.

Film Coefficients

The film coefficients, h, can be determined from the following relationship:

$$Nu = 0.11 \ (Re^a)^{0.686}Pr^{0.4}$$

where: $a = 0.961$ for whole milk and 0.964 for skimmilk (Peeples and Eastham 1962)

These values approach one as the product is condensed. The viscosity values are shown in Figs. 15A, B, C.

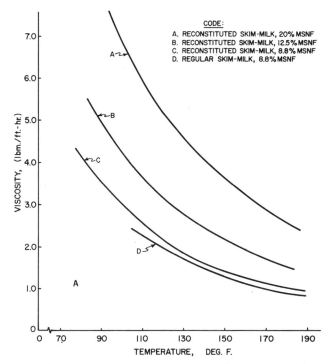

FIG. 15A.   VISCOSITY OF SKIMMILK AS EFFECTED BY MILK-
SOLID-NOT-FAT CONTENT (PEEPLES)

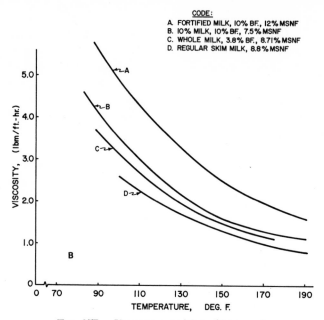

CODE:
A. FORTIFIED MILK, 10% BF., 12% MSNF
B. 10% MILK, 10% BF., 7.5% MSNF
C. WHOLE MILK, 3.8% BF., 8.71% MSNF
D. REGULAR SKIM MILK, 8.8% MSNF

FIG. 15B.   VISCOSITY OF MILK (PEEPLES)

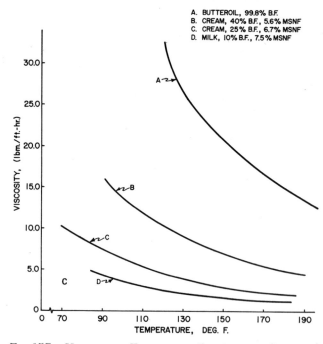

A. BUTTEROIL, 99.8% B.F.
B. CREAM, 40% B.F., 5.6% MSNF
C. CREAM, 25% B.F., 6.7% MSNF
D. MILK, 10% B.F., 7.5% MSNF

FIG. 15C.   VISCOSITY AS EFFECTED BY FAT CONTENT (PEEPLES)

$$\text{Nu} = \frac{hD}{k}$$

$$\text{Re} = \frac{DV\rho}{\mu}$$

$$\text{Pr} = \frac{c}{k}\mu$$

where

k = thermal conductivity, B.t.u. per hr. sq. ft. °F. (Table 7)
h = film coefficient, B.t.u. per hr. sq. ft. °F.
D = diameter, or equivalent, ft.
V = velocity of flow, ft. per hr.
ρ = density, lb. per cu. ft.
c = specific heat, B.t.u. per lb. °F. (Table 8)
μ = viscosity, lb. per ft. hr. (Re)

$$\frac{\text{ft. sec.}}{\text{lb.}} = \frac{\text{lb.}}{\text{ft. hr. (3600)}}$$

TABLE 7

THERMAL CONDUCTIVITY, k, OF SOME DAIRY PRODUCTS
K cal. per m h °C. or B.t.u. per Hr. ft^2 °F. per ft.

	Temperature	
	20 °C.	80 °C.
Whole milk	0.473 (0.32)	0.528 (0.35)
Concentrated milk (ratio 1.9)	0.418 (0.28)	0.486 (0.32)
Concentrated milk (ratio 2.54)	0.392 (0.26)	0.457 (0.30)
Skimmilk	0.464 (0.31)	0.546 (0.36)
Whey	0.465 (0.31)	0.551 (0.36)
Water		0.57 (0.38)

Source: Peeples (1962) from a Russian publication by Govorkov.

TABLE 8

SPECIFIC HEAT, c, OF SOME DAIRY PRODUCTS
B.t.u. per lb. °F. or cal. per gm. °C.

	Specific Heat
Whole milk	0.935
Skimmilk	0.950
Evaporated milk	0.920
Condensed milk	0.940
Ice cream mix	0.800

Source: Scott (1964).

The film coefficient of the product, h_2, also changes as evaporation occurs. The film thickness increases and the viscosity increases with concentration, causing a decrease in the rate of heat transfer.

For good heat transfer of a falling film the Reynolds (Re) number should exceed 2000. For a falling film:

$$\mathrm{Re} \ = \ \frac{4w}{\pi\mu'D_0}$$

where

w = rate of evaporation, lb. per sec.
μ' = viscosity, lb. per ft. sec. (centipoise/1488)
D_0 = tube diameter, ft.

Temperature Difference

The difference in temperature between the product and the steam provides the driving force for transfer of heat to vaporize moisture from the product. A vacuum is commonly used for vaporization of milk and other food products. With a given steam temperature, the temperature difference, represented by Δt, increases as the vacuum in the evaporator is increased and the boiling point is decreased. The vacuum provides a method of obtaining more rapid heat transfer than does atmospheric boiling with a given steam temperature. In addition, the vacuum aids in removal of vapor bubbles formed in the product.

The Δt may vary throughout the system. Either an average temperature (Δt) difference between the mediums or a log mean temperature difference (Δt_m) might be utilized.

$$\Delta t \ = \ t_{st} - t_{pr} \qquad \Delta t_m \ = \ \frac{\Delta t_1 - \Delta t_2}{\ln \dfrac{\Delta t_1}{\Delta t_2}}$$

where:

t_{st} = temperature steam, °F.
t_{pr} = temperature product, °F.
Δt_1 = greatest temperature difference
Δt_2 = least temperature difference

with the above temperature differences taken at opposite ends of the heat exchanger.

The difference in temperature changes along the elevation of a vertical tube. In the evaporator, there is an increase in boiling point of the product at the bottom of the tube due to the head in fluid and hydrostatic pressure exerted which causes an increase in boiling point of the product and thereby a slight decrease in temperature difference (Fig. 16). Also, the velocity variations in the tubes with liquid flowing or liquid changing to vapor or vapor flowing provide considerable variation in the values of U and Δt. The effect of hydrostatic head becomes more important as Δt becomes smaller and as the vacuum is higher. With a surface at a boiling point of 170°F., there is an increase in boiling point of about 25°F. for each 10 ft. of head; at 195°F., 170°F.; and 212°F., 10°F. increase (Fig. 17).

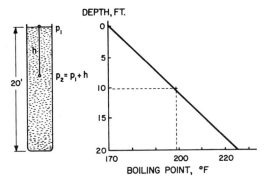

FIG. 16. EFFECT OF HYDROSTATIC PRESSURE ON BOILING POINT

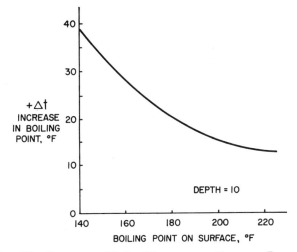

FIG. 17. EFFECT OF 10 FT. HYDROSTATIC HEAD ON BOILING
POINT

The temperature difference between steam and product is affected by
the change in boiling point of the product. The boiling point of the
product is higher than the boiling point of water at the same pressure.
(See previous discussion on the theory involved). The boiling point is
directly related to the quantity of dissolved material and inversely proportional to the molecular weight of the substance. Materials such as
milk, having a high molecular weight, would have a boiling point rise
of less than 1°F. above water at the beginning of the evaporation process,
but increase more than 1°F. after the total solids are increased.

Velocity of Liquid Flow

The velocity of liquid flow is very important in considering the transfer of heat from steam to a boiling liquid. As the velocity is increased, the rate of heat transfer increases. The velocity selected must consider the additional pumping costs versus the smaller heat transfer surface needed with the higher velocity. Liquid velocities are normally from 4 to 5 ft. per sec. by natural convection or 12 to 15 ft. per sec. by forced convection. Steam jets or air jets may be used to increase the turbulence and heat transfer rate. The rate of heat transfer is also dependent upon the diameter of the tubes and, in general, increases as $1/D^{0.2}$. D represents the diameter of the tube which is usually about 1 to 2 in. In film evaporators, there is a considerable increase in velocity and turbulence as the liquid changes to a vapor. Vapor moves at about 50 to 175 ft. per sec. as it leaves the tube.

Liquid Discharge

The relationship between liquid discharge from an evaporator is represented by a Taylor series expansion (Moore and Hesler 1963):

$$\% \text{ removed} = (1 - e^{-t/R})$$

where:

t = time, min.
R = ratio of holding volume to discharge

For an ideal evaporator, to reduce heat effects on the product, it is desirable to have no circulation (single pass) of the product. With an evaporator with a product volume of 10 gal., a feed of 2 g.p.m., and a discharge of 1 g.p.m., the retention time would be 10 min. Mathematically, at the end of 10 min. only 65% of the original capacity of the vessel has been removed. After 20 and 46 min., 87 and 99% were removed, respectively.

Agitated Film Evaporator

Effective heat transfer for vaporization, without damage to the product, is related to a short time for a particular temperature, and a turbulent thin film next to the heating surface. The agitated film evaporator provides an approach to meeting these conditions (Figs. 18A and B). The entering product is moved to the outer edge of the chamber by centrifugal force of the rotating blades. The chamber is heated around the circumference. Both the vapor and condensed product move vertically downward and are removed and separated by conventional means. The unit is particularly suited to products of high viscosity.

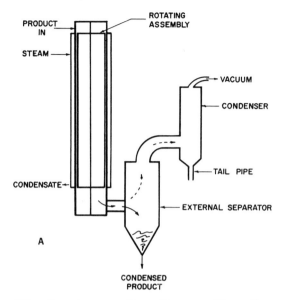

FIG. 18A. THE AGITATED FILM VERTICAL HEAT TRANSFER
UNIT FOR EVAPORATING WATER FROM PRODUCT

Roto-Vak—A trade name of Buflovac, Blaw-Knox.

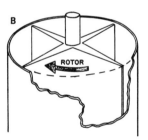

FIG. 18B. ROTATING ASSEMBLY

The rotating rotor agitates the downflowing film of liquid into
turbulent action.

Swept or Scraped Surface Evaporator

The scraped surface heat exchanger, such as used for the ice cream freezer or product heating, can be utilized as an evaporator. Low pressure steam surrounds the chamber. A vacuum can be maintained in the chamber. The scrapers or knives assist in moving high viscosity products from the heating surface. A product of 70 to 80% solids can be obtained.

Centrifugal Evaporator (Centri-Therm)

The centrifugal equipment consists of a stack of rotating cone elements which preceded the development of the expanding flow evaporator. Each cone element is made up of stainless steel walls with a hollow section into which steam is supplied through the hollow spindle which supports the elements. The product to be condensed is sprayed onto the heated surfaces of the cone elements. The centrifugal force developed by the rotation which provides up to 200 gm. spreads the liquid in a thin layer over the inner surface. The product is about 0.004 in. thick over the surface. The centrifugal force keeps the lower surface inside the cone element free from condensate.

The conditions at the heating surface are such to provide good heat transfer and to assist in vaporization. The condensate is discharged immediately by the centrifugal force from the steam heated surface. The thin product layer and the turbulence of the product provides rapid heat transfer and little resistance to heat transfer due to vapor. The concentrate accumulates around the outer diameter of the cone and is displaced upward where it is removed. Evaporating vapors are removed under a vacuum through a large side outlet. The steam condensate is collected under the steam jacket. A standard unit provides a vaporization of 1800 lb. per hr. of water with concentration up to 35% solids. The unit can be cleaned-in-place.

Evaporator with Conical Heating Surfaces

A recent design utilizes heating surfaces comprised of inverted nested stainless steel cones (Fig. 19) known as an expanding flow evaporator. Alternate passages are connected with the product inlet with the other passages connected with the steam inlet. Steam enters at A and rises, then flows toward the center, then down through every other passage between the cones and leaves as condensate at B. The product to be evaporated enters at C and is distributed by nozzles into the passages not occupied by the steam. The product is heated to boiling as it moves

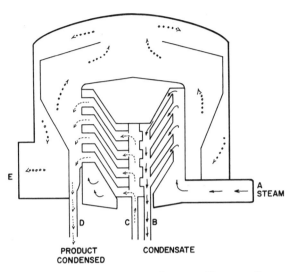

FIG. 19. AN EVAPORATOR WITH CONICAL HEATING SURFACES
(ALFA-LAVAL)

A. Steam enters.
B. Steam condensate leaves.
C. Product to be evaporated enters.
D. Condensed product leaves.
E. Vapor from product leaves.

up and out over the heat surface. The vapor and droplets of the condensed product move into the shell surrounding the heating cones when the droplets move down, D. The vapor moves up, with the separation aided by a deflector, and leaves at E. The vapor leaving at E can be directed to the next effect or condenser.

Countercurrent flow of heat exchange mediums result with a larger cross section utilized for the larger volume of product and steam. A double-effect unit with a thermo-compressor (11,000 lb. per hr.) of skimmilk uses about 0.39 lb. steam per pound of water evaporated.

VAPOR RECOMPRESSION

Vapor recompression offers a method of obtaining increased efficiencies in utilization of energy similar to a double-effect evaporator. Energy is added to the vapor removed from the evaporator. Energy is added by a mechanical compressor or a steam jet to provide mechanical vapor recompression or thermal steam jet vapor recompression (known as a thermo-compressor). Consider 1 lb. of saturated steam leaving a vacuum pan at 6 p.s.i.a. or 170°F. The steam or vapor will contain 1134

B.t.u. per lb. of total heat, of which 996 B.t.u. per lb. are latent heat. The objective of vapor recompression is to increase the pressure and thereby the temperature of the vapor to a point above the condensation temperature so that the latent heat may be removed. In a vapor recompression system, the increase in temperature for a mechanical compressor is about 30°F. which is equivalent to increasing the pressure about 5 p.s.i. and requires an addition of about 10 B.t.u. per lb. to the vapor. The compressed steam can be fed to the heating coils and used for heating the product (Fig. 20).

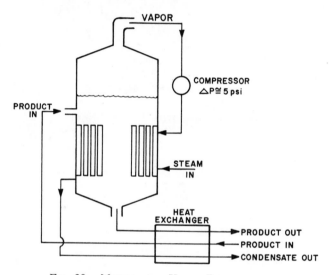

FIG. 20. MECHANICAL VAPOR RECOMPRESSOR

Another approach is to mix incoming high pressure steam with the low pressure vapor being removed from the vacuum pan or evaporator (Figs. 21A and B). This provides a low pressure steam which has a higher temperature of condensation than the low pressure vapor being removed and which can be used for heating the product.

The compressors used for mechanical compression may be either reciprocating positive or centrifugal units. Rotary positive displacement units may be more economical for small capacities.

REFRIGERATION CYCLE LOW TEMPERATURE EVAPORATORS

The refrigeration cycle consists of a compressor, evaporator, condenser, and refrigerant. The refrigerant gas is increased in density at the compressor, after which the gas is condensed to a liquid by removal of heat

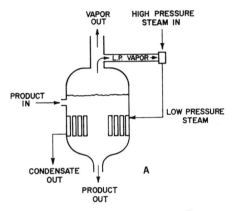

FIG. 21A. THERMAL STEAM JET VAPOR RECOMPRESSION

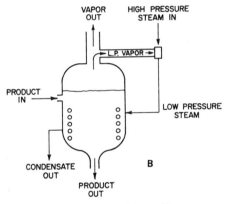

FIG. 21B. THERMAL STEAM JET VAPOR RECOMPRESSION WITH
SPIRAL TUBULAR HEATERS

at the condenser, is metered to the evaporator where it gains heat and the cycle is repeated. The heat given off by the refrigerant at the condenser is used for heating the product in a product evaporator where moisture is removed. A vacuum is pulled on the system, using the steam jet or mechanical vacuum pump. The vapors removed are directed to coils which are cold (the refrigerant evaporator of the refrigeration system) and the vapors condense out of the product system and are removed by a condensate pump.

The refrigeration cycle evaporator is particularly useful for removing water from heat sensitive products, in the temperature range of from 40° to 100°F. Food products so handled need to be pasteurized. From

60 to 100 hp. per 1000 lb. per hr. of water evaporated are required for operating the electrical equipment for a refrigeration cycle evaporator.

PRINCIPLES AND PRACTICES IN OPERATION OF EVAPORATOR OR VACUUM PAN FOR MILK AND MILK PRODUCTS

(1) The unit may be operated either as a batch system or as a continuous system.

(2) Unit should be sanitized before product is placed into the container.

(3) Milk should be placed over the heating coils before steam is moved into the coil to prevent scorching of the product. Otherwise, the product would scorch should it splash on a preheated steam coil.

(4) The product is held at a uniform level in the evaporator. The rate of moving milk into the evaporator should be such that water volume removed will be replaced.

(5) Too rapid boiling will cause an increase in entrainment. A rule-of-thumb for design is to consider approximately 30 lb. per hr. sq. ft. surface area. High pressure steam should be avoided because with a higher temperature of condensation the product might be damaged. Units are normally designed for 5 p.s.i. steam pressure or less.

(6) Economy of cooling water to a water condenser should be checked. No more water should be used than is necessary to condense the vapors being removed.

(7) A measure of the economy of water use is the temperature difference between the cooling water and condenser discharge water, which should be about 5°F.

(8) Air leaks in the system may cause a fluctuation in temperature in the evaporator or a large air leak may cause boiling to stop. Air leaks may occur around valves, fittings, joints, covers, and observation ports.

(9) A vacuum pan or single-effect evaporator for milk is normally operated at a temperature of 130° to 140°F. or 25 in. mercury vacuum.

(10) To stop the evaporator follow proper order of events: (a) turn off the steam; (b) turn off the water to the condenser; (c) stop the vacuum pump; and (d) open the relief valve to the vacuum chamber.

(11) Dry saturated steam is more desirable for vacuum pan operation than wet or superheated steam.

(12) Rule-of-thumb utility requirements for evaporation are as follows: 0.14 lb. of steam per pound of milk for forewarming; 1.1 to 1.2 lb. of steam per pound of water evaporated for single-effect operation; 0.2 to 0.7 lb. of steam per pound of water vapor for steam injector operation; and 13 to 18 lb. of water per pound of water vapor for condenser operation.

(13) Decreased steam costs can be obtained with multiple-effect evaporators. Multiple-effect evaporators can be obtained by using two or more evaporators, a thermal compression, or mechanical recompression of vapors from the product. The additional cost of equipment and operation for a multiple-effect evaporation must be balanced against the savings in fuel.

(14) Commonly accepted overall heat transfer values range from 200 to 600 B.t.u. per hr. sq. ft. °F. Forced circulation systems may go to a U-value of 2000. Manufacturers' literature should be consulted regarding specific evaporators.

Each installation for evaporation is especially designed according to the needs of the processor. The above values are presented only to give an indication of the steam, water, and other utility requirements, and each installation must be carefully figured on the basis of the operation, volume, labor requirements, number of effects, and specific products.

REFERENCES

BADGER, W. L. 1926. Heat Transfer and Evaporation. Chemical Catalog Co., Inc., New York.

BROWN, G. C., et al. 1950. Unit Operations. John Wiley & Sons, New York.

CHARLESWORTH, D. H., and MARSHALL, W. R., JR. 1960. Evaporation from drops containing dissolved solids. Am. Inst. Chem. Eng. J. 6, No. 1, 9–23.

CLARKE, R. J. 1957. Process Engineering in the Food Industries. Philosophical Library, New York.

COULSON, J. M., and RICHARDSON, J. F. 1955. Chemical Engineering 2, McGraw-Hill Book Co., New York.

FARRALL, A. W. 1953. Dairy Engineering. John Wiley & Sons, New York.

HODGMAN, C. D. (Editor). 1962. Handbook of Chemistry and Physics. Chemical Rubber Pub. Co., Cleveland, Ohio.

HUNZIKER, O. F. 1949. Condensed Milk and Milk Powder. Published by the author, LaGrange, Ill.

JENNESS, R., and PATTON, S. 1959. Principles of Dairy Chemistry. John Wiley & Sons, New York.

KERN, D. Q. 1950. Process Heat Transfer. McGraw-Hill Book Co., New York.

LINDSEY, E. 1953. Evaporation. Chem. Eng. 60, Part 1, No. 4, 227–240.

McCABE, W. L., and SMITH, J. C. 1956. Unit Operations of Chemical Engineering. McGraw-Hill Book Co., New York.

MOORE, J. G., and HESLER, W. E. 1963. Evaporation of heat sensitive materials. Chem. Eng. Prog. 59, No. 2, 87–92.

OTHMER, D. F. 1940. Correlating vapor pressure and latent heat data, a new plot. Ind. Eng. Chem. 32, 841–856.

PARKER, M. E., HARVEY, E. H., and STATELER, E. S. 1954. Elements of Food Engineering. Vol. 3, Reinhold Publishing Corp., New York.

PEEPLES, M. L., and EASTHAM, J. 1962. Simplified Nusselt's type equation for describing some of the heat transfer characteristics of fluid dairy products. J. Dairy Sci. 45, 286.

PEEPLES, M. L. 1962. Forced convection heat transfer characteristics of fluid milk products. A Review. J. Dairy Sci. *45*, 297–302.

PERRY, J. H. 1950. Chemical Engineers' Handbook. McGraw-Hill Book Co., New York.

RANZ, W. E., and MARSHALL, W. R., JR. 1952. Evaporation from drops— Part I. Chem. Eng. Progr. *48*, No. 3, 141–146.

RANZ, W. E., and MARSHALL, W. R., JR. 1952. Evaporation from drops— Part II. Chem. Eng. Progr. *48*, No. 4, 173–180.

RICHMOND, H. D. 1920. Dairy Chemistry. Charles Griffin and Co., London.

RIEGEL, E. R. 1944. Chemical Machinery. Reinhold Publishing Corp., New York.

SCOTT, R. 1964. Evaporators and evaporation. Dairy Inds. *29*, 749–754.

CHAPTER 3

Drum or Roller Driers and Miscellaneous Methods of Drying

INTRODUCTION

Milk or milk products can be dried in a thin film on an internally steam heated rotating drum. The product is removed from the exterior of the drum by a "doctor" blade. The process is usually called drum drying, or in the dairy industry, more often called "roller drying." The process is also known as "film drying." On a drum drier, the product enters as a liquid and leaves as a solid. Drum drying requires less space and is more economical than the spray drier for small volumes. The major disadvantages of drum drying are that the dry product may have a scorched flavor and solubility is much lower because of protein denaturization.

CLASSIFICATION OF DRUM DRIERS

Drum driers may be classified according to (Fig. 22):

(1) *Number* of hollow drums—constructed as (a) single drum, (b) twin drum, or (c) double drum units.

(2) *Pressure* surrounding the product—(a) atmospheric, or (b) vacuum.

(3) *Direction* of turning of a twin or double drum unit: (a) the drums may turn up at the center and away at the top as in a twin drum unit, or (b) the drums may turn at the center and together at the top, as in a double drum unit.

(4) *Method of placing* product on the surface of drum: (a) trough or reservoir above for top feed, (b) spray or splash feed, (c) sump below for dip feeding, or (d) trough below for pan feeding.

(5) *Method of obtaining vacuum* for a vacuum drum drier unit—use of (a) steam ejector, or (b) vacuum pump.

(6) *Material* of construction: (a) steel, (b) alloy steel, (c) stainless steel, (d) cast iron, (e) chrome, or nickle-plated steel. Cast iron is usually used. The wear is excessive on stainless steel drums. The metal used for the knife should be softer than the drum.

The double drum atmospheric drier is most commonly used in the dairy industry (Fig. 23). Vacuum drum driers are essentially the same as atmospheric units except that the drums are enclosed so that a vacuum

47

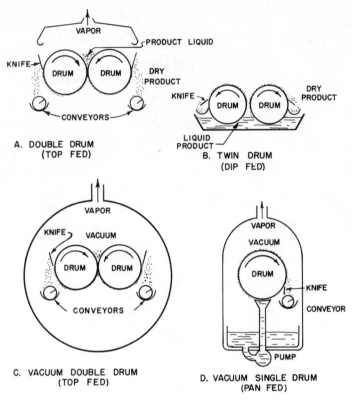

FIG. 22.   TYPES OF DRUM DRIERS IN THE DAIRY INDUSTRY

can be pulled on the product during drying. The single drum with top feed is more commonly used for vacuum. A thicker film is obtained with the top feed.

FLOW OF PRODUCT

The product may be placed in its natural form or condensed in a vacuum pan or evaporator before it is fed to the drum drier. Milk is usually precondensed for single drum units. The product is usually preheated and placed in a reservoir between the upper portion of the drums. Other devices may be used to provide a thin film over the turning drums. The doctor blade, a sharp hard flexible knife, scrapes the dried material from the drum. The blade sits at an angle of 15 to 30° with the surface. The film of dry milk forms a continuous sheet from the knife to the auger trough which is about level with the bottom of the drum. The auger for each drum discharges the product into elevators, then to the hammer mill

Courtesy of Blaw-Knox Company

FIG. 23.   DRUM DRIER AND ACCESSORIES

which pulverizes the product, after which it may be sized. After sizing, the dried product is packaged and stored, or moved to market. These preceding steps are carried out in one continuous operation. In some vacuum units, the product is accumulated in a tub or tray and removed periodically first from one side of an enclosed chamber (separate from the drying chamber) to the other side, while maintaining a vacuum on the drying unit.

VAPOR REMOVAL

Water vapor above the drier has a lower density than the air surrounding the unit, and will rise. A hood must be placed over the drums with a stack for vapors to move out of the area. The stack must have adequate capacity and its hood should be sanitary and must be properly located and sized to collect the vapors. The hood must extend beyond the ends and the midline of the drums. The lower edge of the hood should be formed into a trough to drain away moisture which may accumulate because of condensation, particularly in cold weather.

DRUMS

Drum Description

The atmospheric double drum drier is most commonly used for milk drying. Single drum and twin drum driers are more commonly used for chemicals. The vacuum drum drier is used for foods and milk. The inside of the drum is heated with steam with a low pressure. For vacuum drum drying, the volume between the drum and the housing surrounding the drum, is maintained at 27 to 29 in. Hg vacuum. The product temperature approaches the temperature of the steam heating medium.

The drums used for a drier are 24 to 48 in. in diameter, and up to 12 ft. in length. Drums must be carefully machined, inside and outside, otherwise a difference in thickness will alter heat transfer and drying will not be uniform. The speed of the drums is adjustable usually being from 6 to 24 rpm.; however, the range of speed may be from 1 to 36 r.p.m. A 32 by 120 in. double roll unit would have a 15 hp. motor. The speed of the drums is important as it affects (a) the thickness of film, and (b) the time the product is on the roll. The speed of the drums may be varied according to the concentration of the product and the dryness desired. Both drums must turn at the same speed.

The product is removed after $3/4$ to $7/8$ of a revolution of the drum has taken place. The product is in contact with the drum for about 3 sec. or less.

One drum of a double drum drier is mounted on a stationary bearing. The other drum is mounted on a bearing which can be moved to provide the desired clearance between drums.

The spacing between drums of a double cylinder drier is about 0.02 to 0.04 in. when the drums are cold (Hillman and Warren 1959). The level of milk between the drums affects the capacity of the unit and is designated by the distance above the centerline of the drums. The product is contained over the drums with end plates.

Heat Transfer Through Drum

Steam carries heat into the system. The heat moves through the steam film (inside the drum), the metal, the product film (outside the drum), into the product. The drums must be of uniform thickness to provide uniform heat transfer. From 1.2 to 1.3 lb. of steam per pound of water evaporated are required.

An excellent analysis of heat transfer through the drum and evaporation of moisture for a drum drier is presented by Hougen (1940). Drying may occur in either (a) constant rate zone, or (b) falling rate zone. The constant rate occurs primarily as the moisture is removed from the liquid surface and is carried away by air moving over the surface of the liquid sump above the drums. The falling rate portion of the drying occurs primarily on the surface of the drums.

Uniform Film

A single drum drier may be equipped with up to five spreader rolls to increase the uniformity of thickness of the film on the drier. The concentrated product is fed onto the drum next to the spreader rolls. As the drum turns, a thin film is formed, and an additional film then forms on the next spreader roll. The unit was developed by Buflovak to dry mashed potatoes, but offers possibilities for drying of other viscous products.

EVAPORATION

Constant Rate of Evaporation

Although a specific particle may be considered as having moisture removed in the constant and falling rate zones, the rate of moisture removal by a drum drier is essentially a constant rate of water evaporation as the product is continually fed between the drums, and the dried product removed.

During the constant rate period of drying the evaporation rate of a drum drier is found by:

$$R = 2.45 V^{0.8} \Delta p, \text{ lb. per hr. sq. ft.}$$

where:

R = evaporation rate from surface, lb. per hr. sq. ft.
V = velocity. ft. per sec.
$\Delta p = (p_s - p_a)$, atm.
$(p_s - p_a)$ = (saturation vapor pressure − vapor pressure of air), atmosphere

The rate of drying, given in pounds of water removed per hour per square foot, is related to the velocity of airflow over the surface of the liquid sump, and the vapor pressure difference between the saturated surface and the surrounding atmosphere. As the liquid level between the drums is increased there is more moisture loss. The evaporation from the liquid above a double drum roller is about 30 to 45% of the total water evaporated during drum drying.

Coefficient of Evaporation

The equivalent coefficient of evaporation from the surface of the drum can be calculated and is based on (Hougen 1940):

$$h_e = \frac{2.45 V^{0.8} \Delta p L}{\Delta t_f} \cong 54$$

(The example applies to a 4 ft. diameter drum, steam heated from the inside at 200°F. with atmospheric air moving over the liquid surface, 40% relative humidity at 300 f.p.m., and with the product at 100°F.)

where:

h_e = the evaporation heat transfer coefficient equivalent, B.t.u. per hr. sq. ft. °F.
Δp = 0.514 atm.
L = latent heat, 988 B.t.u. per lb.
V = velocity, 5 ft. per sec.
Δt_f = film temperature difference, assumed at 83 °F.

The amount of heat transferred through the drum is represented by:

$$\frac{Q}{A\theta} = \frac{t_s - t_a}{\dfrac{1}{h_s} + \dfrac{1}{h_m} + \dfrac{1}{h_p} + \dfrac{1}{h_c + h_r + h_e}} \quad \frac{\text{B.t.u.}}{\text{ft.}^2 \text{ hr.}}$$

h_s = equivalent film coefficient of steam, B.t.u. per hr. sq. ft. °F. $\cong$ 800
h_m = equivalent film coefficient of metal, B.t.u. per hr. sq. ft. °F. $\cong$ 1000
h_p = equivalent film coefficient of product, B.t.u. per hr. sq. ft. °F. $\cong$ 800
h_c = convection coefficient $\cong$ 1.1
h_e = radiation coefficient $\cong$ 1.5
h_e = evaporation film coefficient equivalent $\cong$ 54
t_s, t_a = temperature of steam and air, respectively $\cong$ 200; 100

The preceding equation can be used to determine approximately the number of square feet of drum surface or heat transfer surface area required to provide a desired heat transfer in B.t.u. per hour. Also, the equation can be used to determine the amount of heat transfer in B.t.u. per hour which can be transferred through a specific drum area.

For quick estimates the following rules-of-thumb are helpful. Unconcentrated milk can be dried on a drum drier with 18.5 lb. per hr. of water evaporated per sq. ft. of drum. If the product is concentrated or contains sugar there is a lower rate of water evaporation, but more dry product is produced. The amount of dry product is normally about 1 to 6 lb. per hr. per sq. ft., but may be as high as 10 lb. per hr. per sq. ft. of surface area. The overall heat transfer coefficient, U, is approximately 360 B.t.u. per hr. sq. ft. °F., based on the area in contact with milk. A U-value of 220 B.t.u. per sq. ft. hr. °F. is considered reasonable for a double drum center feed unit. The area is determined from the line at which the film is formed to the scraper blade. The area will normally account for approximately $^3/_4$ to $^7/_8$ of the circumference of the drum. The temperature difference through the metal is between 35° and 65°F. and is used as a basis for estimating the heat transfer (Van Marle 1938). The dry product leaving the drum is within 3° to 5°F. of the drum temperature.

MOISTURE REMOVAL

Moisture is removed by circulation of air which carries away the moisture above the unit. A hood above the unit directs the water vapors from the drier. A product entering the reservoir above a double drum with 9% solids would leave at about 12 to 16% total solids. A hood should be placed over the sump or reservoir which is between and above the drums. Adequate airflow must move over the surface of the drums to carry away moisture. An excessive amount of airflow should be avoided since this causes low efficiency due to rapid cooling of the drum surfaces and of the product. In a vacuum drum drier, moisture must be removed from the chamber in much the same procedure as described in the discussion of evaporators.

STEAM FLOW

Steam is fed into the center of the drum at one end of the shaft through the hub. Steam pressures up to 90 p.s.i. are utilized with dry saturated steam up to 300°F. Superheated steam should be avoided. As heat is removed from steam the vapor condenses to a liquid. The liquid condensate moves to the bottom of the drum and must be removed by a pump or siphon. Flooding of the inside of the drum with condensate reduces the rate of heat transfer. It is essential that condensate be removed as rapidly as it accumulates in the bottom of the drum (Fig. 24).

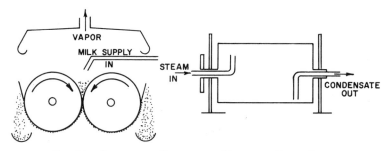

FIG. 24. STEAM AND CONDENSATE FLOW FOR DRUM DRIER

FACTORS AFFECTING PRODUCTION FROM A DRUM DRIER

In a recent report by Hillman and Warren (1959) and an earlier investigation by Harcourt (1938) the variables affecting the rate of drying of a 28 by 60 in. double drum atmospheric drier were evaluated (Table 9). Changes were made from the following standard conditions: steam pressure, 60 p.s.i.; drum speed, 15 r.p.m.; drum gap, 0.04 in.; milk level, 10 in. above the centerline of the drums; milk feed temperature, 160°F.; and milk inlet with 3.5% fat and 12% total solids. These results were obtained:

(1) **Milk Feed Temperature.**—As the milk feed temperature is increased from 120° to 160°F., the rate of drying increases proportionally

at about 2.2% for each 10°F. increase in feed temperature with little increase above 160°F.

(2) **Height of Milk.**—Increasing the height of milk over the drum from 10 to 12 in. gives an increase in drying rate of 10% (98 gal. per hr. ± 6 gal. per hr.).

(3) **Drum Gap.**—The roller gap should be adjusted between 0.020 and 0.043 in. If the drums are farther apart, leakage will occur. The thickness of the film on the drum is directly related to the distance between the two drums.

(4) **Drum Speed.**—Only a slight increase in drying rate occurs as the drum speed was increased from 12 to 19 r.p.m. As the speed is increased

TABLE 9

EFFECT OF INCREASING VARIABLES ON PERFORMANCE OF A DOUBLE DRUM ATMOSPHERIC DRIER

	Variables Increased			
	Steam Pressure	Feed Temperature	Drum, r.p.m.	Distance Between Drums
Film thickness	Increase	Increase	Decrease	Increase
Evaporation between drums	Increase	Decrease	Increase	Decrease
Evaporation on drum	Increase	Increase	Decrease	Increase
Total heat transfer	Increase	No change	No change	No change
Moisture content of powder	Decrease	Decrease	Increase	Increase
Production of dried product	Increase	Increase	Increase	Increase

Source: Harcourt (1938) and Drazga and Eskew (1962).

the film becomes thinner so that the production in both pounds of water evaporated and pounds of dry product produced remain approximately the same with only slight increases in drying rate as the speed is increased.

(5) **Steam Pressure.**—With an increase in steam pressure, the temperature is increased and thereby the rate of drying increased. Too high a steam pressure or superheated steam must be avoided because scorching of the film will result. Increasing the steam pressure from 55 to 65 p.s.i. (302° to 311°F.) increases the production of dry product by approximately 10%.

Approximately the same results were reported by Combs and Hubbard (1932) with the additional observation that the level of the product should be maintained as high as possible without spillage. They also reported that if the speed is increased the steam pressure (and temperature) must be increased to obtain a dry product regardless of inlet temperature of the skimmilk. Higher speeds can be used with less difficulty for concentrated milk than whole milk or skimmilk. To obtain the greatest capacity from a particular drum drier, it is desirable to use concentrated products.

Speeds up to 36 r.p.m. (for a 24 in. diameter drum) were utilized by Combs and Hubbard.

The product may be damaged and scorched (Dickson 1954) if there is uneven milk supply, incomplete removal of film, imperfect roller alignment, rough roller, too high a temperature of the product caused by too high a steam temperature or too slow a drum speed.

OPERATION AND MAINTENANCE

Several items must be checked periodically to assure maintenance of quality of the product at sufficient, rated capacities or output. The following important factors must be considered (Davis 1955; Dickson 1954):

(1) *Drums* must be properly *aligned,* particularly for double drum or twin drum units and have identical characteristics of speed, heat transfer, wear, etc.

(2) *Knives* must be *reground* regularly (approximately every 100 hr.) and be of uniform sharpness.

(3) *Knives* must be *flexible,* machined on both edges, have uniform thickness, and be easily adjusted. A uniform knife pressure against the drum must be maintained. The knife pressure against the drum is adjusted by screws at each end of the knife and by numerous screws along the length of the knife for local adjustment. Excessive pressure of blade on the drum increases the energy to operate the drum and danger of metallic shavings in the product.

(4) *Drum* surface must be kept *smooth.* It may be necessary to resurface the drums after 1000 to 3000 hr. of operation. Smoothing the surface may be done with emery cloth or by wet sanding by attaching a sanding bar to the scraper knife. Pits in the drum surface fill with milk, escape the blade and flake off gradually, causing scorched particles. The drum should receive a film of oil or paraffin wax when not used regularly to prevent rust.

(5) *Condensate* must be promptly removed from inside the drum.

(6) Drum must be *vented* of air to assure that all of the interior heat transfer surface of the drum can be utilized for steam.

(7) *Processing* operations before drum drying consist primarily of removal of moisture, normally 2:1 concentration to 16 to 18% solids, in an evaporator or vacuum pan. Clarification helps to some extent (Davis 1955). Homogenization before drying on a drum does not affect the rate of drying, but will help quality (Hunziker 1949) for whole milk.

(8) *Starting* the drier is accomplished by (a) lifting the knives, (b) starting the drums, (c) turning on the steam, and (d) placing the product on the drums. The very first powder is not of as high a quality as the

remainder because of the tendency to be either over or under dried. The drums should be set in motion when the heat is applied by gradually turning on the steam to prevent warping.

(9) *Preheating* of product is desirable, especially if intended for bakery use.

(10) *End plates* or end drums, which are spring loaded and contain the product above double drums, must be properly adjusted to prevent (a) leakage and (b) accumulation of solids.

(11) The level of milk in the reservoir must be uniform and holes kept open in the distribution tubing. Change in the milk level affects the film thickness on the drum and thus the steam requirements.

(12) *High moisture* in the product is due to low temperature, thick film, high total solids, and fast speed (rpm.).

FREEZE DRYING

Freeze drying consists of (a) freezing the product, (b) supplying heat so moisture is removed without passing through the liquid phase, or as is usual, by (c) maintaining a vacuum in the vaporizing chamber. The vapors are removed before they reach the vacuum pump.

Rapid freezing is desirable to provide formation of small ice crystals. Small crystals provide least change in the properties of the product and a product which will reconstitute easily. Low temperature processing is desired to provide a product with minimum change. The frozen product is placed in a chamber with high vacuum to move the water molecules from the solid to the vapor state without going through the liquid state, known as sublimation. The heat or sublimation must be provided to vaporize the ice. The heat of vaporization is equal to the sum of the latent heat of fusion, sensible heat of liquid, and latent heat of evaporation. Moving heat to the product in the vacuum chamber is a major problem. Heating can be done by radiation, conduction, or convection.

Freezing can be done in a separate or the same chamber in which drying is done. Freezing is done to $-20°F$. or lower. Multiple chambers are used for commercial operations with the complete drying process in each chamber. A vacuum approaching 1 in. Hg abs. is used. The air and vapor are passed over a refrigerated coil to condense the vapor. The quantity of heat supplied must be limited to avoid damage to the product, particularly as the conduction properties change with drying. Electric heating platens, which can be decreased in temperature as drying progresses, are usually used for batch processes.

Freeze drying costs about 5 to 10 times conventional drying for foods (estimated at 10 cents per lb. of water removed). This estimate is based upon 1965 figures.

VACUUM DRYING

One method of vacuum drying is to meter nitrogen into concentrated whole milk after homogenization. The product is then cooled to 35°F. in a scraped surface heat exchanger. The product is placed on a solid stainless steel horizontal belt 12 in. wide. It goes through first a heating drum, then a cooling drum, each 2 ft. diameter for heating and cooling. Heat is supplied by 19 banks of 2 kw. radiant heaters. A vacuum of about 50 mm. Hg abs. is supplied. Scraper blades remove the product from the belt (Anon. 1963).

FOAM MAT DRYING

The product to be dried is first stabilized. Air is fed to the product in a closed mixer where a porous product is formed. The air-product mixture is extruded onto perforated drying trays. The tray of foam passes over air jets cratering the foam to increase drying surface. The trays then move into and upward in the drying chamber in the same direction as the air moves. Air at 220°F. with a velocity of 330 to 400 f.p.m. is used, traveling co-currently with the tray. At the top section of the drier, after the warm, drying air is emitted, cool air enters to cool the product. The dry product is produced in $1^{1}/_{2}$ to 15 min. and is porous and readily soluble (Lawler 1962).

A foam film up to 40 mil thick is placed on a stainless steel belt. The product is exposed a minute and does not exceed 170°F. The system can produce 1.0 to 1.5 lb. per hr. of dry product per sq. ft. of belt (Anon. 1965).

VORTEX METHOD

The vortex method has been used in the laboratory for drying casein. The vertical drying cylinder has three overflow holes through which the dried product is discharged. The cylinder has a conical base and contains a horizontal sieve, which supports the column of product and a dividing wall. The raw casein is fed in at the top; the drying air is fed in from the base causing intensive agitation and rapid drying of the particles. The air enters at 265°F. and casein column is heated to about 113°F. The drier used 1.9 lb. steam per pound of water evaporated, as compared to 3 to 5 lb. per lb. for table driers (Cesul et al. 1963).

REFERENCES

ANON. 1943. A method of smoothing dryer rolls. Butter and Cheese J. 34, No. 4, 24.

ANON. 1963. Continuous vacuum drying of whole milk. Dairy Eng. 80, No. 1, 26.

ANON. 1965. Higher quality drying at a lower cost. Food Proc. 26, No. 1, 92, 94.

CESUL, J., CIBROWSKI, J., and FILIPKOWSKI, S. 1963. Drying of casein by the vortex method. Milchwissenschaft 8, No. 7, 325–333 (In German) Reviewed in Dairy Sci. Abs. 25, No. 11, 464.

CLARKE, R. J. 1957. Process Engineering in the Food Industries. Philosophical Library, New York.

COMBS, W. B., and HUBBARD, E. F. 1932. Some factors influencing the capacity of the atmospheric drum dryer. J. Dairy Sci. 15, 147–154.

DAVIS, J. G. 1955. A Dictionary of Dairying. (Rev.) Leonard Hill, London.

DICKSON, R. M. 1954. The prevention of charred particles in roller dried milk. Aust. J. Dairy Tech. 9, 162–165.

DRAZGA, F. H., and ESKEW, R. K. 1962. Observations on drum drying mashed potatoes. Food Technol. 16, No. 12, 103–105.

HARCOURT, G. N. 1938. Effective drum drying by present-day methods. Chem. Met. Eng. 45, No. 4, 179–182.

HILLMAN, H. C., and WARREN, R. J. 1959. Rate of production of roller-process milk powder: Effect of variations of operating conditions. XV Int. Dairy Congress 4, 2141–2145. London.

HOUGEN, O. A. 1940. Typical dryer calculations. Chem. Met. Eng. 47, No. 1, 15–17.

HUNZIKER, O. F. 1949. Condensed Milk and Milk Powder. Published by the author, LaGrange, Ill.

LAWLER, F. K. 1962. Foam-mat drying goes to work. Food Eng. 34, No. 2, 68–69.

MOORE, J. G., and SAMUEL, O. C. 1964. Advances in drum drying process. Food Processing 25, No. 7, 58–61.

SCOTT, A. W. 1932. The engineering aspects of the condensing and drying of milk. Hannah Dairy Research Institute Bull. 4 (Scotland).

VAN MARLE, D. J. 1938. Drum drying. Ind. Eng. Chem. 30, 1006–1008.

WOODCOCK, A. H., and TESSIER, H. 1943. A laboratory spray dryer. Canadian J. Research 21, Sec. A, No. 9, 75–78.

CHAPTER 4

Spray Drying

INTRODUCTION

In spray drying the objective is to remove moisture from a liquid product to form a powder of solid material. The purposes of drying are to remove the moisture, to reduce the cost of transportation, to improve storage of the product, and to provide a product which can be utilized for many food manufacturing operations. The use of the spray drier has increased in recent years and it is the most important method of drying liquids to solids for milk and milk products. The spray drier is flexible and can be used for many different food products. The system, incorporating various components of equipment, is complicated, even though the principle of operation is simple.

GENERAL DESCRIPTION

The spray drier utilizes a product which is first condensed in a vacuum pan or an evaporator. The product is then atomized inside a drying chamber of a drier. The drying functions include: moving the air, cleaning the air, heating the air, atomizing the liquid, mixing the liquid in the hot air, removing the dry material from the air, additional drying of the product, cooling the product, and pulverizing and sizing the product. Different methods of atomization may be used, but the most common procedure for the milk industry in the United States is to force the product by high pressure pump through a nozzle to assist in the break-up of the liquid. As the atomized product is introduced into the drying chamber, heated air is forced through the chamber. The air furnishes heat for the evaporation of the moisture and the air is a carrier for moisture to be removed from the drier. The air may be forced through the drier by either a pressure or suction system. After drying, the product and air must be separated. The product is then cooled and packaged. Controls maintain the proper adjustment of the variables involved in drying. Additional operations may be utilized to provide a product that will dissolve rapidly. The capacity of commercial driers may vary from 500 to 5000 lb. per hr. and higher of dried product, with 2000 to 3000 lb. per hr. being common in 1965. Proper management of a drier operation is important to provide good quality solid products from good quality liquid products.

CLASSIFICATION

Spray driers may be classified according to:

(1) Method of atomizing spray material: (a) high pressure nozzles; (b) centrifugal spinning discs; (c) two fluid systems—air, steam.

(2) Method of furnishing heat: (a) steam; (b) gas; (c) fuel oil; (d) electricity.

(3) Method of heating air: (a) direct—gas or fuel oil; (b) indirect—utilizing heat exchanger plates or coils.

(4) Position of drying chamber: (a) vertical (Fig. 25); (b) horizontal (Fig. 26).

(5) Number of drying chambers: (a) one drying chamber (Figs. 27A and B); (b) two drying chambers—main and secondary drying chamber (Fig. 38 p. 81).

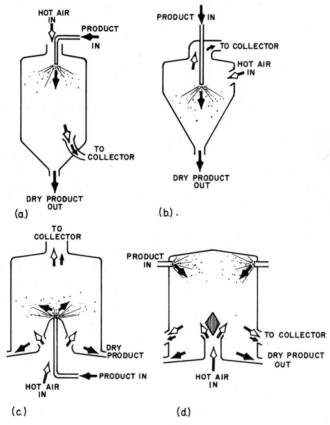

FIG. 25. VERTICAL CHAMBER DRIERS

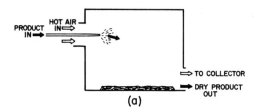

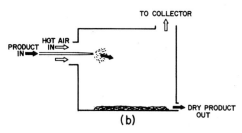

FIG. 26. HORIZONTAL CHAMBER DRIERS

(6) Direction of airflow in relation to product flow: (a) countercurrent; (b) parallel; (c) right-angle.

(7) Pressure in drier: (a) atmospheric (usually a very slight pressure); (b) vacuum.

(8) Method of separation of powder from air: (a) cyclone; (b) multicyclone; (c) bag filter.

(9) Treatment and movement of air: (a) recirculation of air; (b) dehydration of air; (c) conventional—atmospheric air used and exhausted after use.

(10) Removal of powder from drying chamber: (a) conveyor; (b) vibrator; (c) sweep conveyor; (d) air conveyed to cyclone.

(11) Method of heat transfer: (a) convection; (b) radiation,

(12) Kind of atmosphere in drying chamber: (a) nitrogen; (b) air; (c) other gas, usually inert.

(13) Position of fan providing; (a) pressure in chamber; (b) suction in chamber.

(14) Direction of airflow in chamber: (a) updraft; (b) downdraft; (c) horizontal; (d) mixed.

(15) Shape of drying chamber: (a) silo or cylindrical; (b) box-like, square cross-section; (c) tear-drop.

(16) Product being dried: (a) milk; (b) other milk products; (c) egg; (d) other food products; (e) chemicals; (f) detergents.

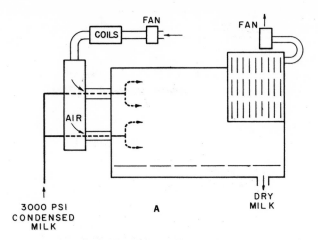

FIG. 27A. HENSLEY SPRAY DRIER

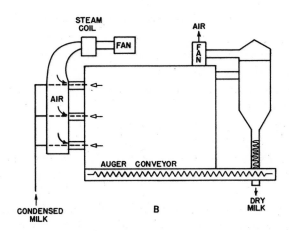

FIG. 27B. BLAW-KNOX SPRAY DRIER

CHARACTERISTICS OF SPRAY DRIERS

The acceptance of the spray drier is attributable to its favorable characteristics, particularly for food products obtained and the economics of operation. These characteristics are:

(1) continuous operation; (2) little labor required to operate; (3) handle many different products; (4) heat contact to product is short during drying and removal (quality is more likely to be preserved as overheating is less likely); (5) thermal efficiency is low compared to other types of driers; (6) operation dependent on large surface area of product which is obtained by different methods of atomizing; (7) separation of air and product after drying is important; (8) unit is easy to clean if operating

properly; (9) economy of moisture removal improved by condensing product before drying; (10) product properties and quality may be effectively controlled; (11) normally, product does not contact drier surfaces until dry. (However, sticking of the dried or semi-dried material to the surface of the drier is a problem if the heat source fails and the product continues to enter the drier.)

PARTS OF A SPRAY DRIER

The essential parts of a spray drier provide for: (1) moving, filtering, and heating of the air; (2) atomizing of heated product; (3) mixing of heated air with atomized product for heating and for carrying the moisture from the product; (4) separating of air and large particles, usually inside the drier, and the removal of large particles; (5) separating of air and small particles, usually external to the drier.

Associated with these basic parts are observation ports, instruments, and controls to provide for proper inspection and operation.

HEATING AIR

Air for drying is filtered and heated before it passes through the atomized product. Filtering is normally done by mechanical means. Future developments will probably involve more thorough filtering of the air.

Air may be heated with an indirect heater, such as steam; a direct-fired indirect heater, which burns the fuel with the heat transferred across the metal surface to the air; or with a direct-fired unit, in which the products of combustion enter the drier. Most spray driers have been of the indirect type, but many new installations are made with direct-fired units. Radiators, with steam at 50 to 100 p.s.i., have been quite common as a means of heating.

Direct-fired indirect heaters may burn liquid, solid, or gaseous fuels. Fuels for the direct-fired units are limited to gas and light oil (No. 2) to avoid soot formation. The direct-fired units are more efficient due to less heat loss during heat transfer. With a direct-fired unit, the selection of fuel is based primarily on the cost and the effect of combustion products of the fuel on the drying product. In laboratory units and in areas where electricity is competitive in cost, electrical heating may be utilized. Adequate heat transfer surface areas must be provided for heating the required quantity of air to the proper temperature.

The area of the heat exchanger surface must be determined for each installation, inasmuch as the overall heat transfer coefficients vary considerably with the velocity of the air and the design of the heat exchanger or radiator. *U*-values for steam radiators for heating air vary from 0.5 to 10 B.t.u. per hr. sq. ft. °F.

Air is heated to about 300° to 500°F. for drying milk products. The relative humidity of the drying air is quite low, even if the ambient air is of a high relative humidity. Air heated from 80° to 300°F. and originally at 100% R.H. will have a relative humidity after heating of only 3 to 4%. However, air with a high relative humidity before heating does not have the same capacity to dry as low relative humidity air, although the difference is very small. The products of combustion for a direct-fired unit cause an increase in relative humidity of the drying air.

Interest has developed recently in using radiant heating to heat the air and the product by placing gas or electric infrared heating units opposite, but facing the atomizing units in the drier. As high a temperature as is consistent with obtaining a desirable product is used. The air may be forced through the drier by placing the fan ahead of the heater, providing a pressure system. If the fan is located at the discharge of the drier, the drier operates under a slight vacuum. Some driers use both fans with the neutral pressure zone approximately in the middle of the drying chamber.

Flow

The air must be properly and uniformly directed into and through the drying chamber. Otherwise, the heat in the air will not be utilized efficiently and a partially dried product may accumulate on the inside edge of the drier. Air straightening vanes are often used at bends in the ducts to direct the air into the drying chamber and to reduce pressure drops. Entrances to the drying chamber must be properly designed to permit and provide uniform airflow with a minimum loss of pressure drop. Each installation must be analyzed from the standpoint of airflow patterns and friction loss during airflow.

In a plant operation, a Niro spray drier was used for six months with air and milk streams flowing in the same direction followed by six months with flow of air and product in opposite directions. With flow in opposite directions as compared to parallel flow, there was (a) less tendency of the powder to stick to the drier, and (b) more particle aggregation with better wettability which was slightly less soluble.

The horizontal drier is in common use in the United States today. These units usually use cocurrent flow of air and product. Centrifugal atomizers are not used in horizontal units.

The vertical drier is more flexible than the horizontal drier. Spiral flow of air rather than straight flow permits a longer residue time thus permitting a unit wider and shorter than for straight-line flow.

The efficiency of a drier can be improved by using the heat in the air discharged from the drying chamber by preheating the incoming product and/or the incoming air to the drier using an indirect heat exchanger.

Steam Coil vs. Direct Flame Heating

Heating air for the drier is usually done with either steam for indirect heating using heating coils, or gas for direct flame heating. It is easier to control a uniform low temperature with steam. Steam provides a heat source which is relatively nonhazardous. Higher efficiency is obtained with a higher temperature of inlet air, direct heating with gas providing a more efficient operation than indirect heating. With direct heating, the products of combustion enter the heated air stream providing approximately 1 gal. of water for each gallon or equivalent of fuel.

Fuels

The heating value of fuels is the quantity of heat released by burning a unit weight or unit volume of fuel. From 30 to 80% of the energy in the fuel may do useful work in a heater or drier. The heating values (high or gross) for coal are from 11,000 to 14,000 B.t.u. per lb.; liquid fuels, 18,000 to 20,000 B.t.u. per lb.; gaseous fuels 500 to 1000 B.t.u. per cu. ft.; and fibrous fuels (wood, bagasse agricultural by-products) 8000 to 11,000 B.t.u. per lb.

ATOMIZATION

The purpose of atomization is to obtain many small particles with a large surface area, preferably uniform in size, generally ranging from 50 to 150μ in diameter. Uniform particles provide: (a) superior instantizing product, (b) reduced product losses, (c) less over and under drying, and (d) more efficient drying. Large drops are more difficult to dry and require a longer time or a higher temperature, or both. The large surface area provides easy transfer of heat to the droplet and transfer of moisture away from the droplet. The atomization of 1 gal. of product to 50μ in diameter will produce 200,000 sq. in. of surface area.

The type and efficiency of atomization affects the desired design (size, air temperatures, exposure time, evaporation rate, and efficiency). The atomization also affects the product properties, such as air content, moisture, bulk density, particle size (range and average), and reconstitutability.

The pattern produced by the atomizer must be directed so that the particles will be dried before hitting the surface of the drying chamber. Otherwise, an accumulation of partially dried product will occur on the drier. There are three major methods of atomizing: pressure nozzle, spinning disc, and pressure.

Pressure Nozzle

Pressure spray nozzles include the swirl nozzle (called whizzer or centrifugal pressure nozzle), the solid cone spray nozzle and the fan nozzle. The pressure spray nozzle is most common in the United States for milk and food product atomization for spray drying. Pressures from 1500 to 5000 p.s.i. are used. A high pressure pump such as the three or five piston homogenizer pump is commonly used. For products which are preheated to a high temperature, it is important that a pump be used which does not operate at too high a speed to avoid cavitation in the pump cylinders. The spray from a nozzle may range in shape from a flat sheet to a 120° cone. An equation for the energy for atomization of a swirl type pressure nozzle is given by Marshall (1954):

$$E = 19.2 \ Q(\Delta P)$$

where:

E = energy or power required, ft. lb. per min.
Q = flow, g.p.m.
ΔP = pressure difference, p.s.i.

Note that the equation is general and does not include a factor for viscosity which might be quite important for many milk and food products.

The principle in pressure atomization is to rapidly apply enough energy to overcome the surface tension of the product. At high pressures the surface tension is the controlling factor in atomization (Marshall and Seltzer 1950) while at low pressure, viscosity is considered to be the controlling factor. As the atomized product moves through the air, or vice versa, friction of the air assists in atomization.

The following describes the pressure spray nozzle:

(1) As the viscosity increases, the cone angle decreases. For example, when increasing from 1 to 100 centipoise viscosity, with a 0.036 in. diameter orifice, the cone angle decreases from 60° to 42° in a nearly linear relationship (McIrvine 1952). Also, the capacity may increase or decrease at a given pressure with a change in viscosity, depending upon the formation of the air core.

(2) Large core angles provide soft sprays and do not penetrate as far in the drier as hard sprays (Marshall 1954).

(3) With an increase in pressure there will be only a small increase in core angle (McIrvine 1952).

(4) As the tangential velocity increases, the core angle increases, being 55° at 10 ft. per sec. and 76° at 46 ft. per sec. for a 0.031 in. diameter orifice (Tate and Marshall 1953).

(5) The drop size (a) will change inversely to the square root of the pressure, (b) will increase approximately as the square root of the viscosity, (c) will increase approximately directly as the surface tension, and (d) will increase approximately directly with the density (Marshall 1954).

The pressure nozzle with a swirl chamber (Fig. 28) and the centrifugal pressure nozzle, with a grooved core insert to impart a spin or circular motion, are popular for drying of milk and milk products (Fig. 29). A conical spray pattern usually results, made up of small droplets with a hollow spherical core. The pressure nozzle is made of two parts, the core which is identified by a number by most manufacturers, and an orifice, also designiated by a number (Table 10). One large nozzle or several small nozzles may be used. Nozzles made of hardened stainless steel will

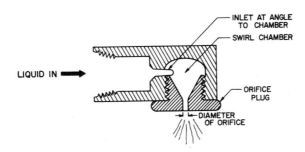

FIG. 28. PRESSURE NOZZLE WITH A SWIRL CHAMBER (SPRAYING SYSTEMS CO.)

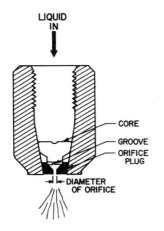

FIG. 29. CENTRIFUGAL PRESSURE NOZZLE WITH GROOVED CORE INSERT (SPRAYING SYSTEMS CO.)

TABLE 10

CAPACITY AND ANGLE OF PRESSURE NOZZLES OPERATING AT 4000 P.S.I.

Orifice No	Diameter, In.	Core insert No.	Capacity, water Lb. per Hr.	Grooves No.	Spray Angle, Deg.
80	0.0135	10	66.6	2	56
76	0.02	16		2	
72	0.025	16	148.3	2	70
68	0.031	20		2	
65	0.035	20	452	4	68
60	0.040	17	533	4	73
55	0.052	21	702	4	75
52	0.0635	27	1400	4	75

Source: Spraying Systems Co.

show wear after 15 hr. if a small orifice is involved. By using tungsten carbide, the nozzles will have as much as 25 to 30 times the life of hardened steel.

Strainers or filters usually precede the spray nozzle to prevent clogging and excessive wear of the nozzle. Nozzle capacities are usually rated on the basis of flow of water. The output of various food products which have a higher viscosity would be a little less and it is suggested that a conversion factor for 40 and 24% solids be 70 and 80% of the flow of water, respectively (Seltzer and Settelmeyer 1949).

When several nozzles are used, they must be arranged so that the spray patterns do not overlap. If spray patterns overlap the droplets may combine and uneven or difficulties in drying result.

A recent development is the use of sonics or ultrasonics to vibrate the nozzle to increase the uniformity of the droplets from the spray nozzle. There are no published technical research reports on this subject at present, but manufacturers report considerable possibilities for this application of ultrasonics.

A sonic field is claimed to provide a means of obtaining uniform droplets to improve spray drying. The sonic field can be obtained with compressed air. A constant frequency sonic field is provided in the vicinity of a nozzle emitting the product. The sound waves assist in forming droplets of desired uniform diameters. The particle size is a function of the flow rate, pressure, orifice size, and resonator position. The nominal operating frequency is 10 kc. for flow rates up to 40 p.p.m. which requires a flow of 130 std. c.f.m. of gas and a pressure of 90 p.s.i.g.

Centrifugal Spinning Disc

The spinning disc is particularly useful for viscous materials and for materials in a suspension. The spinning disc is used more for the drying of milk in Europe and England than in the United States

Various types of centrifugal spinning discs consist of a radial-vaned disc (vanes placed between two discs), multiple discs (three or more discs), and a bowl or hemispherically shaped liquid chamber through which the product moves (Fig. 30). The product may be ejected from the spinning disc over a lip or through a slot, hole, or other opening. The disc atomizer permits considerable variation in capacity of from ±25% of the design capacity. The pattern produced by the disc is in the shape

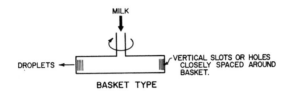

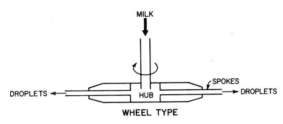

Fig. 30.   Centrifugal Disc Atomizers (Spraying Systems Co.)

of an umbrella, although for very fine droplets a mist of cloud is formed. The drop size produced by the disc atomizer decreases with an increase in speed up to about 400 ft. per sec. for the product coming from the atomizer. The centrifugal unit is used in a vertical drying chamber.

Energy is used to rotate the nozzle, often at high speeds instead of putting energy in a pump for converting pressure energy to velocity as is done for the pressure nozzle. Small diameter discs of approximately 2 in. may be driven at 50,000 r.p.m. and a large 30 in. diameter disc may run at as low as 3500 rpm. Peripheral speeds from 250 to 600 ft. per sec. are involved with capacities up to 60,000 lb. per hr. (Belcher *et al* 1963). An equation for determining the power requirements for driving the disc atomizer is (Adler and Marshall 1951):

$$P = 1.25 \times 10^{-11} \ WN^2 \left(D^2 - \frac{D_0{}^2}{2} \right)$$

where:

P = power, kw. hr.
W = flow rate, lb. per min.

N = speed of disc, r.p.m.
D = disc diameter, in.
D_0 = diameter of disc where product is fed, in.

At speeds above 4000 r.p.m., additional energy is required for moving air so that the equation must be used with judgment.

Conflicting reports have been given on the comparative power requirements for the centrifugal disc as compared to the pressure nozzle for atomization. Marshall and Seltzer (1950) report that the power requirements for the rotating disc is about twice that for the pressure nozzle. Marshall (1954) reports that for comparable conditions approximately the same amount of power is required for discs or nozzles.

Centrifugal vs. High Pressure Atomizer

The specific weight of dry milk prepared with a centrifugal disc atomizer is dependent on the specific gravity of the concentrated milk, which is not true for dry milk prepared with high pressure atomizers (Blaauw 1960). The decrease in air content of powders with increasing solids of the concentrated milk is much more pronounced with the centrifugal as compared to the high pressure atomizer. The organoleptic properties of dry milk produced by the two methods remain the same when gas packed and stored for one year at 60° to 70°F. Bulk density of powdered milk can be varied greatly by varying the pump and product pressure.

Pneumatic or Two-Fluid Atomization

The fluid atomization is produced by passing compressed air or steam over an opening leading to the liquid, usually a pipe. Energy released from the fluid vaporizes the liquid. The unit operates similar to a perfume atomizer or sprayer. The principle is usually used for small flow rates, such as mentioned above, or for laboratory sprayer systems, and is particularly useful for providing uniform small droplets of 15μ or less.

Units may be classified as external mixing two-fluid systems or internal mixing two-fluid systems (Fig. 31). A pneumatic unit is available in which air or steam is passed through a venturi when a liquid stream is injected into it, thereby causing droplet formation.

The two-fluid system is inexpensive in initial cost, but expensive to operate. It is particularly useful for highly viscous materials. During atomization, the droplets cool because of the expansion of the air or steam. Air and steam can transfer more energy per pound of fluid than is obtained by the increase in pressure of the liquid in the pressure nozzle system.

The major disadvantage of the pneumatic atomizer is the large amount of energy required. In general, 2 to 3 times as much energy is required

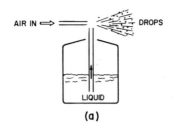

(a)

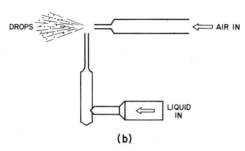

(b)

FIG. 31. TWO FLUID ATOMIZERS (SPRAYING SYSTEMS CO.)

for the two-fluid or pneumatic atomizer as with the pressure spray nozzle. The following equation relates the energy requirements:

$$E = W_a RT \ln P_1/P_2$$

where

E = energy, ft. lb. per lb.
W_a = weight of air used to atomize 1 lb. of liquid, lb.
R = gas constant, 53.3 for air; or 1544 divided by the molecular weight of the gas
T = absolute temperature, °R = °F. temperature + 460°
P_1 = initial air pressure, p.s.i.a.
P_2 = final gas pressure, p.s.i.a.

For $P_1/P_2 = 3.0$, $E = 32,000$ ft. lb. per pound assuming 1 lb. of air per pound of liquid.

A particular two-fluid nozzle consists of a supply conduit tipped with

TABLE 11

POWER TO ATOMIZE 1 G.P.M. OF WATER TO 50 μDROPS

Methods	Power	Conditions
Theoretical	5 ft. lb. per min.	Based on surface tension and area produced
Pressure nozzle	80,000	4000 p.s.i.
Rotating disc	160,000	6 in. disc at 15,000 r.p.m.
Two fluid	200,000	Using high pressure air

a spray jet which is emitted within a circular air supply. The air nozzle is equipped with an electrical resistance heating element to provide accurate control of the air temperature contacting the product (Anon. 1958B).

Comparison of Methods of Atomization

The amount of energy required to produce droplets is considerably higher than the theoretical calculated value. For example, to atomize 1 g.p.m. of water to 50μ drops requires only 5 ft. lb./min. theoretically, compared to 80,000 ft. lb/min. with the pressure nozzle (Table 11). This tremendous difference is attributed to energy for overcoming the friction of the air and for dispersing the droplets inside the drier, in addition to the small energy required for producing new surface area by overcoming surface tension. The heating effect of the liquid during atomization is a measure of the efficiency of formation of droplets. The greater the temperature increase the lower the efficiency of atomization.

It is difficult to determine which is the best method of atomization. The following criteria have been suggested by Marshall and Seltzer (1950). They suggest that methods of atomization be compared on the basis of (a) capacity in terms of square feet of surface area per pound of liquid in a minute, (b) power on the basis of square foot of surface area produced, (c) distribution of the drop size, and (d) weight flow distribution curve.

By analyzing each of these methods individually and in combination, for the particular application, one of the methods of atomization may be superior. It is impossible to say that one method is superior for a majority of applications.

DROPLET

Size Designation

Little information is available, particularly for milk and food products, designating the droplet size specification for spray drying. Data on changes of the droplet as it proceeds from the nozzle to the dried particles are practically nonexistent. A whole new field of investigation awaits the researcher who will develop means of obtaining data on droplet distribution, amalgamation, size, area, temperature, and volume as these variables change with time and as the droplet proceeds from the atomizer as a liquid until the powder is formed.

A frequency distribution curve relates the number or percentage of drops of various diameters, surface areas, volumes, or weights. A normal

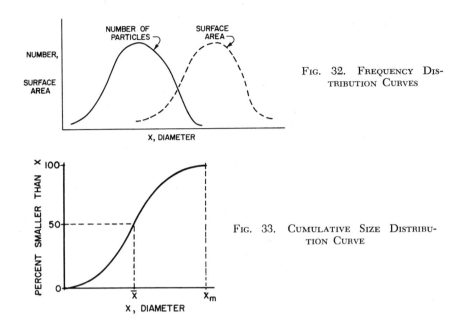

Fig. 32. Frequency Distribution Curves

Fig. 33. Cumulative Size Distribution Curve

distribution would provide a standard bell-shaped curve (Fig. 32). The cumulative distribution representation is also used considerably (Fig. 33). One hundred per cent of the drops would be the maximum size, X_m or less. The median diameter $\overline{X}$ is the point where 50% of the droplets (or some other variable) is included, and divides the spray into two equal portions. Table 12 presents a recommended manner of designation and use of mean diameter which can be expressed by mathematical and statistical relationships (Marshall 1954). Most data presented pertaining to spray drying of milk and milk products are based on the mean diameter and do not include frequency distribution or the standard deviation from the mean diameter. Considerable work has been done on the pressure atomization of fuels where information is also incomplete (De Juhasz 1959).

TABLE 12

TABLE OF MEAN DIAMETERS

Mean Diameter	Symbol	Field of Application
Linear	$\overline{X}_a$	Evaporation
Surface	$\overline{X}_s$	Absorption (area involved)
Volume	$\overline{X}_v$	Distribution of mass in a spray
Surface diameter	$\overline{X}_{s\text{-}d}$	Adsorption
Volume, diameter	$\overline{X}_{v\text{-}d}$	Evaporation
Volume, surface (Sauter Method)	$\overline{X}_{vs}$	Mass transfer

Source: Mugele and Evans, (1951).

Evaporation from Droplet in Drier

In the evaporator water is removed from the product from within the container and through the surface of the product. The condensed product is forced to nozzles, atomized and dispersed in a hot air stream of the drier. Water is then removed from the droplets of the product which are 40 to 70% water, with the remainder as dissolved solids. As the product changes essentially from a liquid at entrance to a moist solid and then to a dry solid the rate of drying also changes.

The rate of drying can be classified into two zones: The constant rate period and the falling rate period, separated with a transitional period (Fig. 34). The constant rate period of drying is characterized by a uni-

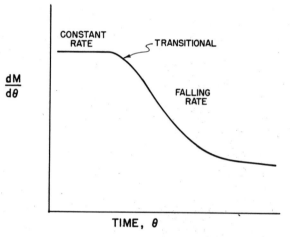

FIG. 34. THEORETICAL ZONES OF DRYING

form loss of moisture from the product similar to evaporation from a water droplet. The rate is controlled by the temperature and speed of airflow over the droplet and continues until the free water surface ceases. During the constant rate period the drying rate varies with the velocity at $V^{0.8}$. After the constant rate period, a crust or solid phase forms over the droplet. The crust offers resistance to the flow of moisture from the product, with the resistance increasing as the water-vapor phase recedes to the center of the droplet. The falling rate of drying begins at the moisture content at which the solid phase is formed which varies with the material, temperature, and air velocity, and has not been evaluated for the variables involved. For coffee extract the change from liquid to solid appears to occur at 1.2 lb. water per pound dry solid (55%, w.b.)

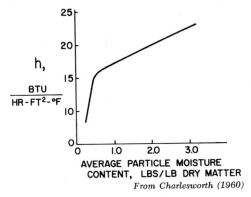

From Charlesworth (1960)

Fig. 35. Change in Heat Transfer Coefficient with Moisture Content for Drying of Coffee Extract

as shown by the change in h-value (Fig. 35) (Charlesworth and Marshall 1960). Below 40% moisture content the remainder of the water was bound and had a higher latent heat, which may be due to a sharp retreat of the liquid vapor interface.

The testing of most spray driers is done with water. Most reported information other than with water is with chemicals. In practice the drying rate of material with dissolved solids will be less than with water. Drying of a droplet consists of two simultaneous processes. First, the heat must be transferred from the air into and through the droplet; and secondly, the moisture must be transferred to the surface and evaporated from the product and carried away by the surrounding air.

HEAT TRANSFER

The unit of measure of the heat transfer from the air through the film into the droplet is the h-value, known as surface or film coefficient of heat transfer, B.t.u. per hr. sq. ft. °F. The quantity of heat transfer can be obtained by:

$$Q = hA(\Delta t)$$

where:

A = area of surface of the droplet, sq. ft.
Δt = difference in temperature between droplet and surrounding air outside of the film, °F.

"The drop temperature can be estimated from a psychrometric chart at the point where the adiabatic saturation line drawn through the drying air condition crosses the curve for humidity over a saturated solution of the nonvolatile material" (Ranz and Marshall 1952, Part II).

The film coefficient, h, can be calculated from

$$h = k/D(2.0 + 0.54 \ \mathrm{Re}^{1/2})$$

or

$$h = 1.6 \ (1 + 0.3 \ (\mathrm{Pr})^{1/3} \ (\mathrm{Re})^{1/3})$$

where:

k = conductivity coefficient, B.t.u. ft. per hr. sq. ft. °F.
D = diameter of drop, ft.
Re = Reynolds number = $DV\rho/\mu$

with

V = relative velocity of drop and air, ft. per sec.
ρ = density of air, lb. per cu. ft.
μ = viscosity of air, centipoise/1488 (Table 13); lb. per ft. sec.

TABLE 13

VISCOSITY OF AIR AND WATER VAPOR

Temperature, °C.	Temperature, °F.	Poises
Air		
0	32.0	171×10^{-6}
18	64.4	183×10^{-6}
40	104.0	196×10^{-6}
74	165.2	210×10^{-6}
100	212.0	220×10^{-6}
229	444.2	264×10^{-6}
334	633.2	312×10^{-6}
357	692.6	318×10^{-6}
409	768.2	341×10^{-6}
Water vapor		
0	32.0	90×10^{-6}
100	212.0	132×10^{-6}

Other dimensionless numbers commonly utilized for expressing relationship in heat transfer are Nu (Nusselt number) and Pr (Prandtl number), which follow:

$$\mathrm{Nu} = hD/k \ \text{(Nusselt number)}$$

$$\mathrm{Pr} = c\mu/k \ \text{(Prandtl number)}$$

The relationship of the Nusselt number and the Prandtl number and Reynolds number are given in Fig. 36. Along the ordinate are plotted values of the Nusselt number from which it is a common practice to calculate h (from hD/k) which can be determined if values for D, V, ρ, c, Nu, and k are known.

Corresponding to heat transfer coefficient is the mass transfer coefficient, k_g, which can be utilized for determining the rate of mass transfer, water or vapor, in pounds per hour.

$$R = k_g A(\Delta p)$$

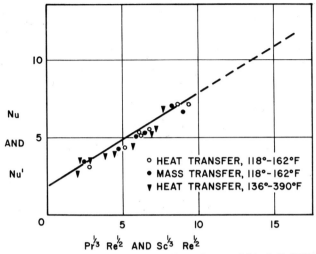

From Ranz and Marshall (1952)

FIG. 36. HEAT AND MASS TRANSFER RATES OF WATER DROPS
EVAPORATING

where:

R = rate of mass transfer, lb. per hr.
k_g = mass transfer coefficient, lb. per hr. sq. ft. unit partial pressure
A = area, ft.^{2}
Δp = dimensionless partial pressure ratio

$$\frac{p_{vi} - p_a}{p_{vi} - p_s} = \frac{\text{vapor pressure at interface} - \text{pressure of diffusing vapor}}{\text{vapor pressure at interface} - \text{vapor pressure of drying air}}$$

Inasmuch as heat and mass transfer take place simultaneously these values can be plotted as shown in Fig. 36. However, different dimensionless numbers are used to represent mass transfer, such as Schmidt number, Sc, and modified Nusselt number, Nu'. Dimensionless values utilized in analyzing mass transfer are as follows:

$$\text{Nu}' = \frac{k_g MDP_f}{(\text{diff})(\rho)} \quad \text{(modified Nusselt number)}$$

$$\text{Sc} = \frac{\mu}{\rho(\text{diff})} \quad \text{(Schmidt number)}$$

$$\text{Re} = \frac{DV\rho}{\mu} \quad \text{(Reynolds number)}$$

where:

M = average molecular weight of gas mixture in transfer path
D = diameter of droplet, ft.
P_f = average value of pressure difference across transfer path
diff = diffusivity of vapor in air, at 300 °F., diff = 1.425 sq. ft. per hr. at 32 °F., diff = 0.853 sq. ft. per hr. μ and ρ previously defined, for air

Thus, when using Fig. 36 for mass transfer, the modified Nusselt number, Nu', versus Sc and Re, and for heat transfer, Nu versus Pr and Re, are used.

A water droplet will dry or evaporate quite rapidly in a spray drier. The active evaporation in a drier occurs in 4 to 5 sec. As the drop diameter increases and as the temperature of air surrounding the droplet decreases, the length of time for complete evaporation increases. To

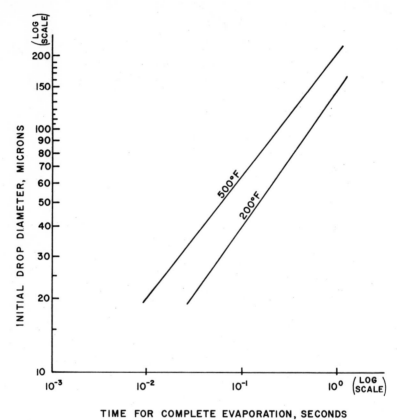

TIME FOR COMPLETE EVAPORATION, SECONDS

From Marshall and Seltzer (1950)

Fig. 37. Time for Evaporation of Water Drops

evaporate the water in a 100μ droplet at 500°F. takes approximately 0.3 sec. (Fig. 37). The time the droplet and the resulting dry product is in the drier is much greater than the theoretical time to dry a water droplet. The droplet and resulting dry particle is in the spray drier from 10 to 30 sec. The time may be much greater if the product is not immediately removed from the drier.

AIR AND AIR CHANGES

Air moves into and through the drier at a velocity of from 1200 to 2000 f.p.m. If the droplet is not given an initial velocity it will be carried by the high velocity air. If, on the other hand, the initial velocity is greater than the velocity of the air, the droplet velocity will decrease. The greater the velocity, the more likely the dry product will be carried from the drier. Air moving at 1 ft. per sec. or 60 f.p.m. will carry 100μ or smaller particles. The velocity which will support the particular size of particle is known as the terminal velocity (Table 14). The terminal velocity increases with an increase in diameter of product.

TABLE 14

TERMINAL VELOCITY OF PRODUCT WITH SPECIFIC GRAVITY OF
1.0 BASED ON STOKES LAW WITH AIR AT $155°F$.

Diameter, μ	Terminal Velocity, Ft. per Sec.
10	0.008
30	0.086
60	0.31
90	0.70
120	1.24

The proper quantity of air must be supplied to furnish heat for vaporization and to provide moisture carrying capacity for removing the vapor. For making preliminary calculations, the following equation may be used:

$$wc(t_1 - t_2) = q = Ua(\Delta t_m)V_d$$

where:

w = mass flow rate of the air, lb. per hr.
c = specific heat or humid heat capacity, B.t.u. per lb. $°F$.
t_1 = inlet air temperature of the drier, $°F$.
t_2 = outlet air temperature of the drier, $°F$.
q = total rate of heat transfer which may be calculated on basis of water to be evaporated, B.t.u. per hr.
Ua = volumetric heat transfer coefficient, with values of 9 to 17 B.t.u. per hr. cu. ft. $°F$. (Tsao, Schen, and Tai 1948).
Δt_m = mean temperature difference of product throughout drier, $°F$.
V_d = volume of drier, cu. ft.

The value of U as reported above of 9 to 17 B.t.u. per hr. cu. ft. $°F$. is for a small experimental unit. These relationships used can be based only on experience, knowing the temperatures involved and quantity of product to be dried. Also, the quantity of air can be determined from psychrometric calculations (see Chapter 5).

One pound of air heated to $300°F$. and exhausted at $175°F$. will give up 29.8 B.t.u. and evaporate 0.030 lb. of water; from $500°F$. to $175°F$. will evaporate 0.078 lb. of water.

The air supply to the drier may come from outside or inside the plant. With either source, the air should be filtered. An outside air supply next to roads, drives, or other conditions where dust or contamination may be present should be avoided. Inside sources might provide a more uniform temperature the year around, but the practice makes it difficult to maintain uniform room temperature. It is usually easier to supply heat as needed to outside air to provide a uniform air temperature to the drier.

For a vertical spray drier economy can be achieved by using warm air from the highest point in the building such as the top of the main drier penthouse. Using this procedure not only improves heating efficiency, but tends to maintain a more desirable temperature in the drying room.

The thermal efficiency of air utilization in the drier is enhanced by (1) high inlet temperature, (2) low outlet temperature, (3) high efficiency of heat exchanger for heating air, (4) use of outgoing exhaust air to heat incoming air, (5) insulated drier body to reduce heat losses, and (6) lack of air leaks. About 6 to 8% heat loss results from radiation from an insulated drier ($U = 0.5$) when operated at full capacity. Based on steam entering heater, the drier is about 40% efficient. Based on heat entering the drier, the drier is 55 to 60% efficient.

The use of gas engines to drive the supply and exhaust fans has proved satisfactory where natural gas is available. The heat from the radiator cooling water and from the exhaust gases is used to preheat incoming air. Also, the gas engine might be sized to drive an electric generator in addition to the fans. The electricity generated can be used to drive smaller motors in the drying system or in the plant.

The steam or vapor formed by the evaporation of 1% of water from the product is 17 times the volume of the particle. The first 1% moisture evaporated surrounds the product with a layer of cool gas equal to about five times the radius of the particle (Fogler and Kleinschmidt 1938). The evaporation of 5% of the water from the water particle will lower the temperature 50°F.

Dehumidifying incoming air provides a method of increasing the moisture carrying capacity of the incoming air. The cost of dehumidifying air is more expensive than supplying additional heat to carry the moisture. Unless the additional heat, or longer time involved by using more air, give a damaging effect on the product, dehumidifying the air is not economical.

Some units accomplish drying in both the main drier and the secondary drier (Fig. 38). For nonfat dry milk, the product leaves the main drier at a moisture content of 6 to 7% and the secondary drier at 3 to 4%. The secondary drier is furnished with fresh room air which is heated in the

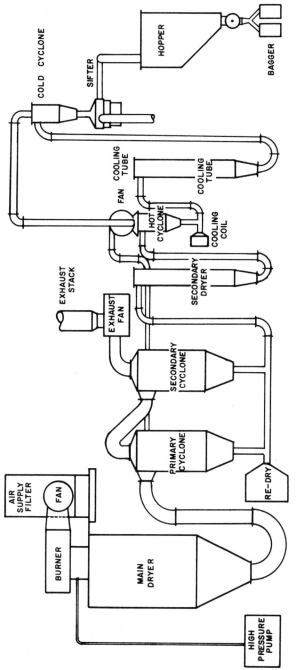

Fig. 38. Spray Drying System with Two Drying Chambers

redry burner. The thermal efficiency of the unit is increased by using this procedure. The drying capacity of heated room air is greater than the air leaving the main drier. Thus, the drying unit is operated with less air per pound of powder than if all the drying is done in the main drying chamber.

SEPARATION OF AIR AND POWDER

As the product is dried it is necessary to separate the dried product from the air. Without special design features the product will be carried by the moist air from the drier. It is necessary to remove the particles (a) to get a maximum yield from the drier by saving all powder product, and (b) to avoid air pollution surrounding the drying plant.

Location

The powder may be separated from the air primarily either (a) inside the drier (internal) (Fig. 39), or (b) outside the drier (external) (Fig. 40). In both cases it is necessary to use an additional device or component outside the drier to remove the fines or small particles which will not normally settle out in the drier. External separation devices are used on all driers.

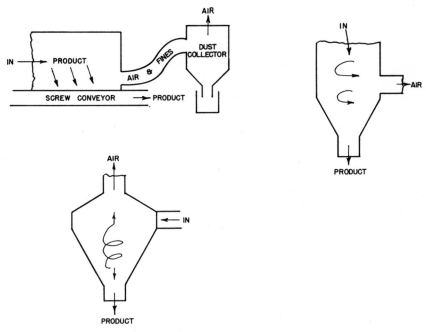

FIG. 39. SEPARATION OF DRY MILK IN DRIER

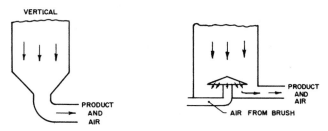

FIG. 40. AIR AND PRODUCT ARE REMOVED TOGETHER

Methods

The product which is separated from the air in the drier can be removed from the drier by: (1) an air brush, by which air from outside the drier, either at room temperature or conditioned to a lower temperature, is used to direct a jet of air to move the product from the bottom of the drier; (2) a rake or broom which is pulled across the bottom of the drier; (3) a conveyor, flight or auger type; (4) a gravity system.

Vibrators are often attached to the sides of the drier to prevent or reduce sticking of the powder and to move the product rapidly from the drier.

Variables

The variables affecting separation of powder from the air are particle size, concentration, nature of the material, and quantity of the product. In a drier, where most of the powder is removed internally, the quantity of air handled by the separation system will be the same as for external separation, but the quantity of product will be much less. The product nature or characteristics such as fat, moisture, cohesion, and friction greatly affect the efficiency of separation or collection. The efficiency of collection is designated as the ratio of the output divided by input times 100% with the product at the same moisture content at the two locations. Some manufacturers rate the efficiency of collection or separation on the basis of pounds of dry matter input. If all of a product at three per cent moisture is collected, an efficiency of 103% would be claimed using this method of rating.

Types

Three types of air-product separators are commonly used for milk driers. The cyclone or multicyclone, which is a type of inertial separator, is most commonly used. Cloth collectors or bag filters have been used for many years. The material may be cotton, wool, or plastic. Approxi-

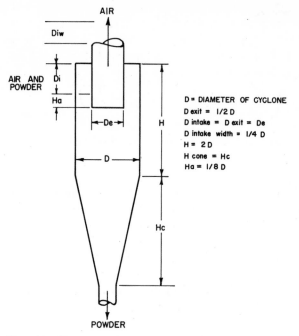

FIG. 41. EXTERNAL SEPARATION OF AIR AND POWDER

mately $^1/_3$ to $^1/_2$ sq. ft. of bag or cloth surface per cubic feet per minute of airflow is provided for external separation. Losses generally range from 0.2 to 0.5%. The wet scrubber or liquid device may be used where the fines passed through other separation devices are removed from the exhaust air and returned to the incoming product for redrying. From 1 to 16 g.p.m. of milk are used for each 1000 c.f.m. of air for a wet scrubber. The wet scrubber provides for a high recovery of product and heat from the exhaust air. The disadvantages are the possibility of product deterioration and contamination.

Other methods of separation not now commonly used in the milk industry to separate fine solids from the air stream consist of the use of electrostatic, sonic or ultrasonic, electrical, and packed beds of granular or fibrous materials.

CYCLONE SEPARATORS

The cyclonic separator is most commonly used for removing the dry product from the air. Air at a high velocity moves into a cylinder or cone which has a much larger cross-section than the entering duct. The velocity of the air is decreased in the cone, thus permitting the settling of the solids. The velocity of the air decreases near the wall of the cylinder or cone and the product falls by gravity and is removed from the bottom (Fig. 41). Cyclones may be used for storage of product before packaging, to provide for a more efficient packaging operation.

Centrifugal Force

The centrifugal force acting on the particle for removing the product from the air is:

$$CF = m \times a = \frac{w}{g} \times a = \frac{wV^2}{g\,r}$$

where:

CF = force on the particle, lb.
w = weight of particle, lb.
g = acceleration due to gravity, 32.2 ft. per sec.^{2}
V = velocity of particle, ft. per sec.
r = radius, ft.; i.e., the radius from the center of the cyclone to the particle position or the radius of rotation in ft.

The centrifugal force on the product is exerted toward the edge of the cyclone. The weight of the particle is effective vertically in the direction of the outlet.

Arrangement

Cyclones can be used individually or in combination to provide multi-cyclone units. Units of 96 tubes or cones have been used to get high efficiency of separation in which cyclones were arranged in two elements, each with 48 cyclones in rows of six and each mounted at a 45° angle (Anon. 1958A). However, the trend is toward fewer, but larger cyclones to remove fines from air.

Velocity

High air velocity is needed to separate small diameter, lightweight materials from air. A high air velocity is obtained by small diameter cyclones, several of which may be placed in parallel. Cyclones can be placed in series with a small diameter high velocity unit at the end to remove small particles in the stream. With a higher velocity there is a greater pressure drop and thus it is more costly to remove the product. The cost of obtaining the separation must be balanced against the value of the product obtained, assuming that minimum requirements are met from the standpoint of loss of product and contamination of the environmental atmosphere.

Efficiency

The efficiency of separation with a cyclone unit is based on product, cyclone design, and on the size of particle to be removed, but losses range from 0.5 to 3 and average one per cent. The cyclone is normally used for separation of material between 5 and 200 μ. As the size of particle decreases the efficiency of the cyclone decreases. A properly designed cyclone will remove 99% of the solids larger than 30 μ (Brown *et al.* 1950), 98% of material larger than 20μ, 90% of material larger than 10 μ, but only 50% of the material smaller than 5 μ. One can see particles of about 10 μ and larger with the eye.

A qualitative evaluation of the efficiency of the separator in operation can be obtained by placing a receptacle filled with water on the roof of a building under the outlet. Collection should proceed at intervals for about one month. The receptacle will catch particles over 1 μ in diameter.

The following characteristics identify the cyclone separator:

(1) The separation is a function of the difference in diameter and density of the product and air with the larger values giving more efficient separation.

(2) The velocity of the inlet is one of the main factors in controlling efficiency as it is related to the cyclone diameter. Velocities may approach 100 ft. per sec.

(3) A smaller diameter cyclone at a fixed pressure drop will have a higher efficiency than a larger diameter one.

(4) The diameter of the inlet is usually about one-fourth of the diameter of the cyclone.

(5) The collection efficiency can be increased by reducing the air outlet diameter, which at the same time will increase the pressure drop. A major problem in operation for a unit with these design characteristics is the leaking of air in at the outlet of the unit causing vacuum. The design must incorporate a change in cross-section entering the cyclone which is gradual.

(6) A smaller inlet and outlet result in separation of smaller particles. The practical limit is set by the permissible and economical pressure drop of the system.

(7) The pressure drop to the cyclone decreases as powder or dust is incorporated in the air.

(8) The pressure drop of the system decreases as the dust concentration increases for the same airflow in cubic feet per minute.

EXPLOSION

Explosion caused by concentration of organic materials in the atmosphere depends upon the increase in temperature of the particles to the ignition point. The possibility of explosion usually is designated on the basis of concentration of the product in the air. It is generally considered that a concentration of organic dust of 5 gm. per cu. ft. or less is safe from the standpoint of explosion. Adequate ventilation must be provided to maintain the concentration at a low level.

PARTICLE

The size of particles is designated by microns or mesh. A micron is 0.001 mm. Mesh refers to the number of screen openings per lineal inch. The opening also depends upon the wire size used in making the mesh material. The Tyler sieve, the U.S. Scale sieve and the B.S.S. (British Standard) use different sizes of wires for making up the screen. Small particles will often stick together forming a larger apparent diameter of particle.

A 100-mesh Tyler sieve has an opening of 0.0058 in. or 0.147 mm. which is equivalent to 147 μ. The 400-mesh opening is equivalent to 38 μ diameter. For smaller sizes a microscopic analysis is made to determine the particle diameter.

Various methods may be used for representing the distribution of weight or distribution of size of particles involved. Excellent references covering these topics have been written by Lapple (1944–1946).

COOLING THE POWDER

The dried product should be removed from the drier as quickly as possible after it is produced to minimize the effect of heat damage on the product. The product and air may be removed together from the drier and separated outside of the drier to reduce heat effect.

Product cooling is done to prevent clumping, sticking, and heat damage to the product. Prolonged heating causes staleness in nonfat dry milk. Prolonged heating causes the fat to melt and move to the surface of whole milk powder. With more of the fat on the surface of the powder the product will not keep as well in storage. Warm powder will hold the heat for some time in a bulk container, thus increasing the heat damage. The thermal conductivity, or k-value, is estimated at 0.03 B.t.u. per hr. sq. ft. °F. per ft., which is considerably lower than most food particles and very similar to insulation materials.

Some cooling of the product will take place in the drier when using an air brush supplied with cool air to remove the dried product from the sides and bottom of the drier.

The three principles of cooling powder outside the drier involve:

(1) Conduction cooling in which the product is cooled when moving through a water jacketed screw conveyor.

(2) Convecton cooling by using room air or refrigerated air to cool to 100°F. or by moving conditioned air over the product or through the conveyor handling the product.

(3) Radiation cooling by placing a cold evaporator surface in view of the warm product offers another possibility. This method has not been exploited by the dry milk industry.

The outlet of a cyclone separator can be surrounded by a chamber through which cold air is moved to cool the product. The material moves on to an entrainment separator for separation of solids and air. A vibratory conveyor for moving dry milk permits cooling as the product moves through the surrounding air (Anon. 1957).

Oxygen can be removed more easily from warm powder than cold, when an inert gas, such as nitrogen, is used for packaging under vacuum.

The amount of heat to be removed is given in B.t.u. by multiplying the specific heat by the weight, in pounds times the difference in temperature, °F. A specific heat of approximately 0.25 B.t.u. per °F. lb. may be used for dry powder. The density of drum dried milk is 0.3 to 0.5 gm. per ml. and for spray dried is 0.5 to 0.6 gm. per ml. (Hunziker 1949).

Some drying occurs in the cooling process. About one-third to one-half of the heat removed in the cooling process can be considered to be used for vaporization of water. Thus 2000 B.t.u. will remove about 1 lb. of water.

EFFECTIVENESS OF SPRAY DRIER

Two terms are used to describe the operation or effectiveness of spray driers. The evaporative capacity is the pounds of water evaporated per unit of time under standard operating conditions. The thermal efficiency is the per cent of total heat utilized for heating and evaporating the water.

$$\text{Thermal efficiency} = \frac{(1 - R/100)(t_1 - t_2)}{t_1 - t_0}$$

where:

R = radiation loss, per cent of total temperature drop in drier
t_0 = atmospheric air temperature, °F.
t_1 = temperature to which air is heated, °F.
t_2 = temperature of air leaving drier, °F.

A high efficiency is promoted by a high inlet temperature, a low exhaust temperature, reduced radiation loss, and recovery of heat from exhaust gas beyond the drying zone. The exhaust gas can be recirculated to the drying chamber, used to preheat fresh drying air or can be used to preheat and concentrate the product. In practice, 2.2 to 3.2 lb. of steam are required to evaporate each pound of water in the drier.

MORE ON THEORY

For those interested in a more detailed analysis and mathematical treatment of theory of drying, the reader is referred to Fogler and Kleinschmidt (1938), Marshall (1954), Ranz and Marshall (1952), Fröessling (1938), and Marshall and Seltzer (1950).

PRESSURE DROP

The pressure through a drier increases as the velocity is increased for a particular drier. The pressure drop varies from 5 to 25 in. of water, depending upon the design of the unit. The pressure drop may be calculated on the basis of the drop in each of the components plus the pressure drop in the piping. The components responsible for most of the pressure drop of the system are the filter, heat exchangers, collectors, and the ducts.

FOAM SPRAY DRYING

Common dairy products, skimmilk, whole milk, buttermilk, sweet and sour cream (up to 3:1 fat to SNF) whey and emulsified cheese slurry can be foam spray dried. Foam spray drying can be accomplished by (a)

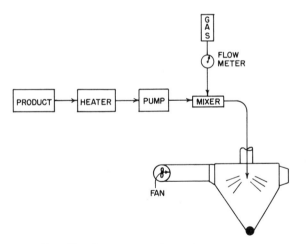

FIG. 42. SYSTEM FOR FOAM SPRAY DRYING

preparing a mixture of gas and liquid ahead of the pump, or (b) forcing the gas into the liquid after the pump, before the atomizing nozzle supplying the drier (Fig. 42) (Hanrahan and Webb 1961). It is necessary to maintain a pressure over the mixture to get appreciable incorporation of the gas in the liquid, if done ahead of the drier. An Oakes mixer is commonly used for this purpose (Oakes 1963). To move gas into the liquid requires that the gas be about 200 p.s.i. pressure above the pump pressure which is commonly at 1000 to 1800 p.s.i. for milk.

Air is commonly used as the added gas for making foam spray nonfat dry milk. Nitrogen is commonly used for making foam spray dried whole milk. The quantity of gas used is from 0.5 to 2.0 std. cu. ft. per gal of product, for 40 to 60% total solids (NF).

Foam spray drying provides a means of using most conventional spray drying equipment for drying liquids up to a maximum of 60% total solids as compared to 50% on a particular drier; for drying special products, such as malted milk and cottage cheese whey; for obtaining an instant type powder, but with characteristics different that the product prepared by agglomerating processes normally used following the spray drier; and provides a procedure for increasing the capacity of conventional equipment.

Foam spray nonfat dry milk has a bulk density of about 0.35 gm. per ml. or less (about the same as agglomerated or conventional instantized powder). The foam spray dried product will withstand considerable handling and can be compressed into smaller volumes without appreciably affecting the dispersibility. The foam spray dried product will remain on top of water for an extended period and needs to be mechanically mixed to form a solution. During reconstitution, considerable foam is formed. A more uniform particle size is obtained with foam spray drying.

The capacity of a conventional drier can be increased appreciably by using a higher total solids in the input. By removing the water ahead of the drier, the cost of operation can be decreased. Approximately twice as much product can be handled by the drier with an in-feed of 60% t.s. as compared to 42.5%.

Providing a liquid milk product at 60% t.s. requires special consideration for heat exchangers and evaporation. In general, larger heat exchanger surfaces or continuous forced flow of the product through the heat exchanger is required. As the hold time increases, the viscosity increases and makes handling difficult and dispersibility decreases. By holding the product, however, one can obtain a product with more crystalline lactose. Holding 52% t.s. condensed nonfat milk at 95°F. for 15 min. provides 7.7% crystalline lactose in the powder; for 35 min., 17.1%; for 60 min., 35%. Conventional agglomerated dry milk has from 5 to 25% crystalline lactose. The greater the percentage of crystallized lactose, the less hygroscopic is the dry product (Webb 1965).

The principle has been especially valuable for producing dry cottage cheese whey. The product was provided to the drier at 130°F. at 1800 p.s.i., but difficulty has been encountered because of sticking to the drier. A product of low density, excellent flow properties, highly hygroscopic, and rapid solubility is obtained using this method. The gas is injected at room temperature at a rate of 1 to $1^1/_2$ c.f.m. with the whey at 45% t.s. being pumped at a rate of 1 g.p.m.

For foam spray drying, with an increase in total solids, it is necessary to increase the rate of injection of compressed gas required for drying. By increasing the rate of injection, it was possible to dry up to 60% t.s. and to substantially increase the drier output.

Dry in Tall Towers Without Added Heat

Towers 200 to 250 ft. high and 50 ft. in diameter can be used for drying with dehumidified air at normal atmospheric temperatures of about 86°F. The product falls through the tower by gravity in about $1^1/_2$ min. The air used for drying is dehydrated to about 3% R.H. by using silica-gel. The air enters at the bottom and rises at $^1/_6$ to 3.0 ft. per sec. leaving the top at nearly 90% R.H. A cyclone separator is used. Up to 13,000 lb. per hr. of water can be removed and the product is dried to two to four per cent (Lang 1963). The product has a rapid solubility and good flavor because there are no high temperature heating effects (Anon. 1963).

Jet Spray Drier

Rapid drying of concentrated milk (35% solids) has been obtained with a high velocity air stream over a jet of product. An experimental unit drying 75 lb. per hr. of condensed product uses 500 lb. per hr. of 600°F. primary air and 4500 lb. per hr. of secondary air to product three to four percent moisture product. An air velocity of 1400 ft. per sec. is used in the drying tube (6 x $^2/_3$ ft.) (Ashmus 1954).

REFERENCES

ADLER, C. R., and MARSHALL, W. R., JR. 1951. Performance of spinning disk atomizers. Part I. Chem. Eng. Progr. 47, 515–522. Part II. Chem. Eng. Progr. 47, 601–608.

AMMON, R. 1956. The manufacture of powdered cream. Int. Dairy Cong. 1, No. 2, 329–331 (In French) Reviewed in Dairy Sci. Abs. 19, 35.

ANDERMATT, C., and SAUTER, K. 1957. Method and equipment for the cooling of spray-dried products. German pat. 961,967 (In German) Reviewed in Dairy Sci. Abs. 19, 910.

ANON. 1957. Cooling powder. Milk Products J. 48, No. 12, 12–13.

ANON. 1958A. New powder collector. Food Eng. 30, No. 4, 131, 133.

ANON. 1958B. Milk powder's improved dispersibility is object of this new spray dryer. Part 2. Food Eng. 30, No. 11, 113.

ANON. 1963. Latest drying techniques. Food Eng. 35, No. 2, 76–77.

ASHMUS, D. H. 1954. Momentum transport in coaxial jets in a high velocity spray dryer. Thesis for M.S., Purdue University.

BELCHER, D. W., SMITH, D. A., and COOK, E. M. 1963. Design and use of spray dryers: Part 1, Principles and applications. Chem. Eng. 70, No. 21, 83–88. Part 2, Design. Chem. Eng. 70, No. 21, 201–208.

BELL, R. W., HANRAHAN, F. P., and WEBB, B. H. 1963. Foam spray drying methods of making readily dispersible nonfat dry milk. J. Dairy Sci. 46, 1352–1356.

BLAAUW, J. 1960. Observations on dried milk obtained by the centrifugal and high pressure spray drying technique. (In Dutch) Misset's Zuivel 66, No. 49, 1123–1127.

BRADFORD, P., and BRIGGS, S. W. 1963. Equipment for food industry-3, jet spray drying. Chem. Eng. Progr. 59, No. 3, 76–80.

BROWN, G. G., et al. 1950. Unit Operations. John Wiley and Sons, New York.

BUCKHAM, J. A., and MOULTON, R. W. 1955. Factors affecting gas recirculation and particle expansion in spray drying. Chem. Eng. Progr. 51, 126–133.

BULLOCK, D. H., HAMILTON, M. O., and IRVINE, D. M. 1963. Manufacture of spray-dried cheese. Food in Canada 23, No. 3, 26–30.

CHARLESWORTH, D. H., and MARSHALL, W. R., JR. 1960. Evaporation from drops containing dissolved solids. Am. Instit. Chem. Eng. J. 6, No. 1, 9–23.

DALLA VALLE, J. M. 1948. Micromeritics, The Technology of Fine Particles. Pitman Pub. Corp., New York.

DE JUHASZ, K. J. (Editor) 1959. Spray Literature Abstracts. Am. Soc. Mech. Eng., New York.

FOGLER, B. B., and KLEINSCHMIDT, R. V. 1938. Spray drying. Ind. Eng. Chem. 30, 1372–1384.

FORTMAN, W. K. 1962. Apparatus for the acoustic treatment of liquids. Off. Gazette, U. S. Pat. Office 785, 1332. December 25 (U. S. Pat. 3,070,313).

FORTMAN, W. 1961. Spraying with sound waves. Chem. Week 89, 95–96.

FRÖESSLING, N. 1938. Gerlands Beitr. Geophys. 52, 170 (as quoted in Marshall 1954).

HANRAHAN, F. P., and WEBB, B. H. 1961. USDA develops foam-spray drying. Food Eng. 33, No. 8, 37–38.

HENDERSON, S. M., and PERRY, R. L. 1955. Agricultural Process Engineering. John Wiley and Sons, New York.

HUNZIKER, O. F. 1949. Condensed Milk and Milk Powder. Published by author, LaGrange, Ill.

LANG, FRANCIS 1963. No-heat drying plant in Switzerland. Dairy Eng. 80, 410–411.

LAPPLE, C. E. 1944. Mist and dust collection. Heating, Piping and Air Conditioning, 16, No. 7, 410–414; No. 8, 464–466; No. 10, 578–581; Mist and dust collection in industries and buildings. Heating, Piping, and Air Conditioning 16, No. 11, 635–640. 17, No. 11, 611–615, 1945. 18, No. 2, 108–113, 1946.

LAPPLE, C. E. 1954. Elements of dust and mist collection. Chem. Eng. Progr. 50, 283–287.

MARSHALL, W. R., JR. 1954. Atomization and spray drying. New York: Am. Inst. Chem. Eng. Monograph Series 50, No. 2.

MARSHALL, W. R., JR., and SELTZER, EDWARD. 1950. Principles of spray drying. I. Fundamentals of spray-dryer operation. Chem. Eng. Progr. 46,

501–508. II. Elements of spray dryer design. Chem. Eng. Progr. *46*, 575–584.

McIrvine, J. D. 1952. Atomization of viscous liquids with centrifugal pressure nozzle. Thesis for M.S., Department of Chemical Engineering, University of Wisconsin.

Mugele, R. A., and Evans, H. D. 1951. Droplet size distribution in sprays. Ind. Eng. Chem. *43*, Part 1, 1317–1324.

Nickerson, T. A., Coulter, S. T., and Jenness, R. 1952. Some properties of freeze-dried milk. J. Dairy Sci. *35*, 77–85.

Oakes, E. T., *et al.* 1963. Preparation for soluble milk powder. U.S. Patent 3,072,486.

Perry, John S. 1950. Chemical Engineers' Handbook. 3rd Edition. McGraw-Hill Book Co., New York.

Ranz, W. E., and Marshall, W. R., Jr. 1952. Evaporation for drops. I. Chem. Eng. Progr. *48*, 141–146. II. Chem. Eng. Progr. *48*, 173–180.

Salkeld, J. N. 1962. Parallel and counter-current rotation of milk and air streams in a spray drier. Aust. J. Dairy Tech. *17*, 28–29.

Scott, A. W. 1932. The engineering aspects of the condensing and drying of milk. Scotland: Hannah Dairy Research Inst. Bull. *4*.

Seltzer, E., and Settelmeyer, J. T. 1949. Spray drying of foods. Academic Press, New York, Advances in Food Research, *2*, 399–520.

Shepherd, C. D., and Lapfle, C. E. 1939. Flow pattern and pressure drop in cyclone dust collectors. Ind. Eng. Chem. *31*, 972–984.

Sjenitzer, F. 1952. Spray drying. Chem. Eng. Sci. *1*, No. 3, 101–117.

Tate, R. W., and Marshall, W. R., Jr. 1953. Atomization by centrifugal pressure nozzles. Chem. Eng. Progr. *49*, Part I: 169–174; No. 5, Part II: 226–234.

Terrett, J. P. 1960. Air heating for modern dryers. Proceedings of Milk Concentrate Conf. *4*, 50–54, Michigan State University, March.

Tsao, P. H., Schen, F., and Tai, H. 1948. Engineering reports. National Tsing Hua Univ. *4*, 115.

Webb, B. 1965. Foam spray drying. Address before the Am. Dry Milk Inst., Chicago, April 21.

CHAPTER 5

Instrumentation and Control

VACUUM PAN AND EVAPORATOR

Objective

The objective in evaporator operation is to produce (a·) a product of uniform quality, and (b) a specified product at maximum economy.

More uniformity of product is secured by using instruments and automatic control rather than manual operations. Production frequently can be increased with automatic control, particularly through adequate level control. Several instruments are utilized for automatic control in order to:

(1) Maintain uniform product boiling pressure, and thereby a constant temperature of evaporation in the vacuum pan or evaporator.· Uniform pressure avoids sudden changes in the boiling point of the product which may cause entrainment of liquid in the vapors being removed.

(2) Maintain level of liquid in the vacuum pan or evaporator to obtain maximum heat transfer efficiency and to prevent entrainment of liquid in the vapors.

(3) Maintain uniform pressure on the steam to provide uniform heating without damage to the product.

(4) Remove product when properly concentrated.

Instruments

There are many instruments which may be used to maintain desirable operating conditions. The locations of these sensing and control elements are shown in Figs. 43 and 44 for a vacuum pan and evaporator. These instruments are:

(1) The **absolute pressure controller** maintains the vacuum on the product. The pressure within the pan or evaporator may be controlled by (a) regulating the flow of water to the condenser, (b) operating the pressure creating device, such as a vacuum pump, and (c) operating the airflow into the vacuum pan space so that when air enters the volume, the absolute pressure is increased and the vacuum reduced. The operation is known as air-loading.

The absolute pressure is always measured using a perfect vacuum as the base or zero value. The "absolute" value of pressure is used which is independent of atmospheric conditions. Vacuum and gage pressure

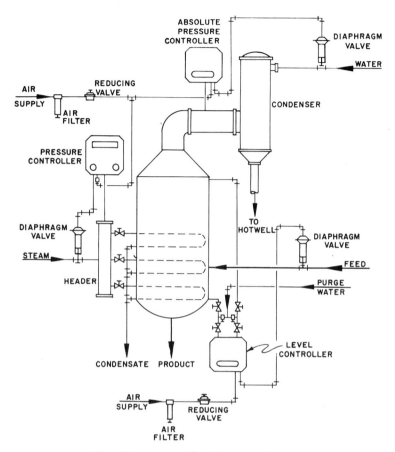

FIG. 43.   VACUUM PAN CONTROL SYSTEM

measuring instruments always measure the pressure or vacuum from the atmospheric pressure as a base.

To convert from gage pressure (above atmospheric pressure) to absolute pressure, the atmospheric pressure existing when the instrument is read is added to the gage pressure reading. For example, at sea level if the nominal value of atmospheric pressure (14.7 p.s.i.a.) existed at the time the gage was read, the absolute pressure (p.s.i.a.) is the gage pressure plus 14.7. Under vacuum conditions (sub-atmospheric conditions) the absolute pressure is the atmospheric pressure minus the vacuum. The higher the vacuum the less is the absolute pressure.

Absolute pressure measuring elements such as the bellows shown in Fig. 45 or the U-tube in Fig. 46 do not need to be compensated for in

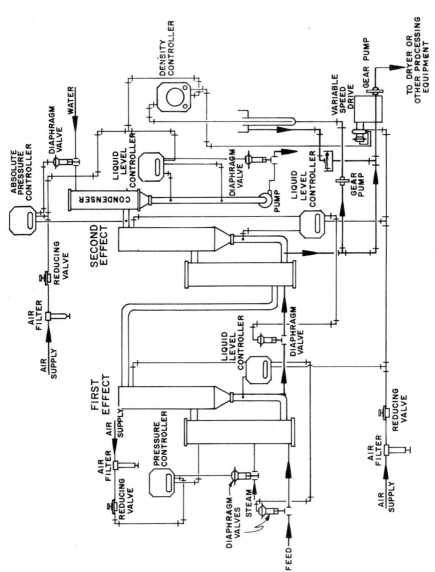

FIG. 44. CONTINUOUS EVAPORATOR CONTROL SYSTEM

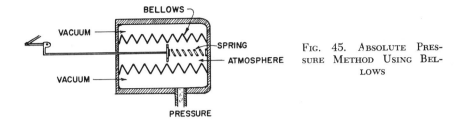

Fig. 45. Absolute Pressure Method Using Bellows

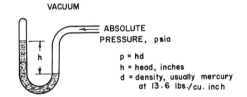

Fig. 46. Absolute Pressure Gage

atmospheric conditions because by design these units measure from a zero base (perfect vacuum). These elements are preferred for evaporator control since the boiling temperature depends on absolute pressure.

Pressure is frequently expressed in different units, equivalent values of which are: 14.7 p.s.i. = 33.9 ft. of water = 29.9 in. of Hg = 76 cm. of Hg = 760 mm. of Hg

Units of pressure must be consistent when calculating the absolute pressure from vacuum or gage pressure readings. Usual units for gage pressure are pounds per square inch gage; for vacuum—inches of mercury (in. Hg); for absolute pressure—pounds per square inch absolute.

(2) The **level controller** regulates the product feed to the vacuum pan or individual level controllers regulate the feed to each effect of a multiple-effect evaporator if the evaporator has natural or forced re-circulation. If the evaporator is of the "single pass" or "straight through" type and therefore has no recirculation, level controllers, if used, merely maintain a level on the suction side of the product withdrawal pumps.

Although an operating evaporator exhibits no clear-cut level (because of the frothing condition of the boiling liquid), there is an effective level which can be measured and controlled.

On manually controlled recirculation evaporators, the level is visually observed in a sight glass and the product feed hand valve is throttled to maintain the level.

For automatic level control, numerous approaches have been used. Some evaporators have built-in level control devices operated by a float or on the basis or differential pressure. However, the most common means of sensing evaporator level for control purposes is with a differential pressure measuring instrument. One side of the differential pressure instrument is connected near the bottom of the evaporator body and the other is connected to the vapor space to provide a reference pressure.

Two types of differential pressure instruments can be used, the "open type" which requires connecting sensing lines between the level taps and the instrument, and the "filled type" which has a sealing metal diaphragm which isolates the product from the instrument proper (Fig. 47). The open type such as aneroid manometers, differential bellows,

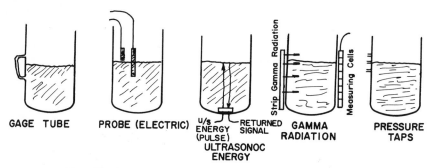

GAGE TUBE   PROBE (ELECTRIC)   U/s RETURNED   GAMMA   PRESSURE
                                            ENERGY SIGNAL   RADIATION   TAPS
                                            (PULSE)
                                            ULTRASONOC
                                            ENERGY

FIG. 47. METHODS OF LEVEL MEASUREMENT AND CONTROL

and differential pressure transmitters require an air or water purge to keep product out of the sensing lines. The "filled type" with a higher initial cost has the advantage of not requiring a purge on the "product side" because the stainless steel diaphragm can generally be mounted directly on the side or bottom of the evaporator body. This also has the advantage of being a more sanitary installation than the open type instrument with its associated purge. A normal range for an evaporator level measuring instrument would be 0 to 100 in. of water. Other devices for level measurement include electrical capacitance air purged pressure taps, load cells for weighing the products, and aneroid (mercuryless) manometers.

Aneroid manometers are commonly used for evaporator control. Brass bellows are provided for measuring the differential pressure developed by the liquid head which activates the instrument. Aneroid brass bellow manometers which operate in the pressure range from 0 to 150 in. of water are used for evaporators (Fig. 48).

(3) The **pressure element and controller** for steam on the heating side of the coils are provided to assure an adequate quantity of steam to carry out the process. A Bourdon tube or spiral element is used for

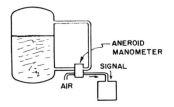

ANEROID MANOMETER
SIGNAL
AIR

FIG. 48. ANEROID MANOMETER

sensing steam pressure. The signal from the element may be used for indicating, or for controlling the pressure by throttling the steam supply valve. A pressure reducing valve from the main steam line to the heating side of the pan or evaporator may also be used. Elements are normally used in the pressure range from 10 to 35 p.s.i.g. for vacuum pans and evaporators (Fig. 49).

(4) The **product concentration** can be very adequately controlled on vacuum pans or recirculating evaporators. Control is accomplished by measuring the final product concentration and varying the discharge rate of the vacuum pan or last effect of the evaporator. In order for the product concentration control system to work properly, however, adequate automatic level control is absolutely necessary.

Three techniques for measuring product concentration for automatic control can be successfully applied. The first is based on density and is accomplished by using a bubble tube-sample column combination with the appropriate pressure measuring instrument. Since the sample column provides a constant height of liquid, the bubble tube back pressure is then a measure of the density. The second approach uses a density transmitter where the sample is run through a constant volume chamber or section of pipe and the weight of this sample is then measured by appropriate means. This is essentially the pycnometer method of

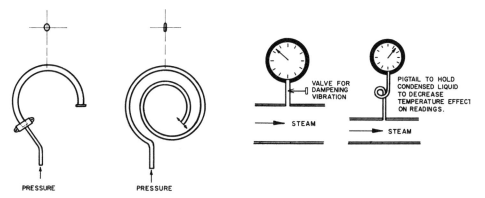

FIG. 49. BOURDON TUBE AND SPIRAL ELEMENT

density measurement. Several instruments are available based on this principle which can practically and accurately measure density for a milk evaporator. The third method is a continuous industrial process refractometer. Such a refractometer is designed to be inserted in a product discharge sample line. It has an electrical output which can easily be amplified and transduced, if desired, to be compatible with conventional pneumatic control instrumentation.

The prime advantage of the latter two product composition measuring approaches over the sample column bubble tube approach, is that they are more easily cleaned-in-place and do not require that air be purged through the product.

If the absolute pressure of the pan or final effect of the evaporator is adequately controlled, temperature compensation of the product concentration measuring system is not required. If the absolute pressure is not controlled, the product temperature for a given composition will vary—temperature compensation is then required.

The specific gravity of an evaporated or condensed (sweetened) product can be calculated by the following equation:

$$\text{S.G.}_{evap} = \frac{100\%}{\dfrac{\% \text{ fat}}{\text{S.G.}_{fat}} + \dfrac{\% \text{ S.N.F.}}{\text{S.G.}_{snf}} + \dfrac{\% \text{ H}_2\text{O}}{\text{S.G.}_{H_2O}} + \dfrac{\% \text{ Sugar}}{\text{S.G.}_{sug}}}$$

at 60°F. the S.G. of fat = 0.93
S.G. of S.N.F. = 1.608
S.G. of water = 1.00
S.G. of sugar = 1.589

The equation shows the effect of various solids on the specific gravity of the mixture. The per cent fat and per cent sugar in the evaporator feed must be constant if density of the final concentration is to have any correlation with the per cent solids-not-fat. This is also true when using refractive index as a measure of concentration.

On single pass evaporators density control by the means previously discussed does not apply since the product flow through the evaporator is controlled from the feed end. Density control can sometimes be achieved on this type of evaporator depending on the specific type of evaporator involved.

The specific gravity is often measured with the Baumé hydrometer. For products heavier than water, a specific gravity of 1 is equivalent to a Baumé reading of 0 and a specific gravity of 1.1 equivalent to 13.18 Baumé degrees. The conversion scale from Baumé to specific gravity is:

$$\text{S.G.} = \frac{145}{145 - \text{B\'e}}$$

Bé = Baumé hydrometer reading, degrees

Another method which shows promise for controlling density is based on the velocity of transmission of sound waves. Sound waves move at different speeds through different materials. Sound waves travel faster through solids than through liquids. The velocity can be correlated with the solids-not-fat and fat in milk products (Winder 1962).

Another reliable means of measuring product concentration is by refractive index. The refractometer is an optical device for determining the total solids based on the refraction of light from a sodium source or a filtered white light. A high degree of accuracy can be achieved. These units can be used for evaluation of a sample or for continuously checking the total solids of the product discharging from the evaporator. The refractive index obtained is temperature dependent so the sample must be cooled to the standard (usually 68°F.) or corrected. To convert the refractive index (RI) to total solids, T (per cent) the following equations are used (Rice and Miscall 1926):

$T = 70 + 444$ (RI $- 1.4658$) for sweetened condensed whole milk
$T = 70 + 393$ (RI $- 1.4698$) for sweetened condensed skimmilk

Refractometers may be installed between evaporators of a multiple effect system to continuously check the rate of evaporation of water.

Turbidity Meter

Evaporated water if not contaminated from the final effect of a multiple effect evaporator can be returned to the boiler or used for washing in a plant. Carry-over of the product into the exhaust vapors can occur if (a) the evaporator is started too quickly, or (b) the level of liquid in the evaporator is too high. In order to reduce the level in the evaporator, it is better to reduce gradually the water to the condenser instead of letting in air to reduce the foam. This practice will reduce carry-over.

Automatic units are available for diverting the flow if the product is contaminated. A turbidity meter can be used for direct divert. If the connection is to the boiler and the return is diverted, it is desirable that other water be directed to the boiler. The contaminated product is often referred to as "cow water."

SPRAY DRIER

The objective in control of a spray drier is to produce a product of high uniform quality without discoloration or undue loss of nutrients.

Although moisture level is critical, the moisture content is maintained indirectly by the rate, temperature, and per cent solids of feed, mass flow rate, and temperature of air. An automatic device is needed for rapidly and continuously determining the moisture content of a milk product so that corrective measures can be taken if the required product moisture content is not obtained.

There are basically three methods of automatic control, all based on directly controlling one or several of the major variables which affect the drying rate. These include: (1) constant product input with inlet air temperature control, (2) constant product input during which the outlet temperature is automatically maintained by control of the air inlet temperature (Fig. 50), and (3) variable product input in which the air inlet and the air outlet temperatures are automatically maintained and the nozzle pressure is regulated from the temperature sensed at the air outlet.

The speed of response for corrective action depends primarily on the design of drier and heating method; and secondarily on the instruments, such as reliability of the sensing elements, the transmitting means of the sensed signal, and the control valve action.

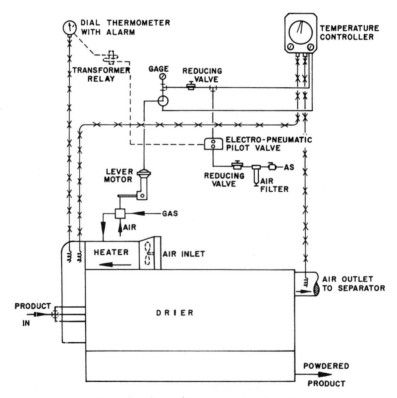

FIG. 50. SPRAY DRIER CONTROL SYSTEM

Since the drying air temperature has the greatest effect on the rate of drying (assuming relatively constant product flow rate and per cent solids), most drier control systems have either inlet or outlet air temperature control. Outlet air temperature control by changing the heat input to the drier is frequently used on driers which have a relatively high air velocity and therefore a minimum temperature lag. On driers with relatively low air velocity, it is preferable to control the inlet air temperature by varying the heat input to the drier.

When inlet air temperature control is used, the outlet air temperature can be controlled by resetting product pressure at the spray nozzles. This additional control tends to compensate for several of the secondary variables which affect drier operation and final product moisture content.

Instruments and Controls

The following instruments and controls are normally incorporated in a spray drier:

(1) **Inlet Air Temperature Controller.**—The sensing element is placed in the air stream after the heater before the drier. These units normally operate in the range of 200° to 500°F. The response for corrective action consists of a steam or gas unit inlet valve.

(2) **Outlet Air Temperature Controllers.**—A temperature sensing device is placed in the air stream leaving the drier. It normally operates in the range of 50° to 300°F. If inlet temperature control is not used, the outlet air temperature controller may provide corrective action by control of heat source through gas or steam. If inlet temperature control is used, the outlet temperature controller resets product pressure.

(3) **Pressure Inlet Pressure Controller.**—The product may be provided to the drier from the evaporator at a constant temperature, but either (a) a constant pressure, or (b) a variable pressure may be used. In the variable pressure design, the pressure is varied by the outlet air temperature controller. The controller normally operates in the range of 0 to 5000 p.s.i.

(4) **Steam or Gas Inlet Valve.**—The valve is controlled by inlet air temperature controller or outlet air temperature controller. The valve is always a variable device if automatic control is used.

(5) **Time Cycle Controller.**—The controller, according to pre-set timing, automatically operates the bag shaker, preventing products from caking onto the bag.

(6) **Temperature Recorder.**—A multipoint temperature recorder is often utilized for obtaining and recording temperatures throughout the process. Temperatures may be obtained of concentrate entering the drier, product leaving the drier, the powder leaving the cooler, inlet air, drier

hot air, redrier discharge air, and cooling inlet air. Several temperatures up to 600°F. may be recorded on a strip chart unit. Temperatures may be recorded in different time intervals, from 2 to 32 sec. intervals. Instruments may be operated with electricity, air (pneumatic), or oil (hydraulic). For air operated instruments, an air supply at 30 p.s.i.g. is required.

Drier Automatic Feed Control

The product capacity to a spray drier can be controlled automatically (Anon. 1964). The capacity may be related to the temperature of the product, concentration of the product, temperature of the drying medium, partial blocking of the nozzles, and temperature and humidity of the atmosphere. The output of the high pressure pump feeding the drier is controlled by the drier exhaust temperature. Controllers, through a series of hydraulic linkages, maintain a preset drier exhaust temperature. The airflow on a weight basis through the drier is kept constant. Generally the air inlet temperature is kept constant.

PSYCHROMETRY

The psychrometric or humidity chart is useful for analyzing the operation of a drier, as well as for developing theoretical aspects of drying (Fig. 51). Data involved in psychrometric calculations are concerned with the external environment of the product. The psychrometric chart can be useful for determining:

(1) **Air flow** in weight or volume per unit time, such as pound per minute, or cubic feet per minute.

(2) **Moisture carrying** capacity of the air entering and leaving the drier, pounds of water per pound of air.

(3) **Quantity of heat** to be added to bring the air to a desired temperature, B.t.u. per pound.

(4) **Vapor pressure** of the air, which can be related to the vapor pressure of the product.

Terminology and Explanations

The psychrometric chart provides the thermodynamic properties of the air and vapor (Fig. 52 between pp. 106 and 107). By knowing two of the three, dry bulb, wet bulb, or dew point temperature, the physical properties of the air can be obtained from the psychrometric chart. Terminology relating to the psychrometric chart include:

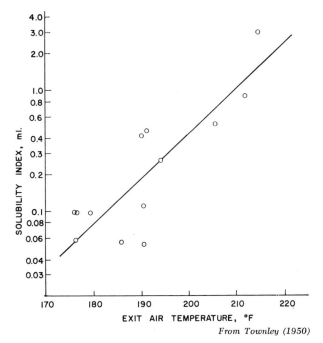

From Townley (1950)

Fig. 51. Relationship Between Solubility Index and Exit Air Temperature

(1) **Dry Bulb Temperature.**—The temperature of the air or product which is not affected by the amount of water vapor in the air as indicated by a thermometer.

(2) **Wet Bulb Temperature.**—The temperature given by a thermometer with its sensing bulb covered with a thin layer of water when dynamic equilibrium exists between heat and mass transfer of the air and water. The wet bulb temperature may be obtained by moving the thermometer bulb, covered by a wick, through the air or by moving air over the wick. An air velocity of at least 900 f.p.m. is required for accurate readings although lower velocities are used and corrected. The difference between the dry bulb and wet bulb temperature is known as the **wet bulb depression.**

(3) **Dew Point Temperature.**—The temperature to which an air-water mixture must be cooled to obtain saturated air. It may also be described as the temperature at which condensation of water vapor will begin if the temperature of the air-water vapor is reduced.

(4) **Humidity.** Humidity is also known as absolute humidity and

specific humidity. It is the weight of water vapor in pounds per pound of dry air (7000 grains = 1 lb.).

(5) **Saturated Air.** Air at a given temperature which contains all of the water vapor it can hold. A decrease in temperature will cause water vapor to be removed from the air.

(6) **Relative Humidity.** The ratio of the partial pressure of water vapor in the air to the vapor pressure of saturated air at the same temperature, usually expressed as a per cent.

(7) **Humid Heat.** The quantity of heat in B.t.u. required to increase the temperature of 1 lb. of air with water vapor through 1°F. In usual terminology, this would be known as the specific heat of 1 lb. of air including the water. The specific heat of dry air is 0.24 and of water vapor is 0.46 B.t.u. per lb. at room temperature and standard pressure.

(8) **Humid Volume.** The total volume in cubic feet of 1 lb. of dry air plus the water vapor it contains at given conditions.

(9) **Vapor Pressure.** The partial pressure exerted by the vapor in the air or in the product usually designated in p.s.i. or mm. Hg (millimeters of mercury).

(10) **Total Enthalpy.** The total heat contained in 1 lb. of air plus the water it contains, usually designated above 32°F. The enthalpy is the sensible heat of dry air, plus the sensible heat of the water vapor, plus the latent heat of the water vapor.

Equations are available for relating and calculating values for the above terms. However, the psychrometric chart is a convenient source of the given relationships. The psychrometric chart is based on atmospheric conditions at sea level.

Psychrometric Relationships

It is helpful to note the following relationships regarding the psychrometric chart:

(1) The dry bulb, wet bulb, and dew point temperatures are equal when the relative humidity is 100%.

(2) The dry bulb temperature is greater than the wet bulb which is greater than the dew point when the relative humidity is less than 100%.

(3) The rate at which heat is transferred from the air to the water is proportional to the wet bulb depression.

(4) The water vapor pressure nearly doubles for each 20°F. increase in temperature.

(5) The density (pound per cubic feet) of saturated air on the basis of dry air is less than the density of dry air at a given temperature.

(6) The difference between the dew point and dry bulb temperature is nearly constant for a given relative humidity.

(7) The latent heat of vaporization increases as the temperature of evaporation decreases.

(8) The dew point of a given air remains the same as the air is heated. Likewise, the vapor pressure remains constant.

Adiabatic Conditions

Adiabatic conditions are usually assumed, at least in the initial stages of calculation, in analyzing drying system. In an adiabatic system, all sensible heat given up by the air is considered to be used for vaporization of water from the material. Thus, the enthalpy or total heat of the air would remain constant after it leaves the heater until it leaves the drier. The line on the psychrometric chart representing the change of conditions as the air passes through the drier is parallel to the wet bulb lines (see Fig. 51).

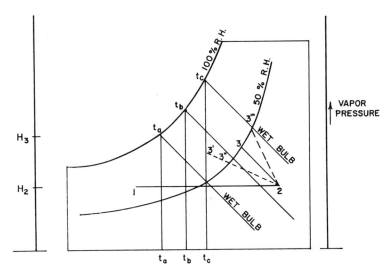

FIG. 53. CHANGE OF AIR CONDITION DURING DRYING

1. Ambient air entering
2. Heated air entering drier
3. Partially saturated air leaving drier

Assuming adiabatic conditions and by referring to Fig. 53, ambient air entering the heater at point 1 is heated to temperature point 2 at a constant vapor pressure. The heated air entering the drier then changes conditions as shown going from point 2 to point 3, parallel to the wet bulb lines. The example represents the air leaving the system at 50% R.H. If the air left in a saturated condition, the 2-3 line would extend to the 100% R.H. line in Fig. 53. The quantity of water removed per pound of dry air is represented by the difference between H_3 and H_2.

In practice, adiabatic conditions do not exist because some heat is lost which is not available for vaporization. Thus, the dotted line from 2 to 3″ would more nearly represent the change of air conditions. If the air left the drier at 50% R.H., a smaller amount of moisture would be carried away, or if the same amount of moisture were removed by the air, the humidity would have to be higher, as shown by point 3 in Fig. 53. If heat was added by the product (which is not likely for spray drying, but could happen if the product entered at a temperature above the air), a line from point 2 moving toward point 3 would be above the solid line (to point 3‴) and more moisture could be carried away with air leaving at 50% R.H. than for the examples cited.

Rate of Drying

With a constant inlet temperature, the outlet temperature, t_3 provides a measure of the rate of drying. The lower the temperature and the higher the relative humidity, the greater the rate of moisture carried away in the exhausted air.

The difference in vapor pressure between the air and the product being dried in the air may be used to determine the rate of drying of the product. The psychrometric chart is useful for determining the vapor pressure of water in the air, v_{pa}. The drying rate, R, during the initial constant rate period of drying is:

$$R = \frac{-dM}{A\,dt}$$

M = moisture content, gm. per gm. of dry matter, $R = 0.00433\,v^{0.8}(v_p - v_{pa})$
R = rate of drying, gm. per hr. sq. cm.
v = velocity of air, m. per sec.
v_p = partial pressure of water at drying surface, mm. Hg
v_{pa} = partial pressure of water in air, mm. Hg
A = drying surface area per unit mass of dry solid, sq. cm./gm.
v_{pa} = (R.H.) (vp_s)
v_{ps} = vapor pressure at saturation, mm. Hg

The equation is difficult to apply because the vapor pressures of the air and of the product are changing and approaching each other. However, the equation can be useful for estimating the rate of drying if the average difference in vapor pressure is known or can be estimated for the product and the air.

The vapor pressure of the air can be calculated by multiplying the relative humidity times the pressure at saturation at a given temperature (Table 15). The vapor pressure of the drying air is constant until heat is given up for drying. The calculations are often useful because for the

TABLE 15

VAPOR PRESSURE OF SATURATED AIR AT VARIOUS TEMPERATURES, v_{ps}
Barometric pressure = 14.7 p.s.i.a. = 0 p.s.i.g. = 760 mm. Hg = 1 atm.

Temperature, °F.	Pressure, p.s.i.a.	Temperature, °F.	Pressure, p.s.i.a.	Temperature, °F.	Pressure, p.s.i.a.	Temperature, °F.	Pressure, p.s.i.a.
32	0.08854	75	0.4298	134	2.4712	218	16.533
33	0.09223	76	0.4443	136	2.6042	220	17.186
34	0.09603	77	0.4593	138	2.7432	222	17.861
35	0.09995	78	0.4747	140	2.8886	224	18.557
36	0.10401	79	0.4906	142	3.0404	226	19.275
37	0.10821	80	0.5069	144	3.1990	228	20.013
38	0.11256	81	0.5237	146	3.365	230	20.780
39	0.11705	82	0.5410	148	3.547	232	21.567
40	0.12170	83	0.5588	150	3.718	234	22.379
41	0.12652	84	0.5771	152	3.906	236	23.217
42	0.13150	85	0.5959	154	4.102	238	24.080
43	0.13665	86	0.6152	156	4.306	240	24.969
44	0.14199	87	0.6351	158	4.519	242	25.884
45	0.14752	88	0.6556	160	4.741	244	26.827
46	0.15323	89	0.6766	162	4.971	246	27.798
47	0.15914	90	0.6982	164	5.212	248	28.797
48	0.16525	91	0.7204	166	5.461	250	29.825
49	0.17157	92	0.7432	168	5.721	252	30.884
50	0.17811	93	0.7666	170	5.992	254	31.973
51	0.18486	94	0.7906	172	6.273	256	33.093
52	0.19182	95	0.8353	174	6.565	258	34.245
53	0.19900	96	0.8407	176	6.868	260	35.429
54	0.20642	97	0.8668	178	7.183	262	36.646
55	0.2141	98	0.8935	180	7.510	264	37.897
56	0.2220	99	0.9210	182	7.850	266	39.182
57	0.2302	100	0.9492	184	8.200	268	40.502
58	0.2386	102	1.0078	186	8.567	270	41.858
59	0.2473	104	1.0695	188	8.946	272	43.252
60	0.2563	106	1.1345	190	9.339	274	44.682
61	0.2655	108	1.2029	192	9.746	276	46.150
62	0.2751	110	1.2748	194	10.168	278	47.657
63	0.2850	112	1.3504	196	10.605	280	49.203
64	0.2951	114	1.4298	198	11.058	282	50.790
65	0.3056	116	1.5130	200	11.526	284	52.416
66	0.3164	118	1.6006	202	12.011	286	54.088
67	0.3276	120	1.6924	204	12.512	288	55.300
68	0.3390	122	1.7888	206	13.031	290	57.556
69	0.3509	124	1.8897	208	14.568	292	59.356
70	0.3631	126	1.9955	210	14.123	294	61.201
71	0.3756	128	2.1064	212	14.696	296	63.091
72	0.3886	130	2.2225	214	15.289	298	65.028
73	0.4019	132	2.3440	216	15.901	300	67.013

Source: Keenan and Keyes, (1936).

higher temperatures involved in spray drying psychrometric charts are difficult to read or do not include the necessary information.

Equilibrium Moisture Content

The partial pressure of water at the drying surface can be determined from the equilibrium moisture content values. Obviously, these values change as the product is being dried. The equilibrium moisture content is the moisture content of a product in a given relative humidity and temperature atmosphere when the rate of moisture loss from the product is equal to the rate of moisture gained by the product from the surrounding air. Thus, at 176°F. and 50% R.H., the equilibrium moisture content is 2.4% wet basis of spray dried product (Table 16). Values of the

TABLE 16

VAPOR PRESSURE AND MOISTURE CONTENT, w.b., OF MILK AND MILK PRODUCTS AT VARIOUS
TEMPERATURES AND RELATIVE HUMIDITIES

Product	Temperature, °F	Relative Humidity, %							
		10	20	30	40	50	60	70	100
Milk powder	50[1]	2.7	3.0	3.4	4.8	7.0	6.5	7.6	
	[2]	0.018	0.036	0.054	0.072	0.089	0.107	0.124	0.178
Full cream	99[1]	2.6	3.3	4.1	4.0	4.5	6.5	7.9	
	[2]	0.09	0.18	0.27	0.37	0.46	0.55	0.64	0.921
Spray dried	176[1]	1.2	1.8	1.6	2.3	2.4	4.3	7.7	
	[2]	0.687	1.37	2.06	2.75	3.43	4.12	4.81	6.868

Source: Morris (1947).
[1] Moisture content, per cent.
[2] Vapor pressure, p.s.i.

equilibrium moisture content of whole milk, powdered milk and milk with different heat treatments are presented in Table 16. For the example cited, the vapor pressure would be equal to 50% times 6.868 p.s.i. the vapor pressure at saturation, or 3.48 p.s.i. Thus, by using 176°F. air, 50% R.H. for drying, the moisture content of the product would approach 2.4% and could not be expected to go lower. Similar relationships hold for the other equilibrium moisture content values.

The equilibrium moisture content of a product when plotted against the relative humidity produces an S-shaped or sigmoid curve (Fig. 54). As the relative humidity increases the moisture content of the product increases. As the temperature of the product is increased at a constant relative humidity, the moisture content of the product decreases.

In tropical areas where the temperature may be 100°F. and the drying air is nearly saturated, difficulty may be encountered in drying milk powder to the desired moisture content at the capacity normally expected. This difficulty can be predicted, knowing that the vapor pressure at 100°F. is 0.949 p.s.i. as compared to 0.507 p.s.i. at 80°F. To dry the product, (a) the temperature of the outlet may be increased, which in effect reduces the equilibrium moisture; (b) the temperature of the inlet may be increased so that more heat is available for vaporization; or (c) the feed rate may be reduced, which is effective because it permits a higher temperature and the air in the drier is at a lower relative humidity.

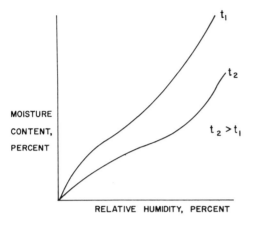

t_1

t_2

MOISTURE CONTENT, PERCENT

$t_2 > t_1$

RELATIVE HUMIDITY, PERCENT

Fig. 54. Equilibrium Moisture Content Relationships

Another approach would be to dehumidify entering air, which is usually not done. Another possibility is to introduce dry air in the final stages of drying through a secondary drier to remove 1 to 2% moisture. This may not be satisfactory because the partially dried product may stick to the drier. The moisture of the powder at the outlet is controlled by the absolute humidity of the outlet air. For a direct-fired unit, from 10 to 15% of the water in the exhaust air may come from the combustion of fuel.

Moisture of Dry Milk

An analysis of moisture content is necessary for quality control of dry milk. The moisture is of prime importance in determining the storability of the powder, the moisture limit being normally below 5%. The moisture content of dry milk is usually designated on a wet basis; that is, for 5% moisture content, there would be 5 gm. of moisture in a 100 gm. wet sample.

Although the term, moisture content, is quite often used, there is no agreement as to what is included as moisture. Water exists in the dry milk product in several forms. Water may be absorbed on the surface of the particles of powder; water may be bound in the crystals of lactose; and water may be imbibed into the colloidal milk protein. Various methods of determining the moisture give different values, primarily because all or part of the water contained in the product may be measured. One can argue that all water in the product should be included in the moisture content. On the other hand, the keeping quality is influenced greatly by the surface moisture. Inasmuch as the time required to make the determination is often critical from the standpoint of processing operations, it seems logical to use a quick method which is reproducible. Such a method will probably be based on the surface moisture content.

The values obtained can be adjusted slightly if the total moisture content is desired for a purpose other than storage. The alpha-lactose hydrate contains about five per cent water of crystallization which would not be measured by some methods of moisture determination such as the toluol method.

Determining Moisture Content

Direct Method.—The moisture content is usually determined by a direct method in which the vapor pressure of the moisture in the product is greater than the surrounding vapor pressure, thus causing moisture to be removed. The loss in weight of the product or the moisture removed can be weighed to determine the per cent moisture. The vapor pressure of the product is normally increased by using heat; either from air, radiation, or from a warm oil which will absorb moisture.

Toluol Distillation.—The toluol distillation method is recommended by U. S. Department of Agriculture (USDA) and the American Dry Milk Institute (ADMI). The toluol distillation method has four characteristics (Thompson and Fleming 1936): (1) enables use of large sizes of samples; (2) each determination can be completed in 1 hr.; (3) it is readily adapted for plant control work; (4) it is as accurate as the vacuum oven method.

Toluol ($C_6H_5CH_3$) is a chemical which has a boiling point of 233°F. Heat is supplied to the toluol, adjusting the amount of heat so that the rate of moisture removal does not exceed four drops per second. The vapor leaving is condensed and collected in a graduated trap. A 50 gm. sample is placed in a 300 ml. Erlenmeyer flask and heated with about 100 ml. of toluol. After 45 min., the quantity of water collected is measured, and then checked again at 60 min., and if the two readings agree within 0.05 ml., the distillation is discontinued and the moisture percentage determined. If, however, the reading is greater than 0.05 ml., heating is continued for another 15 min. and repeated until readings are within the desired limits. The milliliters to the nearest 0.05 of water collected multiplied by two equals the percentage of moisture in the sample.

To eliminate some of the difficulties encountered in using the standard toluol method described above, Kennedy and Stribley (1956) made the following recommendations: (1) use standard taper ground joints—24/40 to avoid possibility of toluol leakage; (2) use a 500 ml. round bottom boiling flask instead of the flat Erlenmeyer flask; (3) use a heating mantle and a variable transformer to control the heat to obtain uniform heating.

The toluol method is rejected by some (Moir 1956) as being too expensive and as dangerous.

Karl Fischer.—The Karl Fischer method utilizes a chemical means for determining the moisture content. Chemical solvents are used which penetrate the cells and aid in quickly removing water from the material. The material must be finely ground. The moisture is extracted with anhydrous methyl alcohol. The Karl Fischer method removes much of the water of hydration. If 50% of the powder was alpha-lactose, the Karl Fischer method would give a moisture content 2.5% higher than by using heating. (See Official Methods of Analysis for more details.)

Air Oven.—An air oven is often used for determining moisture content. One gram of a sample is weighed into a flat bottom metal dish, 5 cm. in diameter, and placed in a vacuum oven at 212°F. with a pressure not above 100 mm. Abs. of Hg. Up to 5 hr. may be required for heating until the sample reaches a constant weight. British technologists recommend 214° to 215°F.

To eliminate the variations when drying with an air oven; because of fluctuation in temperature, air velocity, location of the sample, and to obtain quicker results, the Meihuizen method has been used in the Netherlands. The Meihuizen method uses predried air to avoid some of the variables due to the difference in ovens. Air at 210° ± 2°F. is

TABLE 17

COMPARISON OF THE RESULTS OF DETERMINATIONS OF THE MOISTURE CONTENT OF DRY MILKS

Type of Dried Milk	Moisture Content Found by Method of, %							
	Fischer		Pregl		Meihuizen		Ordinary Drying Oven	
		Mean		Mean		Mean		Mean
Spray dried whole milk	2.49		2.34		2.49		1.96	
	2.51	2.48	2.36	2.37	2.46	2.43	2.01	1.99
	2.47		2.40		2.40		1.97	
	2.43		2.38		2.38		2.02	
Spray dried skimmed milk	3.58		3.57		3.57		3.15	
	3.64	3.63	3.57	3.58	3.52	3.57	3.25	3.30
	3.68		3.60		3.58		3.40	
	3.63		3.56		3.59		3.38	
Roller dried whole milk	2.55		2.46		2.33		2.22	
	2.61	2.59	2.50	2.50	2.34	2.34	2.28	2.30
	2.64		2.50		2.34		2.36	
	2.57		2.52		2.36		2.34	
Roller dried skimmed milk	4.82		4.55		4.79		4.34	
	4.89	4.84	4.68	4.62	4.71	4.74	4.33	4.37
	4.80		4.56		4.77		4.33	
	4.84		4.68		4.69		4.48	
s	0.039		0.042		0.039		0.078	
d			−0.12		−0.12		−0.40	

Source: Kruisheer and Eisses (1956).
s = the standard deviation for the singular determination; d = the mean of the differences between the value for the Fischer method and those obtained by respectively the Pregl, the Meihuizen, and the ordinary drying oven method.
By statistical analysis all these differences appeared to be significant at 5% level.

utilized, which has been passed through silica gel to reduce the moisture content. The sample is dried in about 2 hr. (Kruisheer and Eisses 1956). A comparison of moisture contents of the same sample using different methods is given in Table 17.

Rapid Methods.—Quicker methods are often used in the plant to get a check of the moisture. These determinations should be compared periodically with one of the standard methods mentioned above. The air vacuum oven is often utilized to get more rapid drying with a minimum of oxidation to the product. Heating the sample with an infrared lamp for a specific time, usually 20 min., is one method commonly used. Instant powders appear to present further problems in moisture determinations because of the variation in moisture held by the lactose.

Standard Error.—The standard error obtained by various methods over the range of from two to five per cent moisture content are as follows: the Karl Fischer method ±0.037%; Meihuizen ±0.039; Pregl ±0.042; drying ovens ±0.07.

Selection.—The main two criteria for selecting a method for determining moisture content are: (1) time for making determination; and (2) accuracy required. There are no new rapid methods of making moisture content determinations which are as accurate as those methods which require more time. Cost has not been an important consideration in selection of the methods. An inexpensive, accurate method which would provide the moisture content in a few minutes is needed.

REFERENCES

AMERICAN DRY MILK INSTITUTE. 1962. Standards for grades for dry milk industry, including methods of analysis. Bull. *916,* 221 N. LaSalle Street, Chicago 1, Ill.

ANON. 1962. Progress in process control. Food Eng. *34,* No. 6, 67–83.

ASSOCIATION OF OFFICIAL AGRICULTURAL CHEMISTS. 1960. Official Methods of Analysis. Washington, D.C.

CONSIDINE, D. M. 1956. Process Instruments and Controls. McGraw-Hill Book Co., New York.

HARPER, J. M. 1962. An apparatus for measuring bulk density of dried milk. Food Technol. *16,* No. 9, 144.

HODGMAN, C. E. 1952. Handbook of Chemistry and Physics. Chemical Rubber Publishing Co., Cleveland, O.

KEENAN, J. H., and KEYES, F. G. 1936. Thermodynamic Properties of Steam. John Wiley & Sons, New York.

KENNEDY, J. G., and STRIBLEY, R. C. 1956. A suggested modification of the toluene distillation method for moisture determination in milk powders. J. Dairy Sci. 39, No. 9, 1327.

KITZES, A. S. 1947. Factors influencing the design and operation of a spray drier. Thesis for Ph.D., Dairy Industries Department, University of Minnesota.

KRUISHEER, C. I., and EISSES, J. 1956. Determination of the moisture content of dried milk and other milk products by means of the apparatus of Meihuizen. XIV Int. Dairy Congress III, No. 2, 212–220 (Rome).

KUMETAT, K. 1955. Notes on moisture determination in milk powders. Aust. J. Dairy Technol. *10*, 114–115.

MOIR, G. M. 1956. The analysis of dried milk. XIV Intern. Dairy Congress III, No. 2, 307–316.

MORRIS, T. N. 1947. The Dehydration of Food. D. Van Nostrand Co., New York (also London).

OHMART, P. E. 1956. Radioactive gauging and process control. Proc. Nat'l Dairy Eng. Conf. *4*, 59–60.

RICE, F. E., and MISCALL, J. 1926. Sweet and condensed milk. IV. A refractometer method for determining total solids. J. Dairy Sci. *9*, 140–152.

SIMMONS, T. S. 1964. Automatic feed control. Manufactured Milk Prod. J. *55*, No. 12, 8.

SOROKIN, Y. 1955. Automatic control of the concentration of milk. Mol. Prom. (U.S.S.R.) *16*, 38–39. (Reviewed in Dairy Sci. Abs. *17*, 742.)

TAYLOR INSTRUMENT COMPANY. 1950. Catalog 300-A, June, Rochester 1, New York.

THOMPSON, E. C., and FLEMING, R. S. 1936. The determination of moisture in powdered milk by the toluol distillation method. J. Dairy Sci. *19*, 553–559.

TOWNLEY, V. H. 1950. The operation of a spray drier at high temperatures and under pressure. Thesis for M.S., Dairy Industries Department, University of Minnesota.

WINDER, W. C. 1962. An ultrasonic method for measuring the fat and solids-not-fat content of milk. Proc. Nat'l Dairy Eng. Conf. *10*, 108–111.

CHAPTER 6

Processing, Packaging, and Storage of Evaporated, Condensed, and Sweetened Condensed Milks

INTRODUCTION

Milk concentrated under vacuum to approximately 2:1 in solids, packaged and sterilized in retail size cans is known in the dairy industry as evaporated milk. In recent years a product identical in composition, which has been sterilized by the ultra high temperature method (UHT) and then aseptically packaged in cans, has been marketed as evaporated milk in relatively small volume.

Condensed whole milk and condensed skimmilk are common industry terms that refer to products with increased solids content, achieved by the removal of water under a vacuum. These products may be identified with their solids content, for example, 40% condensed skimmilk. They are packaged in large containers or handled by bulk means for industrial uses in the nonsterile condition.

Sweetened condensed whole milk and sweetened condensed skimmilk have sugar added as a preservative. These products are not sterilized. Sweetened condensed milk may be packaged in retail cans, large containers, or transported in bulk form. Sweetened condensed skimmilk is not packaged in retail size containers.

In the last few years, sterile concentrated and frozen concentrated milks (3:1) have been used to distinguish products processed for reconstitution later in the home for beverage purposes. Efforts have been exerted to keep the "cooked" flavor to a minimum by ultra high temperature sterilization and aseptic packaging. The frozen concentrated milk may or may not be sterilized.

PRODUCTION

Evaporated and condensed milk utilized 4,011,000,000 lb. (3.4%) of 1969 milk production in the United States. In 1960, the utilization was 4.4; 1955, 5.1; 1950, 5.9; and 1945, 7.4%. Following this same trend, production of evaporated milk has declined during the last two decades (Table 18). The major producing areas are the east north central and the east south central (Kentucky and Tennessee) states. More specifically, Ohio was the leader in 1968, with Kentucky, California, Pennsylvania, and

116

Tennessee next in decreasing order. These five states produced half of the evaporated milk.

Condensed whole milk production (bulk) in 1969 was 239,590,000 lb. 15% below 1968. New York produced 29% of the total. Other major states were Maryland, Wisconsin, Pennsylvania, and California.

The amount of sweetened condensed whole milk packaged in cans during 1969 totaled 85,000,000 lb. while bulk product was 62,000,000 lb. Wisconsin manufactured 33,240,000 (50%, 1969) or approximately one-half of the total. The production of bulk sweetened condensed skimmilk was 74,529,000 lb. In this category, Wisconsin led with 53,316,000 lb., and Pennsylvania was second with 5,997,000 lb.

MILK SUPPLY

Generally, manufacturing grade milk is used for evaporated milk, condensed and sweetened condensed whole milk, or skimmilk. Delivery of milk from the farm usually is daily if cans are used and every other day for the farm tank system. Regardless of the method, milk should arrive at the plant in good condition—fresh, sweet (without significant acid development), free of abnormalities and extraneous material. The raw milk should meet the quality and sanitary standards of the Evaporated Milk Association (1962).

When milk is hauled by the tanker system, it should be checked for amount purchased on the farm and amount delivered to the plant. Total net weight may be verified by weighing the tanker, weighing by load cells on storage tanks, or with a meter in the unloading line. Representative samples should be taken for the laboratory analysis. Storage of the raw milk is recommended at 40°F. or less. Processing should take place as soon as practical, because the quality of the raw milk has a direct influence on the quality and storage life of the evaporated or condensed milks produced from it.

EVAPORATED MILK

Heat Stability

Ideally, milk for evaporated milk should have no developed acidity, but the maximum acceptable range extends from 0.18 to 0.22%, depending on initial titratable acidity. Heat stability of the milk is important if the product is sterilized during manufacture. The alcohol test is a common method for testing heat stability of the milk proteins. This method consists of adding 70% alcohol to an equal volume of milk in a test tube (2 ml. of each) and then mixing vigorously. Coagulation indicates the

TABLE 18

THE PRODUCTION OF EVAPORATED, CONDENSED, SWEETENED CONDENSED MILKS, CONDENSED BUTTERMILK, AND CONDENSED WHEY
(add 000 lb.)

	Unsweetened Evaporated Milk		Unsweetened Condensed Milk		Sweetened Condensed Milk			Condensed or Evaporated Buttermilk	Condensed Whey
	Case Goods Whole	Skimmed	Bulk Whole	Bulk Skimmed	Case Goods Whole	Bulk Whole	Bulk Skimmed		
1969	1,413,785	...	239,590	918,754	85,000	62,419	74,529	392	179,818
1968	1,359,958	22,108	284,991	857,043	87,428	67,270	69,254	1,342	158,347
1967	1,493,166	49,994	264,644	871,925	64,376	63,562	62,505	3,054	177,767
1966	1,696,143	10,481	294,921	974,149	128,597	65,154	61,216	2,015	193,628
1965	1,692,974	10,405	324,363	903,168	95,948	64,484	53,551	2,184	150,982
1964	1,888,060	10,393	351,886	820,207	94,623	59,568	57,927	5,245	168,282
1963	1,897,278	11,365	334,012	779,133	78,956	58,704	55,837	6,017	114,235
1962	1,928,834	11,747	354,419	817,613	74,062	51,347	56,742	4,748	85,631
1961	2,117,467	11,995	394,374	803,287	69,837	50,125	52,236	6,564	99,176
1960	2,177,267	11,163	373,234	727,730	67,830	47,771	56,101	8,359	86,960
1955	2,579,831	...	266,495	647,710	33,681	42,124	88,392	62,852	...
1950	2,882,475	...	221,724	523,182	61,973	39,210	210,637	183,470	...
1945	3,776,383	...	128,875	454,443	143,306	77,016	519,188	160,441	...
1940	2,464,668	...	128,017	246,910	61,955	76,138	166,017	111,842	...
1935	1,838,890	...	102,833	164,372	52,985	36,907	133,417	70,543	...

Sources: U.S. Agr. 1944, 1949, 1956, 1961, 1965, 1966, 1967, 1968, 1969, 1970.

milk is susceptible to coagulation during the heat treatment of sterilization.

Milk with substandard stability to the sterilization temperature may be improved with sodium citrate or disodium phosphate at the rate of 1 to 6 oz. per 1000 lb. of milk. Infrequently, the stability is restored by a calcium salt, e.g., calcium chloride.

Manufacturing Procedure

Standardization.—Figure 55 illustrates the steps in the manufacture of evaporated milk. Standardization of the milk fat to the solid-not-fat ratio

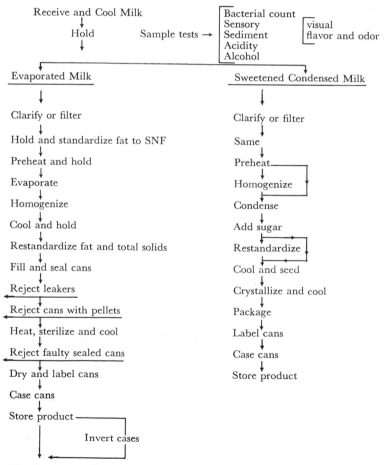

FIG. 55.   SCHEMATIC DIAGRAM OF EVAPORATED MILK AND SWEETENED CONDENSED
MILK PROCESSING

is 1:2.2785 for evaporated milk. Accurate weights and tests of the fat and solids-not-fat are necessary. The Mojonnier method is used in plants for testing fat and total solids. Restandardization to 7.92% fat and 25.95% t.s. after condensing and just before canning is warranted to maintain efficient control and yet not be below the federal standards of 7.9 and 25.9%. A few state standards are slightly less, but none are below 7.7% milk fat and 25.5% T.S. (U.S. Dept. Agr. 1964).

Analysis.—The usual composition of evaporated milk is: fat, 7.9 to 8.0%; t.s., 25.9 to 26.3%; and moisture, 73.6 to 74.1%. The average lactose is 9.0% with protein 6.5% and ash 1.3%. During 1964 and 1965, approximately 84% of the evaporated milk analyzed in the U.S. Dept. Agr. Inspection and Grading Laboratory had a milk fat content of 7.9 to 8.0% and 79% had total solids between 25.9 and 26.3%.

Clarification or Filtering and Preheating.—Clarification of preheated milk may be before or after standardization. Although not essential, it is a common practice for processing good quality evaporated milk. In some plants filtration of the milk is used instead of clarification. Purpose of preheating mainly is to improve stability against coagulation during sterilization, and to maintain a medium viscosity of the finished product. Preheating also increases the temperature of the milk for entry into the evaporator. Later, sterilization will complete the destruction of enzymes and microorganisms which was partially accomplished by the preheating. Temperature and time of preheating should be adjusted to a number of factors such as seasons of the year and feeds, breed of cows, etc. The range of preheating temperature commonly is 200° to 205°F. for 10 to 20 min. or 240° to 260°F. for 1 to 6 min. A combination of preheating in both ranges has been practiced also.

Evaporation of Moisture.—Continuous evaporation follows preheating. The amount of moisture removal from the product is controlled by the determination of the total solids with a Baumé hydrometer, or other instrumentation. The per cent solids and Baumé degrees for several temperatures are given in Fig. 56. A customary practice is to concentrate slightly more than desired in the final product, for greater ease of standardization later. The product is pumped continuously from evaporator to homogenizer.

Homogenization.—Next and important is homogenization. Evaporated milk must be homogenized thoroughly to obtain a uniform fat emulsion and reduce separation of fat to a minimum during storage. A temperature and pressure combination is needed that will reduce the globules to less than 1 μ in diameter. The temperature is usually about 120°F. as the

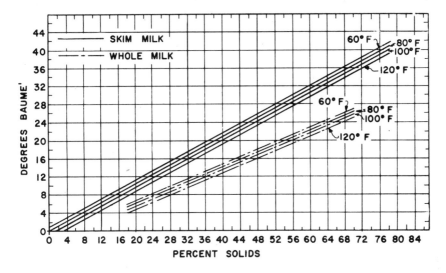

FIG. 56.   SOLIDS VS. BAUMÉ READING AT 60°, 80°, 100°, AND 120°F.

product is removed from the last effect of the evaporator. It may go directly to the homogenizer or receive additional heating. A pressure of approximately 2000 to 3000 p.s.i., in combination with a sufficient temperature, will obtain the desired reduction in size of fat globules.

Restandardization.—After homogenization, the evaporated milk is cooled to 45°F. The product is held in large storage tanks for restandardization of milk fat and/or total solids by the addition of water, skimmilk, condensed skimmilk and/or homogenized cream. Optional fortification with vitamin D concentrate is at the rate of 25 U.S.P. units per fluid oz. Also, the stabilizing salt may be added at this point if pilot tests indicate a need. A laboratory or pilot size sterilizer is used as a guide in determining the amount of stabilizer required. The amount changes with the change of seasons.

Canning.—Packaging of evaporated milk has been in cans of tin coated sheet steel. Popular retail sizes are 14¹/₂ oz. (13 fl. oz.) and 6 oz. (5¹/₃ fl. oz.). The volumetric fillers depend on a uniform temperature and head pressure of product for accurate control of fill. Common fillers have 48 or 60 individual filling cells and each one can be adjusted for volume of evaporated milk into the can. The fill tolerance for a good operation is within one per cent for 6 oz. cans and slightly less for 14¹/₂ oz. cans. One type of machine fills through a small hole in the top. The hole is then mechanically sealed with solder. Another type fills the

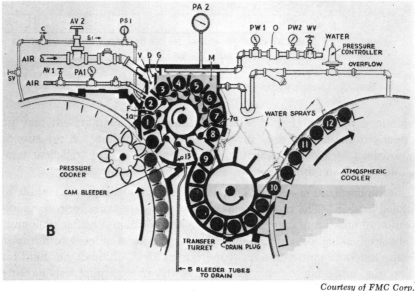

FIG. 57. EVAPORATED MILK STERILIZER AND A CROSS-SECTIONAL DIAGRAM OF THE CAN MOVEMENT

can and the lid is sealed on to it. Leak detectors reject the occasional improperly sealed can. An instrument may be included in the line to automatically reject cans containing solder pellets in the product.

Sterilization.—The cans are conveyed continuously through a retort for sterilization. During forward progress, the evaporated milk is heated rapidly to 240° to 245°F. with hot water and steam; held for 15 min. and cooled with cold water to approximately 80° to 90°F. Both agitation of product by means of can movement, and careful temperature control during sterilizing are necessary to limit burn-on inside the can. Figure 57 shows an evaporated milk sterilizer and a diagram of the passage of the cans through the sterilizer. Lightweight cans are removed automatically. The $6^3/_4$ and 8-lb. cans are usually batch sterilized in a retort, but equipment is now available for continuous sterilization of these two sizes. Cans should be dry before the application of the label by machine. Automatic casers fill the shipping cases and the product is moved to shipping area or storage.

EVAPORATED SKIMMILK

U.S. Department of Agriculture estimated the 1968 production of evaporated skimmilk at 22,108,000 lb. The manufacturing procedure is similar to that of evaporated whole milk. The total solids is standardized to not less than 20% in order to meet the requirements of most states. The minimum range for all states is 18 to 27%. There is no federal standard on composition for evaporated skimmilk. Storage problems are less critical because of the low milk fat (below 0.5%).

STORAGE OF EVAPORATED MILK

Changes in quality of evaporated milk during storage can be delayed by decreasing the temperature. Time also is a contributing factor. The trend in recent years has been to store evaporated milk below room temperature (40° to 60°F.) requiring refrigerated storage. In maintaining freshness the lower the temperature the better, although freezing is detrimental to the body and texture. Webb et al. (1951) observed viscosity and pH decreased while color increased with an increase in storage temperature. Commercial evaporated milk remained acceptable for two years if held below 60°F., but deteriorated rapidly above 70°F. The humidity should be low (below 50%) to inhibit deterioration of the metal cans and labels. Domestic cases should be limited to 12 high when they are stacked in storage. Removal from refrigerated storage should be by a system that prevents condensation of moisture on the containers.

Other details require attention during storage. Inversion of the cases in storage may be necessary to minimize separation of fat and other constituents. Evaporated milks differ to some extent in rate of separation when held under similar conditions. Consequently, a practical policy is to examine representative samples of each lot and turn the cases which have evaporated milk with a small, soft cream layer. If cans are not turned while cream layer is soft, it becomes firm, then tough. In this condition, the milk fat cannot be redispersed by practical methods. The Evaporated Milk Association (Flake 1965) guideline for turning cases in storage is:

60°	3 to 6 months
70°	2 to 3 months
80°	1 to 2 months
90°	after 1 month and every 15 days thereafter

DEFECTS IN EVAPORATED MILK

The principal defects occurring in freshly processed and stored evaporated milk are: (1) cooked flavor; (2) coagulation; (3) discoloration; (4) gelation or thickening; (5) fat separation; (6) staling; (7) protein settling; (8) mineral salt separation; and (9) lack of sterility.

Heat Induced Flavor

The cooked flavor in evaporated milk is attributed to the liberation of sulfhydryl groups. They are caused primarily by the high temperature (240° to 245°F.) and the time (15 min.) required for sterilization. The heat induced effects of flavor are cumulative so the relatively slow heat-up period and cooling time add to the intensity of the cooked flavor. Preheat treatment may contribute, but is relatively unimportant because of the much more pronounced influence of sterilization. Ultra high temperature sterilization of the evaporated milk and aseptic packaging in sterilized cans reduces the cooked flavor substantially, but may introduce other problems such as a higher per cent of spoilage.

Heat Stability

Coagulation during sterilization may be prevented by pretesting the evaporated milk with pilot equipment and ascertaining a satisfactory corrective measure such as optimum preheat treatment for stability during sterilization. More rigid elimination of raw milk with acid development is a consideration, but minor changes in pH of the milk seem to have little effect on heat stability. Other considerations involve stricter application of the alcohol test or other stability tests and elimination of any milk supply that gives an indication of protein coagulation.

The heat stability of milk can be improved by controlling the temperature and time of preheating. Specific lots of milk may differ widely in the effect of similar heatings. But with a heat-up time of 4 sec. and a hold of 25 sec. most milk will have optimum stabilization by preheating between 248° and 284°F. (Webb and Bell 1942). They observed as much as six times more heat stability in some samples preheated by the ultra high temperature method 213.8° to 329.0°F. with a hold of 25 sec. compared to 203°F. for 10 min.

Preheating for maximum stability may require a heat treatment that may induce changes that are less than optimum for other factors such as initial viscosity and age thickening. Consequently, for best results the preheat temperature and time may be a compromise of all factors affected. Test trials with pilot equipment will provide information to help decide the optimum preheating conditions.

Sodium bicarbonate was the first important compound added to milk with excess acidity to improve the heat stability of the evaporated milk. It had the disadvantage of adversely affecting color and flavor. Federal Standards of Identity currently do not allow sodium bicarbonate as an ingredient for evaporated milk.

Later Sommer and Hart (1926) announced the relationship of salt balance to the heat stability of evaporated milk. Generally, an excess of calcium and magnesium occurred in milk with instability. By the addition of a small amount of a citrate or phosphate, stability was increased. Sodium citrate or disodium phosphate are compounds that are used to restore the salt balance.

As the concentration of milk solids increases, the stability to heat decreases. This fact is of little practical significance for the processing of evaporated milk because the milk solids are carefully controlled to the legal minimum. High homogenization pressure lowers heat stability slightly, but stability is greater with increased fat dispersion.

Off-Color

Browning or discoloration is principally attributed to caramelization of lactose and the Maillard reaction. The raw milk supply will influence the color change. Some milks show much more color change from sterilization than others. Normally, a color change in evaporated milk is observed after ten months of storage unless the temperature is below 60°F. The deterioration of color is rapid if the storage is 90°F. or more. Some seasons the milk is more susceptible to this change than others. The method of preheating and sterilizing is a factor influencing color. Short high temperatures are preferable to long and lower temperatures for the least discoloration. Discoloration also increases with the amount of stabilizing salt.

As pH decreases and protein and urea concentrations increase, more discoloration may be expected. The sulfhydryl groups formed during the heat treatments have an inhibiting effect on browning (Jenness and Patton 1959). They also reported sodium bisulfite and hydrogen peroxide reduced browning in milk. Oxygen probably enhances the change by reacting with the sulfhydryl groups.

Gelation

Viscosity of evaporated milks may differ and it is regulated mainly by heat treatment, agitation, storage temperature, and time. Too thin or too viscous a product is undesirable. Therefore, the preheating, and if possible the sterilizing temperature and time, are altered to obtain the desired creamy consistency. Excessive agitation during sterilization can be detrimental to optimum viscosity. Viscosity may decrease the first few weeks of storage and then increase. Low temperature storage delays thickening and gelation.

Vigorous agitation destroys gelation or thickening in evaporated milk. The mechanism of gelation is different from coagulation according to Jenness and Patton (1959). These defects occur more frequently in ultra high temperature sterilized evaporated milk and can be delayed by preheating the evaporated milk for 30 min. at about 190°F. This more severe heating may extend the time before formation of the gel begins. Leviton *et al.* (1962) were successful in delaying gelation in ultra high temperature sterilized concentrate with 36% solids by the addition of 0.05 to 0.15% (based on milk with 12.6% solids) sodium tetraphosphate. Sodium orthophosphate had the opposite effect. As the ratio of protein to other solids-not-fat constituents increases, the chances of gelation are increased. Homogenization especially at high pressures may increase the viscosity slightly.

Milk Fat Separation

One of the principal problems in the storage of evaporated milk is the separation of fat. The defect accelerates as the storage temperature increases (Fig. 58). Any condition or processing treatment that results in below normal viscosity of the evaporated milk will augment fat separation during storage. But adequate homogenization delays fat separation. Routine checks on globule size will assist in controlling the gradual changes that adversely affect homogenization efficiency such as mechanical wear of the homogenizer valves.

Carrageenan is used to retard fat separation. The results have been less than completely successful. Presumably, favorably effects are dependent on associated heat treatments or other factors. Procedure for the use of emulsifiers to delay fat separation should be developed by experimentation with the pilot equipment before making commercial trials.

Off-Flavor During Storage

Stale type of off-flavor development in evaporated milk is augmented by increasing the total heat treatment during manufacture. However,

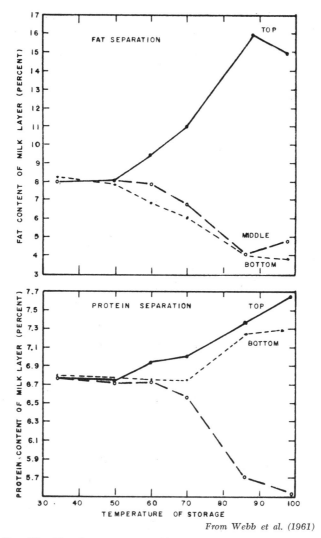

From Webb et al. (1961)

FIG. 58. THE SEPARATION OF FAT AND PROTEIN IN NORMAL
EVAPORATED MILK HELD IN UNDISTURBED STORAGE FOR 757
DAYS AT DIFFERENT TEMPERATURES

The evaporated milk contained 8.05% fat, 6.77% protein and 26.15% total solids.
After storage, fat and protein determinations were made on the contents in the top,
middle and bottom third of each can.

production of the cooked flavor at the same time will mask this off-flavor during the initial stages. Minimizing the severity of heating and decreasing the storage temperature are important means of delaying the occurrence of the stale type of off-flavor in evaporated milk. Much research is needed to reduce the seriousness of this change not only in evaporated, but other sterilized liquid milk products.

Separation of Nonfat Milk Solids

The settling of protein and mineral salts in evaporated milk is reduced by inverting the cans periodically during storage. The case inversion required for separation of fat suffices also for the nonfat solids. By controlling the protein denaturation during preheating and sterilization, and by optimum homogenization, protein settling can be reduced or delayed. A salt sedimentation may form consisting of tricalcium phosphate. This light-greyish separation is rather insoluble and sandy to the touch. It is not a common defect and is observed only after long or unfavorable storage periods. Once formed, this salt does not redissolve by common commercial handling. Salt sedimentation is controlled by lowering storage temperature or shortening the storage time. Calcium citrate crystals were inhibited by the addition of disodium phosphate before sterilization.

Freezing of evaporated milk should be avoided. Fat and solids-not-fat separation result from the destabilizing effect of freezing and thawing. The first evidence of this may be a rough texture. Casein precipitation may become evident by curdled appearance and some crystallized lactose may appear. Freezing may damage the cans by bulging the ends. The freezing point of evaporated milk ranges from 29.59° to 29.50°F.

Bacterial Spoilage

Failure to adequately sterilize evaporated milk by the in-can process is rare with modern equipment. Nonsterility is due almost exclusively to minuscule leaks. A defective container may permit contamination immediately subsequent to sterilization, but instruments in the processing line eliminate practically all these defective containers.

In former years, species within the genus *Bacillus* (spore former) have been identified in many investigations of evaporated milk spoilage. Coagulation may be with or without souring. *B. substilis, B. coagulans, B. cereus,* and *B. calidolactus* are a few of the most common species isolated (Hammer and Babel 1957).

The flat-sour and sweet-curdling types of spoilage are without gas formation and thus are not detected prior to purchase by the user. In some types of spoilage, gas is produced which distends the ends of the can.

These cans are easily identified for removal. Coagulation may occur concomitantly with the development of gas. Anaerobes of the spore forming type of bacteria frequently are the causative organism.

Growth of the spore formers in evaporated milk usually produces several sensory changes, for example, bitterness and fermented odors. If the per cent of spoilage becomes a problem, the first concern is to check the sterilization temperature and time to be sure they are within the intended operating range. Some attention may be given to the general sanitation of all equipment, and to low temperature holding of all unprocessed milk. High spore counts in the milk from the farm could be a contributing factor. While commercial practice has not included a spore count on the farmer's milk, acquiring more information on this type of contamination would be useful for the sterilization of all types of liquid milk products.

MANUFACTURE OF CONDENSED MILK

The equipment (Fig. 59) and procedure for the manufacture of condensed milk or condensed skimmilk is similar to that used for evaporated milk or evaporated skimmilk, with a few exceptions. The condensed milks are not sterilized. Consequently, there is less need for a high preheat treatment to stabilize the milk to prevent coagulation. The ratio of fat to solids-not-fat and total solids may be tailored to customer request, or sold on the basis of the test without standardization. The fat content may be 8 to 12% and the solids-not-fat 20 to 30%. Total solids of condensed skimmilk range from 30 to 36%. These products are packaged in bulk. Usually the storage area is refrigerated and the storage time is limited to a few weeks. In the absence of sterilization; the sanitation of plant and equipment must be adequate to restrict contamination after the heat treatment, otherwise deterioration takes place more rapidly. The temperature and time of heating for pasteurization must be adequate to insure good keeping quality.

SUPERHEATED CONDENSED MILKS

A small amount of condensed milk and condensed skimmilk is superheated. This extra heat treatment imparts qualities that are desired for specific industrial uses primarily in ice cream and baking. The moisture absorption capacity of the milk proteins and the viscosity are increased substantially.

Superheating of condensed milk is accomplished in the vacuum pan after the moisture removal has been completed. The product usually has 28 to 32% T.S. The steam, condenser water, and vacuum pump are shut off, but the vacuum is not released. Steam is distributed directly

FIG. 59. VACUUM PAN USED BY GAIL BORDEN ABOUT 1856 (A) AND A MODERN RECOMPRESSION TRIPLE EFFECT FALLING FILM EVAPORATOR (B)

into the product until the temperature is 180° to 195°F. This temperature is maintained until the condensed milk forms a gel consistency, but without curd particles, usually taking 5 to 15 min. A satisfactory sheen means the direct steam heating can be terminated. The product is cooled by starting the vacuum pump and the flow of condenser water. When the product reaches 125°F. it is removed from the pan and cooled.

Temperature and time to obtain gel formation depends on milk quality, salt balance, preheat temperature and time, degree of concentration, and the other factors affecting heat stability. More rapid superheating may be expected in winter months when cows are fed dry feeds. Considerable skill is required to obtain the correct gel formation without excessive superheating which causes coagulation of the protein and to obtain the other characteristics of a good quality superheated product. If superheating takes too long, the preheat temperature or time may be increased. The reverse is true if the superheating period is too short to control adequately. With a favorable heat stability of the milk and adequate equipment superheating can be accomplished in a vat. However, this procedure is less satisfactory because the cooling period is prolonged in the absence of a vacuum.

SWEETENED CONDENSED WHOLE MILK

Composition

The manufacturing steps of sweetened condensed whole milk are outlined in Fig. 55. The main difference from the manufacture of evaporated milk is the addition of sugar instead of heat sterilization to prolong the keeping quality. Federal specifications for sweetened condensed milk require not less than 8.5% milk fat, 28.0% total milk solids, and sucrose in water ratio of 61.5%. Therefore, the composition of the end product is usually within 8.5 to 9.7% milk fat, 28.5 to 30.5% milk solids (7.3 to 8.0% protein, 10.1 to 11.0% lactose, 1.5 to 1.7% ash), and 43.5 to 44.8% sugar.

Standardization and Preheating

Processing of sweetened condensed whole milk begins with good quality milk. It is standardized (fat to solids-not-fat of 1:2.294), clarified, preheated and may or may not be homogenized (145° to 160°F. and 1500 to 2500 p.s.i.). The preheating is continued up to a temperature of 180° to 200°F. for 5 to 15 min., or to 240° to 300°F. for 30 sec. to 5 min. by the ultra high temperature method. Effective pasteurization must be accomplished during heating to eliminate enzymes (such as lipase) and reduce the microbiological organisms to a low number. Oxidative changes also are delayed by the heat treatment. The exact temperature and time of heating is controlled to provide sufficient viscosity of the product without inducing excessive thickening during subsequent storage.

Concentrating the Solids

Condensing is accomplished by moving the product into a vacuum pan and operating with 25 to 28 in. or more of vacuum and sufficient steam for indirect heating to maintain vigorous boiling action. Condensing is complete when a slightly higher than 2:1 ratio has been attained. The Baumé hydrometer is used to determine progress of condensing. Figure 56 shows the correlation between total solids and degrees by the Baumé without the addition of sugar.

Addition of Sugar

Sugar is added at the end of the condensing step. If added before condensing an increase in viscosity and greater difficulty in evaporating the moisture are encountered. The amount of sugar required to inhibit bacterial growth is 43 to 45%, or approximately 18 to 20 lb. per 100 lb. of condensed milk. Hunziker (1949) advocated a sugar in water concentration of 62.5 to 64.5% as the proper amount for protection against bacterial growth while still avoiding sucrose crystallization. Sucrose in granulated or syrup form must be of good quality. The dry sugar is dissolved in a minimum of water. Liquid sugars should be subjected to a high pasteurizing temperature to destroy the microorganisms before addition to the condensed milk. Care in the determination and addition of correct amounts of sugar for the milk solids is necessary for an efficient operation.

Other sweetening agents such as corn syrup solids, glucose, and dextrose have been used to replace sugar (5 to 25%) for a specific utilization. The disadvantages of these sweetening agents have been their reduced sweetening capacity compared to sucrose and their adverse effects on color and rate of thickening in storage.

After pumping from the vacuum pan the sweetened condensed whole milk is cooled to 85°F. in a tubular cooler or another type of heat exchanger and held in a vat. The product may now be tested for total solids, sugar, and fat and restandardized to within a close tolerance.

Lactose Crystal Formation

Nuclei for crystallization are incorporated into the product by the addition of lactose in very fine powder form; pulverized nonfat dry milk; or sweetened condensed milk from a previous batch. If a powdered material is used, it should be capable of passing through a 200-mesh screen as an indication that the particles are sufficiently small.   The lactose usage rate is about 4 oz. per 1000 lb. of product.   Twice as much nonfat dry milk is required when it is used as the "seed."   A means of assuring uniform dispersion of the seed material is essential.   One procedure consists of blending the "seed" into a small amount of condensed milk and then adding this to the batch during vigorous agitation.   The agitation must continue while crystallization takes place to stimulate the formation of numerous small crystals of lactose rather than fewer, but larger ones.   Fast crystal-

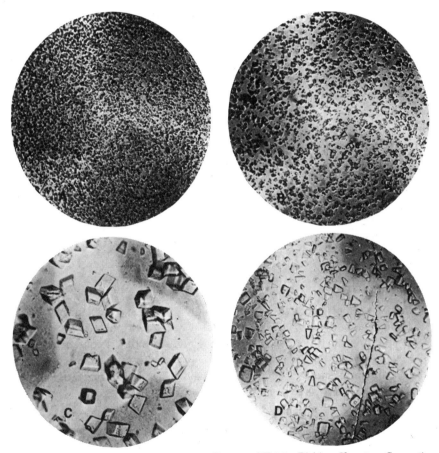

Courtesy of Votator Division, Chemetron Corporation

FIG. 60.   LACTOSE CRYSTALS IN SWEETENED CONDENSED MILK, MAGNIFIED 57.5 ×.
    A.   Seeded and processed through cooler-crystallizer to 50°F.
    B.   Seeded and processed through cooler-crystallizer to 65°F.
    C.   Unseeded and processed through cooler-crystallizer to 50°F.
    D.   Seeded and processed by common vat methods.

lization is conducive to the development of many crystals while slow crystallizing causes the large crystals which are responsible for sandy or gritty body and texture. Proper crystallization produces smooth body and texture as determined by sensory testing (Fig. 60). After the "seed" is uniformly dispersed, cooling of the product may continue slowly to 75°F. This should take approximately an hour, then the cooling is completed to 55° to 65°F. with continued agitation.

The rate of crystal formation is controlled by the amount of agitation, number of nuclei, total solids of the product, temperature, and the viscosity. A temperature decline can reduce the rate of crystallization by increasing viscosity; and at the same time increase the rate by increasing the degree of supersaturation of the lactose. An optimum temperature probably exists for each specific batch.

RAPID COOLING AND SEEDING

A semi-continuous operation is accomplished by continuous condensing and staggering the batch crystallization step. Successful crystallization also has been accomplished by cooling the sweetened condensed product to approximately 95°F., seeding and then rapid-cooling (within a few seconds) the product to 50°F. A heat exchanger with scraping agitation at the product-metal surface has been used for this method. An additional refinement consists of pumping a small portion of the cooled sweetened condensed milk from the outlet of the heat exchanger back to its inlet. The seed is fed into the line ahead of the recycling pump (Bolanowski 1965).

Packaging

The sweetened condensed milk is now ready for packaging. Bulk packaging may be in various size barrels, drums with polyethylene liners, or tin containers. Fillers are used to package sweetened condensed milk (55° to 65°F.) in 15 oz. or other size cans for the retail market. After filling, cans are sealed, labeled, and packed in cases (15 oz. cans packed 24 or 48 per case) for storage and distribution.

Sanitation

Since sweetened condensed milk is not sterilized, special precaution after preheating must be exercised to prevent contamination, particularly of yeast and molds. This requires well cleaned, sanitized equipment and special attention to each step in manufacturing to assure that contamination is not permitted. As a precaution the filling room is usually separate. Emphasis is given to a low level of air-borne microorganisms as well to the sanitation of the filling equipment. Cans and lids are treated with steam or a flame to a high heat and subsequently are protected as they move to the filler and the sealer.

SWEETENED CONDENSED SKIMMILK

The minimum total milk solids for sweetened condensed skimmilk ranges from 18 to 28%, but most states require 24 or 28%. Sugar ranges from 43 to 50%, or generally a trifle higher than for the whole milk product. There are no federal standards for sweetened condensed skimmilk. The procedure and equipment for processing sweetened condensed skimmilk are the same as for sweetened condensed whole milk. Obviously standardization of fat to solids-not-fat is eliminated.

DEFECTS OF SWEETENED CONDENSED MILK

Sweetened condensed milk may have many of the same defects as described for evaporated milk: excessive viscosity, discoloration, chemical, and microbiological changes. Fat separation is a less serious problem in sweetened condensed milk. The intensity of the cooked flavor is usually not objectionable because of the omission of heat sterilization. The chief defects of sweetened condensed milk which warrant attention are: sandiness, sugar separation, and mold growth.

As the lactose crystals increase in size above 10 to 12 μ in the largest dimension, and the number per ml. decreases below 250,000,000; the texture of the product becomes rough or coarse, and then sandy or gritty. Because lactose has less solubility in water than sucrose it will crystallize out of the saturated solution created by the addition of sucrose.

Sandiness

Numerous factors cause sandiness during manufacture and/or storage. Production causes may be one or more of the following: inadequate or improper seeding; product temperature not controlled within desirable limits; lack of sufficient agitation during crystal formation; excessive amount of sucrose added to the condensed milk; high concentration of milk solids; too thin a viscosity and rewarming the cold sweetened condensed milk for packaging. During storage a wide temperature variation may increase the tendency to become sandy. A very low storage temperature such as 32°F. or below is another cause of this problem.

Although the freezing point of sweetened condensed milk is approximately 5°F., the sugar saturation point may be lowered to the point that some sucrose also may crystallize and settle to the bottom of the container. Crystals of sucrose are sufficiently large to be observed by taste and sight.

The prevention of sandiness from lactose or sucrose necessitates satisfactory processing techniques and storage conditions to keep the lactose crystal size small and prevent sucrose crystals from forming.

Viscosity Changes

A standard viscosity of sweetened condensed milk to control sandiness is obtained by an optimum preheat treatment of the milk and to a much lesser extent by homogenization pressure (if used) and temperature adjustments. A cool storage is important too. The addition of 0.05 to 0.15% of sodium tetraphosphate may slow the rate of viscosity change. If sugar is added prior to preheating and condensing the effect of the heat treatment on the protein stability is influenced. A higher concentration of total solids or lower pH increase the rate of thickening during storage.

Sweetened condensed skimmilk with about the same total milk solids as sweetened condensed whole milk will thicken more in the same storage time (Hunziker 1949). According to Webb and Johnson (1965) the viscosity increases logarithmically with storage temperature and arithmetically with storage time.

Color Changes

The cumulative heat effects during processing directly influence the initial color and rate of color change during storage. Browning is increased with an increase in pH especially above 7.0. Reducing sugars, for example glucose and lactose, enhance browning more than sucrose. Hydrogen peroxide reduces the rate of browning. However, discoloration or browning is usually not a problem during the customary storage periods if the temperature is held to 50°F. or below and a good quality milk is processed properly.

Microbiological Changes

Microbiological changes in sweetened condensed milk are not necessarily prevented by the recommended sugar concentration. With increasing contamination in numbers and types of organisms the likelihood of growth and noticeable fermentative deterioration becomes greater.

Mold growth, commonly called mold button, occurs more often than the other types of microbiological changes. Hammer and Babel (1957) indicated one or more species within the genus of *Aspergillus, Catenularia, Penicillium, Cladosporium,* and *Actionomyces* have been isolated from sweetened condensed milk. These molds initiate various color, odor, flavor, and body changes in sweetened condensed milk.

Certain species of yeast and bacteria may produce gas, thicken the product as well as produce objectional sensory changes during growth in sweetened condensed milk. One of these gas producers is the yeast, *Torulopsis lactis-condensi*. Bacterial species in the following genera have grown in sweetened condensed milk: *Micrococcus, Streptococcus, Staphlococcus, Bacillus* and *Mycobacterium*. Coliform organisms have been isolated also from sweetened condensed milk. The heterogeneous contamination of microorganisms produce numerous undesirable flavor changes—sour, cheesy, rancid, musty, fermented, yeasty, fruity, and others.

Control of microbiological deterioration consists of: effective pasteurization, low count sugar, and scrupulous sanitation of surroundings and equipment, including air (during filling and sealing). Sugar content of the aqueous phase of sweetened condensed milk must be not less than 61.5 and preferably 63.0 to 65.0%. Growth of aerobic microorganisms can be inhibited by vacuum packaging of the product.

REFERENCES

BOLANOWSKI, JOHN P. 1965. Controlled crystallization—key to product quality. Food Eng. 37, No. 12, 56–60.

EVAPORATED MILK ASSOCIATION 1962. Evaporated milk industry sanitary standards code and interpretation. Chicago, Ill.

FLAKE, J. C. 1965. Private correspondence. June 28.

HAMMER, B. W., and BABEL, F. J. 1957. Dairy Bacteriology. John Wiley & Sons, New York.

HUNZIKER, O. F. 1949. Condensed Milk and Milk Powder. Published by the author, LaGrange, Ill.

JENNESS, R., and PATTON, S. 1959. Principles of Dairy Chemistry. John Wiley & Sons, New York.

LEVITON, A., PALLANSCH, M. J., and WEBB, B. H. 1962. Effect of phosphate salts on the thickening and gelation of some concentrated milks. 16th Intern. Dairy Congr. Proc. B: Sec. V, 2. 1009-1018 Copenhagen.

SOMMER, H. H., and HART, E. B. 1926. Heat coagulation of evaporated milk. Wis. Agr. Expt. Sta. Res Bull. 67.

U.S. DEPARTMENT OF AGRICULTURE 1944, 1949. Agricultural Statistics 1943, 1948. Washington, D.C.

U.S. DEPARTMENT OF AGRICULTURE 1956, 1961, 1966, 1968, 1970. Production of manufactured dairy products 1955, 1960, 1965, 1967, 1970. Crop Reporting Board, Washington, D.C.

U.S. DEPARTMENT OF AGRICULTURE. 1964. Federal and state standards for the composition of milk products. Handbook 51 (Rev.).

WEBB, B. H., and BELL, R. W. 1942. The effect of high-temperature shorttime forewarming of milk upon the heat stability of its evaporated products. J. Dairy Sci. 25, 301–312.

WEBB, B. H., DEYSHER, E. F., and POTTER, F. E. 1951. Effects of storage temperature on properties of evaporated milk. J. Dairy Sci. 34, 1111–1118.

WEBB, B. H., and JOHNSON, A. H. 1965. Fundamentals of Dairy Chemistry. Avi Publishing Co., Westport, Conn.

CHAPTER 7

Processing, Packaging, and Storage of Nonfat Dry Milk and Dry Whole Milk

INTRODUCTION

Commercial Methods

Milk and skimmilk are dried commercially by the spray, drum, or foam spray methods. Today most milk and skimmilk are dried by the spray method. In the past, the drum dried dairy products constituted a high percentage of the total amount produced. But since World War II decreased economic advantage of processing by drum compared to the spray process, unfavorable government support pricing, and its characteristic scorched flavor and poor solubility due to the severe heat treatment have drastically reduced the volume of drum dried milks. Now only a relatively small amount is manufactured by the drum process.

Foam spray drying, the third method, is new and has caused renewed interest in new products. The product acceptance may be slow for industrial uses because of the low bulk density hence higher packaging costs. Good dispersibility during reconstitution enhances its prospects for home use.

Variation in Milk Production

Seasonal variation in volume of raw milk receipts is a plant problem, especially if surplus Grade A milk from market milk is the chief source of supply. Many milk drying plants have difficulty maintaining plant volume at even 50% of the capacity during the low production months of late summer and fall. For example, the total production of nonfat dry milk is more than twice as high in May and June as in September and October.

A strong effort should be made to convince the producers that uniform production of milk throughout the year is beneficial. Fieldmen and haulers who are well qualified by training with the emphasis on human relations experience and proper personality characteristics are needed to encourage and "sell" producers on the program of uniform yearly milk production. This requires continual education of the supplier and is more effective if augmented by economic rewards to the producers who accomplish specific goals in uniform production.

But the quantity of milk is not the only important factor. Quality of the raw milk (delineated in a subsequent chapter) is an important consideration in producer-plant relations.

Neutralization of Acidity

Fresh, sweet milk contains no lactic acid although mixed herd milk may have a titratable acidity ranging from 0.10 to 0.15% due to milk proteins, carbon dioxide, citrates, and phosphates. Neutralization refers to the practice of reducing the acidity of the fluid product. In the United States neutralization is illegal in the manufacture of dry whole milk and nonfat dry milk for human consumption. If skimmilk has become acid, drying effects a slight decrease in titratable acidity of the reconstituted product.

In countries where neutralization of milk is necessary, and for the neutralization of a high acid product such as whey, food grade alkalies are used. The acidity in these cases should be reduced to a pH of 6.8 (range of 6.6 to 7.0). The specific titratable acidity in this pH is dependent upon the composition of the product and type of alkali that is used. An example of the calculation for acid reduction is:

(a) $A - B = C$
(b) $C \times D = E$
(c) $E \times F = $ lb. of alkali required

A = original acidity (0.25%)
B = desired acidity (0.10%)
C = per cent reduction (0.15%)
D = lb. of product (86,000 lb.)
E = lb. of acid to be neutralized
F = alkali factor (0.9) (0.9 lb. of alkali neutralizes 1 lb. of acid)

Therefore, the addition of 116.1 lb. of alkali will reduce the acidity to 0.10% in the product.

The alkali is dissolved in not less than six times as much water by weight and added to the product with sufficient agitation to prevent excessive localized overneutralization. The product temperature may be 95°F. or less and at least 15 min. is allowed for the reaction before heating. The selection of alkali may be influenced by intended use of the dry product and the processing method. Common alkalies are sodium hydroxide, sodium bicarbonate, sodium carbonate, calcium hydroxide, calcium oxide, magnesuim oxide, and various combinations of the sodium type or of the limes. Sodium bicarbonate causes foaming on a drum so should not be used for the drum drying. A sodium type such as sodium hydroxide is preferred for products requiring a maximum solubility.

·Care should be exercised to minimize the development of defects in the milk or adverse effects of the chemical. Precautions include: (a) the correct neutralizing factor of the alkali, (b) low product temperature and sufficient agitation while adding the alkali, (c) sufficient time for alkali acid reaction before heating, (d) low amount of inert material in alkali, and (e) avoidance of over or underacid reduction by care in weighing or measuring the product, alkali, testing for acidity, and calculations.

MANUFACTURE OF NONFAT DRY MILK BY SPRAY PROCESS

Equipment

The equipment for processing most dry milk products usually includes a separator, preheater, and/or high temperature short time pasteurizer with flow diversion valve, hot well, evaporator, preheater, filter in concentrate line, high pressure pump and other pumps, drier with milk dust collector, cooler, sifter, and packaging equipment. Many variations in equipment and methods are in use. In fact, no two plants are exactly alike in equipment and method.

In the design of equipment layout, a simple forward flow arrangement (Fig. 61) is best for quality and efficiency. The least equipment contact and the faster the forward movement of all the product, the better the quality of the dry milk. Recycling may be detrimental to product quality and should be limited especially for low-heat nonfat dry milk. Elimination of recycling is more difficult for equipment that is used to dry several products compared to single product processing.

Separation

Separation of cream can be with or without preheating the milk. Special cold milk separators are required for the low temperature method. Because fat separation of preheated milk usually is slightly more efficient, suggested temperature is a range of 60° to 90°F. if processing follows immediately. Otherwise if a holding period is involved, cold milk separation is more practical. The most important consideration regardless of method is a high degree of separation efficiency. Not more than 0.05% fat (Babcock method) should remain in the skimmilk.

Preheating for Low-Heat Nonfat Dry Milk

Low-heat nonfat dry milk manufacture requires that heating be carefully controlled to result in a minimum amount of heat induced changes and yet accomplish proper pasteurization. Control of both time and temperature assumes much significance. High temperature short time pasteurization as part of the preheating may be by tubular or plate heaters to not less than 161°F. for at least 15 sec. Higher temperatures or longer holds contribute directly to whey protein denaturation. This index is used as a measure of the cumulative detrimental heat effects during processing of nonfat dry mlik. Not more than 10% denaturation should occur.

Evaporation

Pasteurized skimmilk is condensed continuously in an evaporator (or vacuum pan) to 40 to 48% T.S. The Baumé hydrometer is used to test

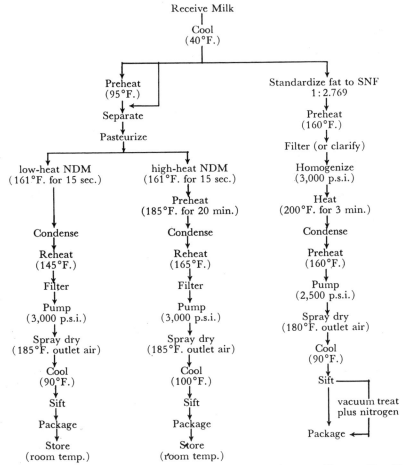

Receive Milk
|
Cool
(40°F.)

Preheat
(95°F.)
|
Separate
|
Pasteurize

Standardize fat to SNF
1 : 2.769
|
Preheat
(160°F.)
|
Filter (or clarify)

low-heat NDM
(161°F. for 15 sec.)

high-heat NDM
(161°F. for 15 sec.)
|
Preheat
(185°F. for 20 min.)

Homogenize
(3,000 p.s.i.)
|
Heat
(200°F. for 3 min.)

Condense
|
Reheat
(145°F.)
|
Filter
|
Pump
(3,000 p.s.i.)
|
Spray dry
(185°F. outlet air)
|
Cool
(90°F.)
|
Sift
|
Package
|
Store
(room temp.)

Condense
|
Reheat
(165°F.)
|
Filter
|
Pump
(3,000 p.s.i.)
|
Spray dry
(185°F. outlet air)
|
Cool
(100°F.)
|
Sift
|
Package
|
Store
(room temp.)

Condense
|
Preheat
(160°F.)
|
Pump
(2,500 p.s.i.)
|
Spray dry
(180°F. outlet air)
|
Cool
(90°F.)
|
Sift
|
vacuum treat
plus nitrogen
|
Package

Fig. 61.   Flow Diagram for Manufacture of Spray Process Nonfat Dry Milk
and Dry Whole Milk

the total solids during condensing. See Fig. 56 for the conversion of Baumé reading to total solids. An in-line refractometer may be used for continuous indication of the total solids. For a low-heat product the evaporator must be designed to operate efficiently with the product temperature below that which causes the denaturation of the proteins. Some types of double and triple effect (or more) evaporators are not suitable for production of low-heat product. Therefore, necessary precaution must be taken to be sure the design and operation of the double effect (or more) evaporator is satisfactory for low-heat product manufacture.

Once the desired total solids is obtained in starting the operation, the condensed skimmilk is pumped continuously from the evaporator to a small balance tank. From there the condensed skimmilk is pumped through a preheater to raise the temperature to 143° to 155°F.

Spray Drying

The high pressure pump, usually of the piston type, forces the hot product through the spray nozzle into mist-like droplets in the drying chamber. The pressure of the pump ranges from about 1000 to 5000 p.s.i. depending on manifold conditions such as nozzle design and size, inlet and outlet air temperature, drying chamber characteristics, particle size, and moisture content desired. Other types of spray nozzles are suitable for low pressure and one utilizes high velocity hot air to achieve atomization of the liquid.

Inlet air is heated by direct flame or steam coils to 250° to 500°F. Drying chamber design and subsequent equipment, as well as climatic conditions and desired moisture in the dry product, influence the inlet and exit air temperatures. The exhaust air temperature is the direct guide in controlling moisture of the product. But changes are achieved through adjustment of the inlet air temperature. Low-heat nonfat milk is dried to 3.0 to 4.0% moisture. Some driers have an auxiliary heater or redrier located after the primary drying chamber. These are useful when the air is humid or for enlarging the total drying capacity of the equipment.

Removal, Cooling, and Sifting

Most milk driers have a continuous removal system (Fig. 62) to immediately separate the dry product from the hot air stream which has been reduced to the range of approximately 175° to 210°F. The dry product should be cooled at once to approximately 90° to 110°F. A few plants still depend upon ambient conditions for cooling. Without cooling equipment, the temperature decline by natural means may be slow. During warm weather and with common warehousing practices of close, high stacks, temperature drop may involve several days with the accompanying product deterioration. Nonfat dry milk packaged too hot may become lumpy due to "heat-caking" and the development of off-flavor and off-color during storage is much more rapid.

A 25-mesh screen with No. 36 wire gage (0.707 mm. sieve opening) is commonly used for sifting nonfat dry milk ahead of packaging. Sifting is needed to remove lumps that occasionally form from a defective spray.

Preheating for High-Heat Nonfat Dry Milk

Except for the total heat treatment before drying the manufacture of high-heat nonfat dry milk is the same as for the low-heat product. In processing high-heat nonfat dry milk, care must be exercised to subject the skimmilk to a time-temperature treatment that will impart good bread baking qualities to the nonfat dry milk. During the heat treatment the whey (serum), protein is denatured. A test for the whey protein nitrogen is used as an index of the suitability of the dry milk for bread. Nonfat dry milk should contain not more than 1.5 mg. per gm. for use in bread formulas.

Courtesy of Marriott Walker Corporation

FIG. 62. INLET AIR, SPRAY NOZZLES, AND REMOVAL APPARATUS IN A FLAT BOTTOM DRIER

The heat treatment usually is applied to skimmilk directly following pasteurization. The general practice is to heat the skimmilk in a hot well to 185° to 190°F. for the equivalent of 15 to 30 min. by a continuous flow of skimmilk in the top and later out the bottom. The inlet distribution should be designed to prevent flow channeling which might result in an insufficient holding time for some of the product.

Imparting the desired heat treatment to skimmilk in the condensed form is risky without adequate instruments to accurately control time and temperature. Deviations that permit overheating quickly result in discoloration and a high solubility index (denatured casein) of the dry product. Therefore, this method is seldom used.

Ultra high temperatures for a short time are being used in some plants to eliminate usage of the hot well. Trials over a period of a year by Hedrick (1960) showed that temperatures of 280° to 300°F. for 30 sec. in the skimmilk prior to condensing gave a satisfactory bake test for the nonfat dry milk; but these results were less satisfactory for overall bread baking quality compared to nonfat dry milk from skimmilk heated to 185°F. for 20 min. The effects of ultra high temperature on the condensed skimmilk would be of interest although literature on this topic is not yet available.

Condensing and Spray Drying

The condensing step parallels the one used for the low-heat product, except the evaporator is operated for maximum efficiency without concern about serum protein denaturation. The reheating of the condensed milk ahead of the high pressure pump may be higher than for low-heat nonfat dry milk. Temperature of condensed milk may be elevated to 165° to 175°F. or higher. A high temperature improves drying efficiency, but is limited by several conditions. The temperature must be maintained below the level where casein denaturation will cause a high solubility index (insoluble sediment when the dry milk is reconstituted). The specific temperature at which the increased solubility index affects the grade of the dry milk is dependent upon heat stability of the milk, the preheat treatment and the time lapse necessary to allow the condensed milk to flow from the heater outlet through the high pressure pump to the spray nozzle. In some plants the temperature is restricted by the build-up of solids on the contact surface of the heater. Since the aim for all plants is a high degree of separation efficiency. Not more than 0.05% fat (Bab-hr.) before cleaning, the maximum temperature is limited.

Moisture Control

The pump pressure ordinarily is within 1500 to 5000 p.s.i. with the outlet air drying temperature usually between 170° and 210°F. The final moisture in the high-heat nonfat dry milk is controlled to approximately three to four per cent. However, the general commercial practice has been to limit moisture to not more than 3.5% if the product will be shipped to warm, humid areas in bags or sold to the Commodity Credit Corporation as surplus.

Orifice Size

The diameter of the orifice in the spray nozzle is selected on the basis of several factors. Drying chambers with a single spray nozzle have a large orifice. The common range in diameter is 0.107 to 0.177 in. Orifice size for the multinozzle chamber usually is between 0.025 and 0.052 in. for nonfat milk drying. Other design characteristics of the drier affect the orifice size. The inlet or outlet air temperature and pump pressure will influence the orifice size selection and spray pattern (Fig. 63). If other conditions are held constant a single large orifice increases the average particle size which in turn increases density and sinkability of the particles compared to drying with several nozzles.

The operator must occasionally observe the spray pattern for a malfunctioning nozzle. A particle may lodge in the nozzle and obstruct the spray pattern. An excessively worn nozzle (Fig. 64) causes an unsatisfactory spraying of the product. Both conditions can cause inadequate drying which may result in milk solids adhering to the drying chamber surfaces, cause lumps and higher moisture in the dry milk.

Intermediate Heat Nonfat Dry Milk

Intermediate heat nonfat dry milk is from skimmilk that has been preheated too much to qualify as low heat, but not enough to meet the high heat requirements. The whey protein nitrogen is more than 1.5 mg. per gm. of nonfat dry milk and less than 6.0 mg. The other processing steps are the same as described for high-heat nonfat dry milk.

Modified Drying Procedure to Improve Dispersibility

To improve the sinkability or dispersibility of nonfat dry milk as an alternative for agglomeration of nonfat dry milk, several processing steps may be modified. Each is changed to affect maximum particle size and uniformity. This involves concentrating the skimmilk to 45 to 50% T.S., a relatively low pressure for spraying, and a larger nozzle orifice. The temperature of the inlet drying air is raised to obtain adequate moisture removal from the larger spray droplets. A system of separating the very small particles either by an air stream or sifting is practiced. Additional variations in procedure such as seeding the condensed skimmilk and partial crystallization of the lactose have been used.

Under regular conditions of drying, the separation of large particles by screening may yield 10 to 40% of the production. The product with improved dispersibility may command a small premium price. But handling of the remaining product may present a disadvantage. Furthermore, the reconstitutability of nonfat dry milk from the modified drying and screen-

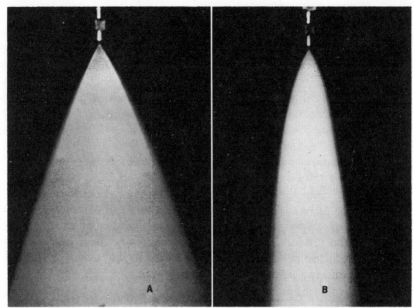

Courtesy of Delavan Manufacturing Company

FIG. 63.   PROFILE OF LOW-PRESSURE SPRAY (A) AND HIGH-PRESSURE SPRAY (B).

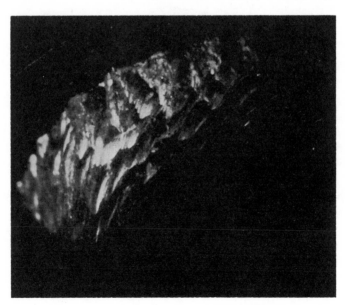

FIG. 64.   EXCESSIVELY WORN NOZZLE ORIFICE

ing method is generally inferior to the agglomerated product. It often is darker in color, higher in moisture and has more insoluble material (denatured protein).

VITAMINS A AND D FORTIFICATION OF SPRAY DRIED NONFAT DRY MILK

Nonfat dry milk contains all the vitamins of whole milk except those that are fat soluble—A, D, E, and K. Of these, vitamins A and D are most frequently associated with milk so their restoration in nonfat dry milk has merit. Nonfat dry milk may be fortified with vitamin A and D by two procedures, the wet and the dry processes. In the wet process the vitamins in a liquid are diluted with bland, edible oil (m.p. 100° to 102°F.) at the rate of 15 lb. of oil to the vitamins A and D concentrate sufficient for 10,000 lb. of milk solids.

It is important that the oil-vitamins must be thoroughly homogenized to delay oxidation. The Commodity Credit Corporation, U.S. Department of Agriculture, requires fortification of not less than 20,000 International Units (I.U.) of vitamin A and 2000 of vitamin D per pound of nonfat dry milk solids. Ten per cent additional I.U. of A and D have been recommended to assure minimum requirement after storage and shipment. The vitamin mixture may be manually added to vats of fluid skimmilk or to the concentrate of known solids content. If the flow of product is reasonably constant, or synchronized, the vitamin mixture may be metered into the product. Sufficient mixing must occur for uniform dispersion before drying.

A dry form of vitamins A and D may be premixed with nonfat dry milk and the premix dry blended into the main lot by batch or continuous procedures. The deterioration of vitamin A in storage is a little slower by the dry blending procedure; however, the vitamin cost and the expense of addition is a trifle higher.

In the selection of the vitamin concentrates for fortification, several considerations are involved in addition to being able to obtain uniform distribution in the nonfat dry milk. The vitamins must not adversely affect the flavor of the nonfat dry milk, either in the fresh condition or after prolonged storage. The vitamin potency should remain stable for the shelf life of the fortified product. A study by Bauernfeind and Parman (1964) (Table 19) indicated that vitamins A and D were stable in nonfat dry milk for one year. The vitamins also should remain biologically available during storage.

The Commodity Credit Corporation approved vitamin A compounds are retinyl palmatate and retinyl acetate. The vitamin D compounds are ergocalciferol and activated dehydrocholesterol.

MANUFACTURE OF NONFAT DRY MILK BY THE
DRUM OR ROLLER PROCESS

The equipment and other facilities for a drum drying operation can be procured for a much lower minimum cost than for spray drying. Consequently, the drum process may be more feasible for a small volume than the spray system. In addition, some industrial users prefer a drum dried nonfat and whole milk.

The skimmilk should be pasteurized and the heating should be continued to approximately 185°F. for 10 min. to be sure of good bake test properties. Removal of moisture by condensing to approximately 2:1 will

TABLE 19

STABILITY OF VITAMINS A AND D IN NONFAT DRY MILK DURING STORAGE

	Vitamin A, I.U.		Vitamin D, I.U.
Sample	75°F.	86°F.	75°F.
Added to control	12,000	12,000	2200
3 months			
Regular	13,500	13,800	...
Gas packed	13,200	13,700	...
5 months			
Regular	13,400	13,700	2100
Gas packed	13,400	13,500	2100
12 months			
Regular	12,700	...	1850
Gas packed	14,000	...	2250

Source: Bauernfeind and Parman (1964).

greatly increase the capacity of the drums. If the total solids is increased above 18%, the film on the roll becomes increasingly difficult to dry satisfactorily. Higher drum temperature or slower speed will cause scorching of the product despite insufficient drying.

The product at 165° to 185°F. is pumped into the reservoir between the two drums. Operating conditions that can be adjusted to regulate capacity of the drum and to control moisture content of the dried product are: steam pressure inside drum, r.p.m. of the drum, temperature of inlet milk, amount of milk in reservoir between two drums, viscosity of milk, distance between the two drums (see Chapter 3). Since the drum speed is normally constant, the principal adjustment is the steam pressure during operation. This should be changed with care to maintain desired moisture, particularly when highest capacity is necessary. Approximately 70 to 90 lb. of steam as indicated by the pressure gage at the inlet of the condensate trap is common. Superheated steam should be avoided to prevent scorching of the product. Milk level between the drums must be uniform for control of the moisture in the dried product. Usually the range in moisture content of drum process nonfat dry milk is 3.0 to 4.0%.

Scorched particles content (brown and black) may be a problem. The U.S. Extra Grade requirement is not more than 22.5 gm. per 25 gm. sample. In addition to the correct steam pressure, the proper pressure of knife against roll and elimination of pits in the drum, thorough cleaning of the complete drum drying unit and accessories each day after operating are necessary precautions to control the amount of scorched particles. Otherwise, under or over dried milk solids residues contaminate the ensuing product. Usually quality difficulties are encountered briefly while starting the drying operation. A common practice is to exclude the initial dry product from the lot until the product is satisfactory in appearance and the drum operation is normal.

Under proper drying conditions a uniform, thin light (in color) sheet of dry milk is scraped from each drum. This film drops into a trough. The screw conveyor breaks up the product into small pieces while being augered to a flaker or hammer mill. Particle size is decreased sufficiently by the mill to permit passage through the 8-mesh screen during the sifting operation. The product is usually packaged in 50- or 100-lb. Kraft bags with a plastic liner reasonably impervious to moisture vapors.

DRY WHOLE MILK

Manufacture by Spray Process

The manufacture of dry whole milk is not much different from nonfat dry milk. Since keeping quality is a restricting factor, care should be taken to obtain the maximum storage life. High quality milk with a low copper and iron content, sanitary plant practices, and a good processing procedure are important influences on keeping quality.

The milk fat is standardized in ratio to the solids-not-fat so the dry product will meet the standard of not less than 26% fat. This requires a fat content of 3.2% assuming the solids-not-fat is 8.86%. Some milk is standardized to have 28% fat in the dry form. Clarification may be conducted either before or after standardization to remove leucocytes and safeguard against the possibility of extraneous material in the product.

Homogenization of the whole milk is common if direct reconstitution of the dry whole milk is contemplated. Without homogenization the fat may churn during agitation while combining with water. Another advantage of homogenization is the improvement of keeping quality. The fat globules, although smaller, are more thoroughly recovered with the protein membrane. A pressure of 2500 to 3500 p.s.i. at 145° to 165°F. provides sufficient homogenization.

Preheating presents a paradox. Low-heat treatment minimizes the cooked flavor in the product, but does not develop antioxidants for delay of oxidation, one of the principal factors in keeping quality. Usually the primary consideration is given to prolonging the shelf life. Preheating also destroys the enzymes. If the lipase enzymes are not destroyed, hydrolytic rancidity will occur in the dry whole milk. Preheating must accomplish pasteurization, thus reducing the viable microorganisms. A beneficial influence on heat stability of the product may occur from the preheat treatment. Heating of the milk also is necessary before it enters the evaporator.

Direct steam preheating the product in the hot well has been replaced by the indirect method. Tubular heaters are currently used in many plants. They eliminate dilution by the steam condensate (6 to 12%) and steam impurities that may be toxic and/or cause off-flavors.

Numerous optimum temperature-time conditions have been used in preheating whole milk. Commercial practice frequently employs the range of 180°F. for 15 min. to 200°F. for 3 min. Heating after concentration has been suggested as preferable, possibly on the basis that a higher percentage of the antioxidants retained. These compounds may be removed along with the moisture during vacuum condensing. When sufficient solids concentration has been attained (35 to 45%) product is continuously removed from the evaporator. The Baumé degrees and corresponding total solids for whole milk are shown in Figure 56.

Another variation in the procedure is to preheat skimmilk to a high temperature necessary for formation of antioxidants and then to condense the skimmilk. Homogenized, pasteurized cream is used to standardize the condensed skimmilk to the desired ratio of milk fat to solids-not-fat. Some operators prefer to homogenize the concentrate after adding the cream.

The temperature of condensed milk after it is pumped from the evaporator is boosted to 145° to 165°F. in a heat exchanger prior to the high pressure pump. The condensed milk is dried with inlet air at 300° to 450°F. and exit air at 165° to 200°F., depending upon drier characteristics. To reduce heat damage during dehydration, and yet obtain the desired moisture, a low exhaust air temperature is preferred.

The dry whole milk should be immediately removed from the hot air stream to maintain better body characteristics and keeping quality. The higher the temperature and the longer the time the product is above the melting point of the fat the greater the amount of free fat that results.

The use of refrigerated air to move the dry whole milk to a cyclone after it leaves the drying chamber is one system of decreasing the temperature. Cooling to room temperature (not below) is preferred. When ambient temperature and humidity are high, the air may be reheated just enough to avoid absorption of moisture by the product due to the dew point. However, other methods of cooling whole milk powder have proved to be satisfactory such as augering across a surface cooled by water or brine in jacketed equipment. The system must be sufficiently protected against the entrance of condensate from the ambient conditions.

A 12-mesh screen is used for sifting dry whole milk. The product should be packaged immediately or held under a vacuum for 7 to 10 days before gas packaging (described later).

Foam Spray Process

Hanrahan and Webb (1961) and Hanrahan et al. (1962) provided the general application of foam spray drying to whole milk. Milk is standardized, preheated, and homogenized the same as for regular spray drying. During evaporation, the solids content may be increased to 50% or above. A gas, preferably nitrogen, is injected into the concentrated milk at 200 (or more) p.s.i. greater than the pump pressure. The gas is distributed into the concentrate by means of a mixing device between the high pressure pump and the nozzle. A regulator and needle valve control the flow of gas into the concentrate coming from the high pressure pump at 1500 to 1800 p.s.i. The usage of gas is at the rate of 0.5 standard cu. ft. per gal. of concentrated whole milk with 50% T.S.

Foam spray dried whole milk has improved dispersibility and approximately one-half the bulk density of the regular spray dried product. The occluded air in the particle causes poor sinkability when product is recombined with water.

Manufacture by Drum Process Dry Whole Milk

Only a small amount of drum dried whole milk has been processed in recent years. The poor keeping quality, unsatisfactory reconstitutability, and scorched flavor are considered to be the chief deterrents. However, for uses demanding a high percentage of free fat such as in certain confections, the drum dried product suffices.

Whole milk which has been standardized at the ratio of 3.5 (fat) to 8.75 (solids-not-fat) will yield a product with 28% fat if dried to two per cent moisture. The standardized milk is pasteurized and may or may not be homogenized. Unhomogenized milk will give a higher free fat in the dried form. The customary homogenization is 2500 to 3000 p.s.i. and the temperature is 145° to 170°F. The preheat treatment is regulated for the intended use, but normally is desirable only for dry milk intended for baking products. The severe heat treatment during drum drying negates any attempt to produce a low-heat product.

Concentration is not necessary, but will increase the drum drier capacity. Total solids increase should be limited to approximately a 20% concentrate. The procedure and precautions for drum drying of whole milk are very similar to those presented for nonfat dry milk by this process.

AGGLOMERATION OR INSTANTIZING

The pioneering research of D. D. Peebles resulted in a commercial process of instantizing. Nation-wide distribution of instantized nonfat dry milk was begun in 1954. The response by retail customers was highly favorable and within a few years instantized nonfat dry milk replaced the regular spray dried product on the retail market.

Purpose

The principal purpose of agglomeration, also called instantizing, is to improve the rate and completeness of the reconstitutability of dry milk products. This process affects wettability, sinkability, dispersibility, and solubility of the particles. However, total solubility is not improved by agglomeration.

Products

In addition to nonfat dry milk, the instantizing process has been applied to dry whole milk and other milk fat containing dry dairy products with only limited improvement in reconstitutability resulting. The milk fat adversely affects wettability because of its hydropholic property. However, agglomeration has been successfully adapted to: flours, starches, dry soups, cocoa products, dry puddings, sweetened flavored milk drinks, stabilizers, and other food products. Some of these products do not have adhesiveness when surface wetted, but are agglomerated by first blending with products that do, such as nonfat dry milk or sugar.

Major Systems

The three major systems of agglomeration in the United States are the Peebles, Cherry-Burrell, and Blaw-Knox. Each varies in equipment and operation details (Hall and Hedrick 1961). The general features in common are: (a) wetting of surface of the particles with steam, atomized water, or a mixture of both, (b) agglomeration whereby the particles collide due to the turbulance and adhere to each other forming clusters, (c) redrying with hot air, and (d) cooling and sizing to eliminate the very large agglomerates and the very small particles.

In the Peebles process (Peebles and Clary 1955), the nonfat dry milk pneumatically enters the agglomerating chamber. The particles are wetted to 10 to 15% moisture in the turbulent air-stream zone and form agglomerates. These fall into the next zone or chamber which redries to approximately 4.0% by means of filtered air at 230° to 250°F. The product is cooled, sized with rollers, screened, and packaged. The fine particles are returned to be recycled through the agglomerating process.

A.R.C.S. system of Cherry-Burrell consists of delivering dry milk at a uniform rate by air, screw, or vibrator to a horizontal tube (Carlson *et al.* 1956). Wetting and agglomeration takes place in the moist rapidly moving air-product mixture which passes into a cyclone. Air returns from the top of the cyclone to recycle and the clusters drop into a filtered hot air-stream at 270° to 300°F. The product is dried by this air while moving into the second cyclone. The air is exhausted from the top of the cyclone and the product clusters descend into a horizontal shaker through which cooling reduces the temperature from 160° to 180° to about 100°F. The sifter removes the fine particles. The clusters pass through sizing rolls and are sifted and packaged. During wetting the moisture of nonfat dry milk increases to 6 to 8% and on redrying declines to 3.5 to 4.5%.

In the Blaw-Knox Instantizer (Moore *et al.* 1964) the dry milk is measured by a rotary-feed valve into a line and fed pneumatically into a small agglomerating tube (see Fig. 65). An alternative system uses a vibrating trough to control product entry into the agglomerating tube. Steam wets the particles to approximately seven per cent moisture as they fall between two jets. Ambient air entering through radial slots in the agglomerating tube maintains the turbulence necessary for the formation of aggregates. The agglomerated product drops onto a conveyor belt for conditioning and transport to vibrating redriers of the deck type. Two or more decks are common. There hot air drys the product to 4.0 to 4.5%. After the large agglomerates pass between sizing rolls the product is screened. Fine particles are reprocessed through the system.

The Niro agglomerator is attached to the bottom of the vertical type drying chamber. Milk is dried to approximately 9% moisture in the drying chamber. A vibrator transports the product from the drier to the inlet of the Niro agglomerator. In the first section agglomeration takes place; in the second section, redrying with hot air; and in the third the product is cooled to room temperature. Very fine mesh screens convey the product through the three sections. Air that passes up through the screens goes out the top of the agglomerator carrying the fine particles. These are returned to the drying chamber.

Factors Affecting Instantizing

Dry milk that is manufactured specifically for agglomeration usually gives the best results. Moisture content and particle size should be as uniform as possible. A minimum of fine particles, less than 20μ in diameter, is

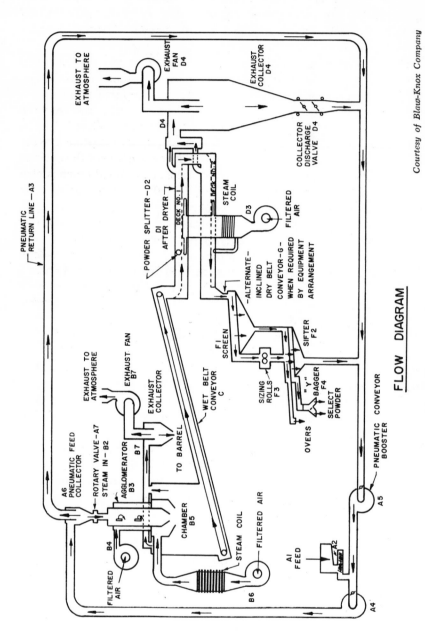

FLOW DIAGRAM

Courtesy of Blaw-Knox Company

FIG. 65. FLOW DIAGRAM OF INSTANTIZING PROCESS

desired with the preferred particle range of 25 to 50μ. Nonfat dry milk for agglomeration should be low in fat content. Low-heat (6 mg. or more of WPN) or medium-heat nonfat dry milk (less than 6 but more than 1.5 mg.) is normally used. High-heat nonfat dry milk will agglomerate satisfactorily, but it shatters much more easily in handling after agglomerating and redrying (Mori and Hedrick 1965).

The success of the instantizing operation depends upon adequate control of each step. The powder distribution into the wetting zone space must be uniform and at a constant rate. Moisture conditions must be uniform in all respects to avoid over or under wetting of particles. Overwetted particles dissolve slowly and too little wetting permits excessive shattering during handling. The air movement has to be stabilized to assure optimum particle collision. Excessive movement causes product adherence to the equipment lining. Control of the redrying air temperature and its flow rate is necessary for adequate moisture removal without heat damage to the agglomerated product.

Well known is the fact that agglomeration lowers the density of dry milks. Usually flavor is not affected, but changes in flavor can occur as a result of agglomeration that are detrimental if the process is not carefully controlled.

PACKAGING DRY MILK PRODUCTS

Packaging of Nonfat Dry Milk

A suitable container for dry milk should be impervious to moisture, light, gases, and insects; should be durable for handling, resistant to corrosion, of low cost; and be relatively easy to fill, seal, handle, and empty. The retail package should have a reclosable opening.

Nonfat dry milk for industrial use and storage may be packaged in barrels, drums, and bags or for retail purposes in metal cans, glass jars, or cartons. Except for the retail market, most of the yearly production goes into 100- or 50-lb. bags. Government purchase under the support program requires a minimum polyethylene liner, 3 mm. thick inside a 6-ply Kraft paper bag. In 1963, the government issued standards for the purchase of nonfat dry milk in bags having a tape over the top seal; and offered a premium price for nonfat dry milk in this bag (Type G) (Fig. 66). In 1966, this type became mandatory. The purpose of this type of bag is to prevent insect infestation of the product after packaging.

Nonfat dry milk in domestic commercial trade is commonly packaged in a 2-mil polyethylene bag inside a 4-ply Kraft paper bag. The outside layer is usually plain, but a crinkled type is available. Freezing, high

Fig. 66.   Type G Tape Sealer in Operation

temperature, and low humidity during storage of bags cause them to become brittle and thus damage more readily in handling.

Manual filling of the bags is most commonly completed by means of a
simple device attached to the sifter.   Automatic bagging equipment is
readily available to dispense the correct weight of powder in one bag before shifting the product flow into the next bag.   Bags are sewn 3 to $3^1/_2$
stitches per inch automatically or by a manually operated sewing machine
suspended and counterbalanced within easy reach of the filling area.
Type G bags require special heat sealing equipment for the protective
tape.   After closing the bags are usually stacked on pallets.   The bag
overhang from the pallet should not be more than 2 in.

Bulk Bins

Portable bulk bins may be used to hold nonfat dry milk a few days prior
to usage in the plant.   These bins also are gaining in usage as a container
for shipment to industrial users.   The bins are fabricated from different
metals or alloys including stainless steel.   Aluminum is most often used by
the dry milk industry, costing about one-half as much and being lighter
than those made from stainless steel.   One common size is about 3.5 by 4
by 6 ft. with a capacity of approximately 74 cu. ft. and a tare weight of

Courtesy of Powell Pressed Steel Company

FIG. 67. BIN FOR HANDLING DRY MILKS IN BULK QUANTITIES

about 225 lb. (Fig. 67). The bin is designed for transfer by a fork lift. The advantages are mainly in reduction of labor and bag costs with possibly a reduction in dust problem.

Automatic conveying equipment is available to fill a group of these bins, each in succession when properly positioned. They are air tight and can be stacked. The bins are emptied into a hopper by elevating and tilting the product flow, or by inversion to allow the product to flow through an iris-type outlet valve. The bins may be easily cleaned by the recirculation spray-ball system.

Retail Carton

Cartons of fiberboard, foil, and plastics have largely supplanted glass and metal as retail containers. Because of the hygroscopic nature of dry milk, the packaging materials must provide a good vapor barrier. Of several types in use, one is a fiberboard carton with an overwrap of foil laminated to paper. Another consists of a fiberboard carton with an inner liner of foil laminated to paper. A polyethylene bag inside the fiberboard carton is used also. Other combinations of layers of polyethylene, foil, and paper either in liner or overwrap, are available for packaging.

Fillers for rapid speed operation are on the market (Fig. 68). Common size packages range from individual use packets of 3.2 oz. (intended for 1 qt. of reconstituted skimmilk) to packages of 9.6 oz. to 5 lb. or larger.

Fig. 68. Combination Dual Tube Bag Machine with Volumetric Filler for Packaging Dry Milks

Courtesy of Triangle Packaging Machinery

Packaging of nonfat dry milk is quite routine. Principal concerns involve keeping machine downtime to a minimum, maintaining the correct net weight within narrow limits, and providing a good seal. The packages are packed in cardboard cases for storage and/or shipment. Coding of each package provides a means of identification for quality control.

Packaging Dry Whole Milk

Oxygen Desorption.—Average production conditions and normal market periods for export and retail sales necessitate gas packaging of dry whole milk to delay oxidative changes. The rapid flavor deterioration of dry whole milk (and other milk fat-containing products) due to oxidation necessitates inhibitory measures. One of these consists of packaging the product with a low oxygen content. The general procedure is to immediately remove oxygen by subjecting the product to 28 in. of vacuum within 24 hr. of drying with final packaging within a few days. The specific procedure selected for gas packaging is governed by final oxygen limit desired in the package, and by the equipment filling rate in relation to oxygen evacuation rate. Oxygen desorption in dry whole milk is slow, principally due to entrapped air in the lactose. The amount of entrapped air is influenced by: content in the concentrate, spraying pressure and orifice size, and temperature and treatment of concentrate.

Less than 2.0% final oxygen in the head space gas of the package is considered satisfactory for most storage conditions. Good quality dry whole milk with a low oxygen content can be expected to withstand room temperature storage for six months or more without an oxidized off-flavor.

Gas Packaging.—To obtain a low level of headspace oxygen in dry whole milk, a double gassing technique is applied. The customary procedure is the collection of filled cans on trays to be conveyed into the vacuum chamber. The air is removed rapidly (60 sec.) with the gage indicator decreasing to 29+ in. of vacuum. After a 2 to 5 min. hold, the pressure is restored with nitrogen to 0.5 to 1.0 p.s.i. above atmospheric pressure. Nitrogen may be replaced with a mixture of nitrogen and carbon dioxide, the latter being restricted to 5 to 20%. After removal from the chamber, the containers are sealed by soldering the 1 to 2 mm. hole in the lid or crimping on the lid. The containers are held for oxygen desorption. When an oxygen equilibrium has been attained in the headspace, usually within a week but at the most ten days, the cans are punctured and the vacuum treatment, pressure restored with nitrogen, and sealing steps are repeated.

Gas packaging of dry whole milk should not be delayed after drying. Otherwise quality deteriorates during the holding period. Warm powder directly from the drier tends to have a more rapid rate of oxygen desorption under vacuum. If the production is not large, dry whole milk may be placed into metal drums and the air exhausted. By holding the product under partial vacuum for oxygen desorption the first gassing step in the package may be eliminated, and yet the final maximum of two per cent oxygen can be attained.

Oxygen Limits.—A good commercial operation using continuous gassing equipment can reduce the oxygen level to approximately 2.5% with a single gassing. This may be satisfactory for many storage conditions, but not 90°F. or above. A maximum limit is 3.0 to 3.5% oxygen in the headspace of the can for a noticeable delay of oxidation. U.S. Standards for grading permit a 2% maximum in Premium and 3% for U.S. Extra Grade dry whole milk.

Oil Impurity.—Water pumped nitrogen (not oil pumped) may be purchased in high pressure drums. It can be prepared at the plant by burning propane gas with controlled conditions. Oxygen impurity is removed by exposure to hot iron shavings.

Oxygen Removal by Reaction.—Another procedure consists of the addition of five per cent hydrogen to the nitrogen used to restore atmospheric pressure after deaeration. A packet containing a catalyst is added to the container after filling with the dry milk. The catalyst may be palladium or platinum. Each of these causes the oxygen to react with the hydrogen forming water, thus effectively removing the oxygen available for oxidative reactions.

Commercial usage and tests have demonstrated that flexible packages as well as metal cans are satisfactory for gas packaging of dry milks containing fat. The flexible packages are more easily damaged by rough handling, but this should not prove a limiting factor in domestic outlets.

Recently an interesting modfication of the flexible package has been developed. The oxygen scavenger is placed in a pouch within the package. Oxygen passes through the wall of the pouch, comes in contact with the scavenger and forms water, which cannot pass back through the pouch wall into the dry product.

MANAGEMENT FACTORS

Considerations in Location of Drying Plants

General.—The proper location of a dry milk plant requires careful, exhaustive study and advanced planning. The high fixed cost of total facilities necessitates a large initial capital investment of approximately $560,000 or more ($1.60 per lb. of milk as a minimum). Operating costs also are relatively high. Consequently, a large volume of raw material is essential. A plant should be assured of a 250,000 lb. daily intake of milk with assured possibilities of an increasing volume as the trend in higher operating expenses continues. Contrary to 15 years ago, daily plant receipts of 1,000,000 lb. of milk are not unique.

Professional Assistance.—The plant should be designed by highly experienced and skilled professionals to obtain proved facilities at reasonable cost. It should be designed and constructed to include latest innovations for maximum operating efficiency consistent with flexibility in all activities from the receiving of the fluid milk to shipment of the processed product. Future expansion must be considered. Reasonably priced real estate for future buildings or additions and parking are part of the original planning.

Other Factors.—To handle large volumes of milk, transportation facilities by truck and possibly railroad must be good. Also an adequate supply of cool water is needed, plus reasonable assurance that ample water will be available in the future including a sufficient quantity for an enlarged production. At present the lowering of the water table has decreased the supply from wells in some areas to the extent that the shortage is serious. Water conservation measures may be considered such as the use of cooling towers, reuse of water removed from milk, etc.

Sewage disposal from the plant can be a serious problem with the possibility of an expensive solution. Conservation and sanitary officials have succeeded in obtaining legislation to prohibit the use of streams. The organic content of sewage from the dry milk plant may be high thus, the use of city sewers has become expensive or is actually prohibited in many areas as population growth overburdens the facilities. A plant's own disposal system has the disadvantage of high initial cost. In view of current problems in many plants one cannot overemphasize the need to have adequate plans for disposal of wastes at a reasonable cost.

Some attention should be given to a suitable labor supply although this factor is less limiting than in former years. The decrease in number of plant employees brought about by more mechanization, and the greater mobility of available workers from other localities tends to offset localized scarcities.

Product Losses in Plant Processing

Actual milk solids losses among dry milk plants vary from 0.5 to 6%, but should be restricted to less than two per cent. Although the tendency may be to place most of the economic significance on the fat losses the solids-not-fat wastes are important also. In computing losses the moisture content of powder should be taken into consideration. To control losses, management must maintain a continual educational program to promote the importance of alertness in all processing activities in the plant. Incentive goals are helpful.

Check List of Product Losses.—A check on weights and tests of the raw milk receipts is desirable. The control list for product losses includes: (a) errors in weights and tests of incoming milk; (b) spoilage, faulty equipment and breakdown delays, especially during the high production season; (c) inefficient separation of milk fat; (d) mistakes in opening valves, improper drainage of lines, etc.; (e) incomplete drainage from vats, pumps, heaters, evaporators and each of the other items of equipment; (f) inaccurate standardization; (g) entrainment losses during evaporation; (h) fine particle losses in drier exhaust air (stack losses); (i) residues in drier; (j) waste and overfillage during packaging; (k) container damage during storage; and (l) insect infestation (qualitative loss rather than quantitative). Drying to a lower moisture content than necessary also represents a form of loss.

Nonfat dry milk yields may vary from 8.5 to 9.2 lb. per 100 lb. of skimmilk in an efficient operation depending upon the total solids of the skimmilk. The yield of whole milk may range from 12.0 to 13.0 lb. per 100 lb. of milk.

Thorough and accurate records are essential for an efficient operation. The quantity of milk handled must be carefully ascertained as well as the total solids of skimmilk and solids-not-fat and fat on whole milk. The

Courtesy of Mojonnier Bros. Company

FIG. 69. APPARATUS TO TEST EXHAUST AIR FOR PRODUCT LOSS

greatest accuracy is obtained by using official tests, as is customary in standardization. But the total solids for whole or skimmilk may be obtained by the use of a lactometer (Watson 1957). In plants receiving whole milk another alternative is to test the fat content and then calculate the solids-not-fat after selecting from several equations the one most appropriate of the local milk supply. Two such equations are:

$$7.01 + (0.400) \times (\% \text{ fat}) = \% \text{ SNF (Jacobson 1936)}$$
$$7.07 + (0.444) \times (\% \text{ fat}) = \% \text{ SNF (Jack } et\ al.\ 1951)$$

The per cent loss for nonfat dry milk should be calculated on the basis of solids purchased and the amount sold, taking into consideration the moisture content of the powder. B.O.D. tests are a good means of checking the total product losses of the plant indicated by the accounting records.

Among the many sources of product losses one of the more serious, but less obvious, in a plant with a cyclone collector system occurs in the exhaust air. Stack loss with cyclones is usually not less than one per cent and can be two to three per cent for nonfat dry milk. Consequently, routine tests should be made. One of several procedures consists of sampling the exhaust air at a point of average velocity. The air is directed into distilled water (Fig. 69). After a specific time, for example 1 hr., the water is tested for total solids and the per cent loss is calculated using drier exhaust air per operating hour. The results can be related to the amount of powder manufactured.

Other Inefficiencies

There are many other inefficiencies in dry milk processing: waste or damage in handling and using supplies, careless use of water and electricity, inefficient production of steam in the boiler operation or fuel cost higher than necessary, failure to use more economical method or equipment in a specific phase of the processing, and excessive labor per pound of product processed. However, labor reduction should be carefully planned. The savings from excessive reduction may be more than offset by failure to maintain equipment properly and make repairs quickly.

In fact, one need fear little contradiction to the statement that few plants have reached the point of diminishing returns in an effort to accomplish greater overall efficiency.

REFERENCES

BAUERNFEIND, J. C., and PARMAN, G. K. 1964. Restoration of nonfat dry milk with vitamins A & D. Food Technol. *18*, No. 2, 52–57.

CARLSON, E., WEIR, R., and ZIEMBA, J. V. 1956. A new way to instantize milk solids. Food Eng. *28*, No. 10, 62–63, 194.

HALL, C. W., and HEDRICK, T. I. 1961. The manufacture of instant milk powder. Dairy Eng. 78, No. 1, 7–11.

HANRAHAN, F. P., and WEBB, B. H. 1961. U.S. Dept. Agr. develops foam-spray drying. Food Eng. 33, No. 8, 37–38.

HANRAHAN, F. P., TAMSMA, A., FOX, K. K., and PALLANSCH, M. J. 1962. Production and properties of spray-dried whole milk foam. J. Dairy Sci. 45, 27–31.

HEDRICK, T. I. 1960. Ultra high temperature processing of dairy products. I. Some characteristics of high-heat nonfat dry milk. Quarterly Bulletin of Mich. Agr. Expt. Sta., Michigan State University 43, 398–406.

JACK, E. L., ROESSLER, E. B., ABBOTT, F. H., and IRWIN, A. W. 1951. Relationship of solids-not-fat to fat in California milk. Cal. Agr. Expt. Sta. Bull. 726.

JACOBSON, M. S. 1936. Butterfat and total solids in New England farmers' milk as delivered to processing plants. J. Dairy Sci. 19, 171–176.

MOORE, J. G., HESLER, W. E., VINCENT, M. W., and DUBBELS, E. C. 1964. Agglomeration of dried materials. Chem. Eng. Progr. 60, No. 5, 63–66.

MORI, K., and HEDRICK, T. I. 1965. Some properties of instantized dry milks. J. Dairy Sci. 48, 253–256.

PEEBLES, D. D., and CLARY, P. D., JR. 1955. Milk treatment process. U.S. Patent 2,710,808.

WATSON, P. D. 1957. Determination of solids in milk by a lactometric method at 102°F. J. Dairy Sci. 40, 394–402.

CHAPTER 8

By-products and Special Products

INTRODUCTION

Numerous special dry dairy products are worthy of consideration in addition to the two principal by-products—buttermilk from the churning of butter and whey from cheese manufacture. These include dry cream, dry ice cream mix, malted milk powder, and dry casein. Other dry products are sodium caseinate, calcium caseinate, dried cheeses (Cheddar and Blue type), nonfat milk-shortening powders, 900- or low-calorie products, dietetic dairy preparations, dry milk products for infants, sweetened chocolate flavored nonfat dry milk, and other flavored dry milk drink products. Products for coffee creaming purposes have been available for more than a decade in dry form. Dry products with high acidity include natural buttermilk, skimmilk, and whey. Grated Italian cheeses are tray and belt dried. Of minor significance are powdered butter, dry yogurt, and cottage cheese which is freeze dried.

SPECIAL PRODUCTS

Special Products Present Drying Problems

The above-mentioned products can be dried with spray drying milk equipment. Some special products require only minor adjustments which may include nozzle orifice and core, inlet and outlet air temperature, and sifting screen. Other special products such as high fat creams necessitate drier adaptation or modification for manufacture of a satisfactory product. Each product should receive careful consideration and perhaps innovation is necessary for optimum processing treatment to develop or preserve desirable characteristics. Further development may be needed to reduce or prevent undesirable changes which might adversely affect the intended utilization or storage.

High milk fat products present a challenge in their efficient removal from most milk driers and subsequent handling. These products are readily susceptible to the development of oxidized and stale off-flavors during storage. The protein in high acid milk products augments drying problems. The acid and heat denatures the casein present which after drying does not redisperse in water. The high acid products discolor readily during spray drying in inverse relation to the final moisture of the dried product.

DRY BUTTERMILKS

Spray and Drum Drying

The method of processing of dry buttermilk by either the spray or drum process is similar to that of nonfat dry milk. Care should be exerted to cool the buttermilk to 40°F. immediately upon drainage from the churn. When intended for human consumption, buttermilk should be stored in stainless steel lined tanks. A common practice is to preheat to 90° to 120°F. and separate the buttermilk to reclaim a portion of the fat. The preheating is continued to 185°F. with a 15-min. hold. After increasing the total solids by condensing to 16% for drum or approximately 40 to 45% for spray process, the concentrate is reheated to 160° to 175°F. and dried to a moisture content of 3.0 to 4.0%. After sifting (12-mesh screen), dry buttermilk is usually packaged in Kraft paper bags with a plastic liner or in a fiber drum with or without a plastic liner.

Storage

The keeping quality of dry buttermilk is relatively short, from 1 to 3 months. The causes have been attributed to the lipid content and to inferior quality from metal contamination and bacterial deterioration prior to drying. However, Ziemer *et al.* (1962) reported on a few samples of commercial dry buttermilk which retained good flavor for more than a year, presumably at room temperature.

Dry High Acid Buttermilk

A modified product known as dry high acid buttermilk is prepared for special uses. One procedure of manufacture consists of pasteurization of the buttermilk at 180°F. for 16 sec. or longer and concentrating to 30% T.S. Then the product temperature is reduced to 115°F. It is inoculated with 1 to 5% *Lactobacillus bulgaricus* and is agitated during incubation. Otherwise homogenization after incubation at a low pressure will give a smooth curd. When the acidity is developed sufficiently, equivalent to approximately 10 to 12% in the dry product, the buttermilk without preheating is spray dried to 3.5 to 4.0% moisture. The dry product should be removed continuously and cooled immediately, and then sifted. Precaution must be observed to control the outlet drying air temperature or excessive browning and a high solubility index will result. In most types of driers, as the moisture content decreases below 3.0% these objectionable characteristics increase. A good quality dry high acid buttermilk is difficult or impossible to manufacture in some driers because of design.

By another procedure, buttermilk is cultured and incubated until the acidity has developed. The high acid product is condensed to 30% T.S. This procedure has the advantage of a higher final acidity. The disadvantages are the large vats required during acid development and the product's sensitivity to the heat applied while condensing.

DRY CREAM PRODUCTS

Regular Spray Dried Cream

Composition.—The term, "dry cream," designates a dried milk product which has a higher fat content than dry whole milk. The usual range for dry cream is 40 to 70% milk fat. The ratio of fat to solids-not-fat in much of the production is too low to be correctly classed as a dry cream. A 18% coffee cream has approximately 72% fat after drying. A dry half-and-half is a more correct reference for much of this type of product. The fat content usually ranges between 50 and 60%; solids-not-fat, 40

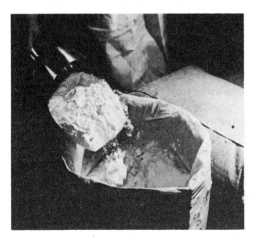

FIG. 70. REGULAR SPRAY DRIED CREAM

to 48%; and the moisture less than 2.0%. Although much of the dry cream can be reconstituted with water to a fat content meeting the legal fat requirement, the solids-not-fat may be higher than is normal for the product.

Dry cream (Fig. 70) has the same advantages as dry whole milk or nonfat dry milk, namely, less perishability, reduced volume for storage and shipment and in general is more economical to distribute and utilize.

Processing Procedures.—Good quality cream should be used in the manufacture of dry cream. A common procedure is to standardize the ratio of milk fat to solids-not-fat, for example, 7.0:2.8 or 1:1 on a solids basis. The cream is pasteurized at 165°F. for 16 sec. and homogenized at 3000 p.s.i. For more resistance to oxidation the heat treatment may be increased by a higher temperature and/or longer hold; heating to 180°F. for 15 min. is common. Cream is concentrated to 40 to 50 % T.S., reheated to 150° to 160°F., and spray dried to 1% moisture. The orifice of the nozzle may be larger and the outlet air temperature lower than used for drying nonfat dry milk.

An alternate procedure consists of preheating the skimmilk, concentrating it and adding pasteurized, homogenized cream. Homogenization may be at 150°F. and a pressure of 2000 p.s.i. on the first stage and 700 p.s.i. on the second stage. Too high a temperature or pressure may cause excessive viscosity.

Immediate continuous removal of dry cream from the drying chamber is certain to result in a better quality product. Dry cream may have a tendency to cling to the sides of the drying chamber in direct relation to the fat content. Special drier design and extra facilities such as vibrators or jets of air are helpful in product removal.

The system for removing the dry cream from the drying chamber should be designed to keep rubbing action to a minimum, especially while the dry cream is hot (fat in liquid state). The rubbing will cause the melted fat to "oil off"—become free fat. Then the globules are no longer covered with membrane. The free fat in the dry product causes an increase in stickiness during handling and a greasy body and texture. The loss of product is higher and more difficulty due to lumpiness is encountered.

Immediate chilling of the product below the melting point of the fat is desirable. Refrigerated air is required to remove the large amount of heat present because of the latent heat involved in changing the milk fat from a liquid to a solid.

Dry cream has a tendency to form lumps during the change of state of the fat. Sifting after the fat solidification will remove the lumps and restore the uniform body and texture characteristics.

Processing and Storage.—Ordinarily dry cream is packed in bags or in drums of preferably small size. Bags cannot be stacked without the increased weight causing proportionally more firmness of body. More free fat develops which in turn causes greater lumpiness.

Compared to dry whole milk, the removal of air from dry cream is relatively easy in gas packing. If packaged in standard tin cans, the oxygen content of the headspace gas can be reduced to 0.75% or less with one deaeration. Oxidation is delayed 12 months or more during storage at 90°F. with not more than 0.75% oxygen in the package.

Storage of dry cream definitely should be below the melting point of its fat to reduce detrimental effects on the body and reduce oxidative changes. At room temperature the keeping quality of air-packed dry cream is limited to 2 to 3 months. Storage in a room cooled to 50° to 60°F. or lower is preferred. Variable storage temperatures increase the lumpiness. Keeping quality and flow properties are improved by the addition of 10 to 20% sucrose prior to dehydration of the product.

Uses.—The industrial uses of dry cream have not been extensive. The chalky flavor, difficulty in maintaining flowability, and prevention of lumping are important reasons for limited sales. Undoubtedly the high cost of dry cream in comparison with other fats is another reason. Dry cream is used in the confection and baking industries as a source of milk solids. It can be reconstituted and standardized with skimmilk for use in making cream and Neufchatel cheeses. The use of dry cream in ice cream having a mild flavor (vanilla) is limited, but the strong flavors such as chocolate will mask the chalkiness or cooked flavors of dry cream.

Foam Spray Dried Cream

Cream that has been standardized and processed as previously described for regular spray drying may be foam spray dried using nitrogen and the same equipment and procedure described for foam spray drying skimmilk. The gas required for a good drying operation is less per pound of solids than for skimmilk. Continuous removal and immediate cooling of the dry product are important for a good quality product. Foam spray dried cream has less heat induced flavors, less "greasy" body and texture, and superior flowability compared to the regular spray dried product.

Foam Spray Dried Sour Cream

Sour cream can be foam spray dried resulting in less adverse effects of the heat during drying than with regular spray drying. The brown discoloration and denaturation of the casein is much less. Consequently, reconstituted foam spray dried sour cream is more nearly like the original product in appearance and in physical-chemical properties than reconstituted regular spray dried product.

Dry Cream Product for Whipping

Regular dry cream when reconstituted does not whip satisfactorily. A special dry cream for whipping by inert gas (nitrous oxide) under pressure was reported by Pyenson and Tracy (1948). Fresh, sweet cream, condensed skimmilk, skimmilk, sugar, emulsifier, antioxidant, and vanilla flavoring were used for drying. Trials revealed the optimum ingredient percentages should be standardized to 30.0 fat, 8.0 solids-not-fat, 5.0 sugar, 0.2 glycerol monestearate and 0.1 pure vanilla concentrate, and 0.03 nordihydroguaiaretic (NDGA) acid or 0.1% gallic acid. NDGA was dissolved in a small amount of butter oil before adding to the mix. Other ingredients were added to the cream with agitation for thorough mixing. The mixture was pasteurized at 160°F. for 30 min. and spray dried to a moisture content of 1% or less using a 0.035-in. orifice and No. 20 core. Smaller orifices decreased drier capacity and product density. Spray pressure of 200 p.s.i. gave optimum whipping properties.

DRY ICE CREAM MIX OR BASE

Production Level

Dry ice cream mix or base has the same advantages characteristic of other dehydrated dairy products, namely, less perishability and lower shipping costs. Most shipments have gone to the Armed Forces and to areas with low milk production. The period of largest production was during and shortly after World War II. Current production is small.

Retail sales have not been encouraging. The ready availability of ice cream at a reasonable price results in little incentive to sacrifice this convenience for the privilege of reconstituting the dry mix with water and freezing it.

Ingredients

The preparation and drying of mixes are a little more involved than for processing the common dairy products. Ingredients are assembled, such as the source of milk fat and solids-not-fat, sugar and stabilizer-emulsifier. Sugar is mainly sucrose although 1 to 25% may be replaced with corn syrup solids or dextrose. Stabilizer-emulsifier combinations that have been used are carrageen, carboxymethylcellulose, gelatin, sodium alginate, and. mono- and di-glycerides. Lecithin has both stabilizing and emulsifying properties. A small percentage of sodium caseinate or buttermilk solids may be used to improve the whippability. A small amount of sodium citrate or similar salts is beneficial.

Standardization to attain proper composition of the dry product is important. Numerous commercial compositions have been manufactured which depend upon ingredients, the intended usage, and the local frozen dessert regulations. The product may be dried with 0 to 100% of the required sugar. A common product contains 20% of the required sugar when dried. The remaining percentage may be dry blended in the form of a finely granulated sugar or the buyer may wish to incorporate the additional sugar.

Dry Ice Cream Mix Composition

Typical composition range of dry mix may be:

	Per Cent
Milk fat	25.0 to 29.0
SNF	25.0 to 30.0
Sugar (full amount)	37.0 to 42.0
Stabilizer-emulsifier	0.5 to 1.2
Moisture	1.0 to 2.5

If the sugar is limited, the following is a suggested range:

	Per Cent
Milk fat	39.70 to 45.20
SNF	39.70 to 46.80
Sugar	7.40 to 8.40
Stabilizer	0.80 to 1.56
Moisture	1.00 to 2.50

Processing Procedure

The mix or base ingredients with 35 to 40% T.S. are assembled in a vat with sufficient agitation for uniformity. The mix is standardized for per cent fat, solid-not-fat, sugar, and total solids. The product is preheated to 150° to 160°F. for homogenization at 2500 p.s.i., heated to 180° to 190°F. for 5 to 10 min., and spray dried to a coarse particle with 2.0 to 2.5% moisture; cooled at once to 90° to 100°F., sifted, and packaged.

Flavorings such as vanilla are usually added by the user when mix is reconstituted at the ratio of 4.25 lb. of mix to 7 lb. of water. If desirable, vanilla in the powder form can be dry blended into the base along with a portion of the sugar or a concentrated vanilla added (less preferable) to the ice cream mix ahead of drying.

Storage

For storage under severe conditions ice cream mix or base should have not more than two per cent moisture. The product should be gas packaged with a two per cent limit on oxygen at equilibrium. Browning, oxidation, and staleness are common defects that develop during storage.

WHEY DRYING

Tremendous quantities of whey are available from cheese manufacture. Only a small proportion of these quantities is dried. The low solids content plus difficulty and expense of drying means a limited monetary return. Low quality and misconception of the nutritional value has deterred usage of dry whey as a food in the past.

The high lactose content of whey solids creates difficulties in the manufacture of dry whey. Heat causes the lactose to become sticky. Caking from the heat or moisture adsorption is particularly objectionable.

Whey and Dry Whey Analysis

Wheys differ widely in composition and characteristics according to the composition of the original milk, method of curd coagulation such as enzyme or acid, type of cheese, and procedure of its manufacture. Whey has approximately 5.8 to 6.5% solids. Of this 4.4 to 5.0% is lactose, 0.7 to 1.0% is protein, and 0.6 to 0.8% is salt. The acidity may vary from 0.11 to 0.75%.

The analysis of dry whey may have the following range:

	Per Cent
Lactose	65.0 to 88.0
Proteins	1.0 to 17.0
Ash	0.7 to 10.0
Fat	0.5 to 2.0
Moisture	2.0 to 15.0
Lactic acid	0.1 to 12.0

Condensing and Drying

Numerous patents and methods have been reported for spray drying whey to obtain a noncaking, flowable product: Simmons, 1930; Eldredge, 1933; Peebles and Manning, 1933, 1937; Chuck, 1935, 1939; Kraft, 1938; Peebles, 1938, 1943; Lavett, 1939, 1940, 1941; Spellacy, 1939; and Supplee, 1939, are a few of the important ones.

In a usual procedure the whey is separated to reclaim the fat if it is not from a skimmilk cheese. Whey with or without neutralization is preheated to 170° to 220°F. for entry into a multiple-effect evaporator. Moisture is removed until the solids have increased to a range of 40 to 70%. The concentrate may be handled by several drying methods and numerous variations of each. One method commonly used is to cool the whey concentrate to 35° to 80°F., seed and crystallize the lactose by holding 24 hr. Then the concentrate is spray dried.

An alternative is to heat the whey after concentrating to 150° to 200°F. and immediately spray dry to 12% moisture. Crystallization is encouraged by a holding period up to 48 hr. Then the product is conveyed to a secondary drier such as a tunnel, shelf, or internal drum to remove approximately 8% more moisture.

Condensed whey with 65 to 70% solids may be dried by directly conveying it to tunnel; here it is distributed onto $1/16$-in. mesh screens, and exposed to moving air at 140° to 190°F.

The drum drying of whey concentrate may be complicated by the formation of a sticky mass. One way of alleviating this problem involves specially designed equipment consisting of two sets of two drums with one set directly above the other. The upper drums increase whey concentrate from 50% solids to 78 to 85%. The mass drops into the trough of the drums below. Material on the lower drums' surface seed the mass for crystallization and rapidly cool the product to 112°F. The product is fed to an internal drum for drying to 4% moisture with air at 130° to 140°F.

In equipment with two sets of similarly designed drums, the lower set of drums removes more of the moisture from the partially dried whey descending from the upper two drums. About half the steam pressure (30 to 35 p.s.i.) is used in the lower set as in the upper set of drums.

Whey for animal feed can be drum or spray dried by mixing of 15 to 50% skimmilk, buttermilk, or certain cereal products (for example, wheat or rye bran), which reduces the problems of high lactose content. The procedure is similar to the dehydration of skimmilk by the drum process.

Foam Spray Drying

Hanrahan and Webb (1961B) introduced foam spray drying of whey. Whey is concentrated to 45 to 50% solids; air or nitrogen at 2000 p.s.i. is injected into this condensed whey at 1800 p.s.i. by a simple mixer (Fig. 71) installed in the line between high pressure pump and spray nozzle.

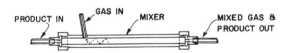

FIG. 71. DEVICE FOR MIXING GAS AND WHEY CONCENTRATE

Gas may be used at the rate of approximately 1.5 c.f.m. per gal. of product. The specific rate depends upon total solids, air drying temperature, desired moisture, and bulk density. Commercial production requires that provisions be made to inhibit a build-up of whey solids on the drying equipment during an operation day.

Packaging

After drying by a drum, tunnel, or internal rotary method the product may be processed through a hammer mill to reduce particle size and may or may not be sifted. The spray processed product usually is sifted. Methods that maximize crystallization of lactose reduce its hygroscopicity. Thus, the tendency to cake in storage is reduced. Dry whey usually is packaged in the conventional bags with plastic liners. Since lactose in the amorphous form is very hygroscopic the liner of the bag must be reasonably impervious to moisture vapors. Methods of production that cause a maximum amount of crystallization reduce the tendency of dry whey to cake in storage.

Compositions.

The analysis of dry whey varies widely due to the composition of the original liquid and the whey processing method. Table 20 shows an average composition.

TABLE 20

ANALYSIS OF NORMAL AND CULTURED DRY WHEY FROM COTTAGE CHEESE

Component	Normal, %	Cultured High Acid, %
Lactose	71.5	68.1
Protein	11.5	11.5
Lactic acid	7.0	10.5
Ash	8.0	7.6
Moisture	2.0	2.3
pH	4.4	3.9

Source: Hanrahan and Webb (1961).

There are numerous commerical procedures which obtain a modified dried whey product. The lactose, protein, and/or minerals can be removed in substantial amounts—the lactose by concentration, crystallization, and centrifugation; the proteins (serum type) by heating to 200°F. for 20 min. and centrifuging with continuous unloading bowl (Pinckney 1960); or the salts by ion selective membrane electrodialysis (Stribley 1965).

The requirements for U.S. Extra Grade dry whey are presented in Chapter 9.

DRY CHEESE PRODUCTS

Introduction

The estimated production of all forms of dried cheese was 22,500,000 lb. in 1969. It consisted of two principal types; the first being the grated and tray or belt dried Italian cheeses, Parmesan was one of the main varieties. The spray dried cheeses, Cheddar the most important, was the second principal type. In this category Blue cheese drying is increasing in importance.

Procedure for Spray Drying

In the manufacture of spray process dry cheese, aged and medium curd, are used. Cheese is comminuted by processing through a food chopper into a jacketed vat with ample agitation for viscous materials. Sufficient water (80° to 90°F.) is added to reduce the total solids to a 35 to 45% slurry. Stabilizer, usually sodium citrate or disodium phosphate, is dissolved in water before addition to the contents of the vat. The range in usage of stabilizer varies from 1.5 to 2.5% of the weight of cheese, depending mainly upon the degree of aging and variety of cheese. A cheese stabilizer promotes an emulsion thus inhibiting oiling off of the milk fat. Sufficient stabilizer should be used to eliminate any graininess which would hamper a smooth body and texture of the slurry. This stabilizing salt should be uniformly dissolved before heating begins, and it should not have a detectable effect on the product flavor.

Cheese coloring is commonly added to meet the market demand. Otherwise the dry product would be light in color. The addition of 0.5% sodium chloride is optional, but accentuates the cheese flavors.

The cheese slurry is heated to pasteurization temperature and held for the required time. Care must be exercised to prevent stringiness of the curd mainly by localized over heating and heating to too high a temperature.

After pasteurization is completed the product is tempered to 140°F. and homogenized at 1500 to 2500 p.s.i. on the first stage and 500 p.s.i. on the second. If an excessively viscous product emerges from the homogenizer, high temperature of homogenization and high total solids generally are the chief causes.

The cheese slurry is spray dried with a relatively large orifice and low pressure to reduce the amount of very small particles. Cheddar cheese is dried to 2.5 to 3.5% by adjustment of the outlet air temperature. The dry cheese is continuously removed from the drying chamber and immediately cooled to 85° to 90°F., but not below the dew point. Twelve-mesh screen is used for sifting prior to packaging in bags or occasionally in fiber drums.

Flavor and Storage Problems

One problem in dry Cheddar cheese manufacture is the loss of cheese flavors from the product during spray drying. Reconstituted Cheddar cheese with the same moisture content as the original cheese prior to dehydration is much milder in flavor. Bradley and Stine (1964) observed that atomization of large particles with as few very small particles as possible resulted in less loss of volatiles, thus more flavor in the dry Cheddar cheese.

The keeping quality of dry Cheddar cheese is relatively short. Fig. 72 illustrates flavor changes during storage. A score of 3.5 or above was considered acceptable by the panel; the Hedonic scale ranged from 0 to 9. Air-packed dry Cheddar cheese should be held at 50°F. or less.

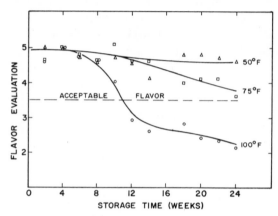

FIG. 72. FLAVOR CHANGES IN AIR PACKED DRIED CHEDDAR
CHEESE DURING STORAGE

Improved Keeping Quality

Gas packaging with nitrogen replacing the oxygen will prolong the keeping quality. Bradley (1964) reported that the addition of 0.05 to 0.08% antioxidant or a combination of antioxidants—butylated hydroxyanisole, butylated hydroxytoluene, propyl gallate, and nordihydroguaiaretic acid, slightly delayed oxidation of spray dried Cheddar cheese, but not of the foam spray dried product.

Dried Cheddar manufactured by the foam spray method retained more cheese flavor than the product spray dried by the conventional method. The body and texture and flowability characteristics also were superior.

Cheese Products and Composition

Modified dry cheese products result from the mixture of cultured (high acid) skimmilk, cultured buttermilk or whey, as a replacement for part or all of the water in preparation of the slurry. The solids of the skimmilk, buttermilk or whey represent 5 to 20% of the dried mixture prior to drying.

Dry Cheddar cheese has the following composition:

	Per Cent
Milk fat	50 to 54
Proteins	36 to 42
Carbohydrates	3 to 4
Ash	3 to 4
Moisture	2 to 5

Dry Grated Cheese

Von Loesecke (1955) delineated a U.S. Dept. Agr. procedure for manufacture of dry grated cheese of hard types in the following: cured cheese is shredded with care to prevent crushing. It is dried initially to 12 to 15% moisture in trays with air at room temperature and low humidity (below 35%). Then the dehydrating air temperature is increased to 145°F. and the drying continued until moisture is 3 to 6%. The product is cooled to 55°F. Dry shreds are reduced in size by grinding and then are packaged.

Infrared Drying of Cheese

Infrared energy can be used for drying of cheese. Cheese is placed about 1 in. deep on a mesh belt conveyor which is 2 ft. wide and about 16 ft. long. The heaters are about 5 in. above the product. The product moves at 6 f.p.m. under the heaters absorbing 68% of the energy. A unit to dry 1500 lb. per hr. by about 15% where 225 lb. of water are removed, uses 680 cu. ft. per hr. of gas or 510,000 B.t.u. (Anon. 1963C).

DRY MILKS WITH FLAVORING MATERIALS

Chocolate Nonfat Dry Milk—Spray Dried

Limited amounts of chocolate flavored nonfat dry milk have been made by spray drying a concentrate of 45% solids composed of skimmilk, cocoa, stabilizer, salt, and a portion of the sugar requirement. The remainder of the sugar is blended into the dry product which is then instantized.

Spray drying a mixture containing regular cocoa or chocolate liquor causes excessive wear on the regular nozzle orifice. The fibrous material in cocoa has an abrasive action. Consequently, special wear resistant orifices should be used. Sapphire or tungsten carbide has the ability to give much longer usage.

Dry Blend

Another procedure in the manufacture of sweetened, chocolate flavored nonfat dry milk is the blending of all ingredients in the dry form, followed by instantizing.

The ingredients for both procedures on the dry product basis are:

	Per Cent
Cocoa processed by Dutch method	10.0 to 15.0
Nonfat dry milk	25.0 to 50.0
Sugar	35.0 to 50.0
Stabilizer	1.0 to 2.0
Salt	0.5
Citrate or phosphate salt	0.5
Moisture	2.0 to 4.0

Product Characteristics

Tests on the product in retail distribution indicated the stabilizers are not completely effective on holding the cocoa particles in suspension in the reliquefied product. The size of sugar particle for dry blending is important in uniformity of appearance, dispersibility characteristics of the product in water, and flowability of the dry product. A medium fine granulation, for example, Bakers Special, imparts optimum qualities for agglomeration. Unfortunately, the terminology of sugar granulation has not been standardized among all sugar manufacturing companies.

A good quality sweetened, chocolate flavored nonfat dry milk has a pleasing flavor blend of milk solids, sugar, and cocoa, disperses readily in hot or cold water, has a minimum of cocoa settling, high density (preferably 0.5 gm. per ml.) and good flowability in the dry form.

Fruit and Other Flavored Dry Dairy Products

Numerous flavored dry milks besides chocolate have been manufactured experimentally. They have not been commercially significant. Some of these flavorings are banana, blueberry, butterscotch, eggnog, strawberry, raspberry, etc. Commercially prepared purées of fruit (without seeds) are convenient source of fruit flavoring.

The drying procedure consists of adding the desired flavoring to skimmilk with 15 to 20% solids. The percentage of flavoring depends upon flavoring concentration and desired flavor strength, but strawberry, raspberry, and banana purées may vary from 75 to 120% of the milk solids with the sugar varying from 40 to 60% of the milk solids. The flavor effect of less than two per cent milk fat in the uncondensed milk is of questionable value because of the masking influence of most flavorings. Consequently, skimmilk or milk having 2.5 to 3.5% fat is used.

After all the ingredients are added to mixing vat and thoroughly dispersed, the mixture is homogenized at 150°F. and 2500 p.s.i. and spray dried to a moisture of 3 to 4%. If dispersibility improvement is desirable, the product may be processed through an instantizer under operating conditions similar to those used for nonfat dry milk.

Chocolate or Milk Crumb

A dry product sometimes called chocolate crumb, is used in various food products particularly for ice cream in some countries. The milk solids-not-fat and fat are standardized and homogenized at 150°F. and 2500 p.s.i. and concentrated. Sugar and cocoa are added in proper ratio to result in 9 parts of milk fat, 22 parts solids-not-fat, 62 parts sugar, and 6 parts cocoa. The mixture is formulated to have 40 to 45% T.S. for drying. It is heated to about 150°F. and is spray dried.

Identical or similar is milk crumb. An example of the composition is:

	Per Cent
Sucrose	52.1
Lactose	13.8
Milk fat	10.9
Chocolate liquor	10.4
Protein	9.8
Ash	1.9
Moisture	1.1

Milk crumb is used as an ingredient of milk chocolate and other confections. The details of processing milk crumb are a trade secret.

COFFEE COLORING PRODUCTS

Private Brands

Many specialty and modified dairy products have been dried experimentally. Some have been sold commercially with success. A few of these have been manufactured by a single company or a small number of companies under the protection of patents or guarded practices. While the production may be a substantial volume of the firm's business, the quantities represent an insignificant amount of the total milk production. Dry products sold under the trade names of *Pream* and *Sprinkle* originally were modified dry milks for direct addition to hot coffee and other beverages. They were used as a replacement for coffee cream in the beverages.

The exact details of manufacture have been confidential company information. In the manufacture of *Pream,* the calcium content of milk was reduced by the base exchange (zeolite) method. *Sprinkle* contained an added complex phosphate. In these products the lactose percentage was increased by addition before drying, dry blending afterwards, or both. The fat content was similar to dry whole milk and the total lactose was higher. These two brands are now manufactured from nondairy products.

Dry Dairy Product

A satisfactory coffee coloring dry dairy product can be processed by using a dispersing agent such as sugar or corn syrup solids or a combination plus a buffering salt—sodium citrate or disodium phosphate. One formula consists of: 100 lb. of condensed skimmilk with 30% T.S.; 7 lb. of sugar; 160 gm. of sodium citrate.

The sugar and sodium citrate are added to the condensed skimmilk. The mixture is thoroughly agitated, pasteurized, and heated to 165°F. for spray drying to a moisture content of approximately 3.0%. The product is continuously removed from the drier, cooled to 90°F., and sifted through a 12-mesh screen.

If a product with 4 to 12% milk fat on the dry basis is desired, the fat to solids-not-fat ratio should be standardized accordingly; the low fat milk is homogenized at 150°F. and 3000 p.s.i. and concentrated to 30% solids. After the addition of the dispersing agent and buffering salt the product is heated to 180°F. for 5 min. It is spray dried and subsequently processed the same as the nonfat product.

If a low sweetening effect is desired corn syrup solids is used to replace the sucrose. Since corn syrup solids vary in moisture absorption, a type with minimum hygroscopicity should be used.

The instantizing process is similar to nonfat dry milk. Too large a percentage of very fine particles decreases the effectiveness of agglomeration and reduces dispersibility. Consequently, these particles are removed by screening with a 100-mesh sieve.

Composition

The composition range of the dry product is:

	Per Cent
Milk fat	1.0 to 12.0
Sugar	8.0 to 25.0
Milk solids-not-fat	50.0 to 80.0
Buffering salt	0.5 to 1.0

Milk fat contributes flavor, but tends to decrease the effectiveness of agglomeration.

Dry Blend

If a large percentage of sugar is used the low fat or nonfat dry milk can be dry blended with the sugar and buffering salt and then instantized for satisfactory dispersibility in hot beverages or other liquids (Fig. 73). However, keeping quality is improved if the product contains fat, by drying the sugar or part of it in the mixture.

FIG. 73. DRY DAIRY WHITENER FOR HOT COFFEE

Dry Nondairy Product

The major dry nondairy coffee coloring products on the market are spray dried blends of corn syrup, vegetable fat, sodium caseinate, stabilizing salt, flavor enhancer, and emulsifier. Other minor ingredients may be included. The Filled Milk Act prevents the use of a dairy product with substitute milk components. The courts have declared that sodium caseinate can be classed legally as a chemical although from milk.

MODIFIED DRY INFANT DAIRY FOODS

For many years modified dairy products have been successfully dried for infant feeding. The volume of milk used for these products is small, but the income is a significant proportion of a few companies total sales. These products are usually modified to approximate more closely human milk and furnish a baby's dietary requirements. Sales of dry infant dairy products have been much more successful as a proprietary product through the drugstores. Consequently, they do not have to comply with the Filled Milk Act.

Modifications

 Common modifications may include: more essential fatty acids; a decrease in fat of 6 to 10%; and an increase in carbohydrates of 25 to 45%. More lactose increases the suppression of harmful bacteria in the intestines. Vitamins, especially vitamin C, may be added. The soluble proteins may be adjusted to resemble more closely human milk. The proteins may be altered by predigestion. A decrease of salts—chlorides, sulfates, and nitrates—inhibits undesirable effects on the kidney.

The fat used in the dry infant products frequently is of vegetable origin or a mixture of vegetable and animal fats. The melting point of the fats used is similar to human milk. In addition to lactose, corn syrup solids may be added to the formula.

The details of the manufacturing procedures are very similar to those presented in Chapter 7 for the processing of dry whole milk powder.

Formula and Analysis

The analysis (Anon. 1963A) of one of the dry infant products was approximately:

	Per Cent		Per Cent
Milk solids	70.0	Carbohydrate	63.0
Sucrose	13.0	Fat	18.0
Vegetable oil	9.0	Protein	14.0
Maltdextrin	5.0	Ash	3.0
Maltose	1.0	Moisture	2.0

CASEIN

Dry Casein Manufacture

Specific details vary tremendously regarding manufacture of casein and the resulting product varies widely in composition and properties. Table 21 provides minimum guidelines for an edible grade casein. In one common method of preparation, pasteurized skimmilk is tempered to a specific temperature within the range of 90° to 115°F. Hydrochloric acid having a specific gravity of 1.2 is diluted with four parts of water and then is continuously fed into the stream of skimmilk using equipment designed to give complete mixing. Sufficient acid is used to lower the pH to 4.6. Contents flow into a vat. After holding 10 to 15 min., the

TABLE 21

GUIDE FOR MANUFACTURE OF EDIBLE CASEIN

Component	Specifications Not More Than
Moisture	9.0%
Protein	85.0%[1]
Ash	2.0%
Fat	1.0%
Sediment	1.0 mg./50 gm. sample
Metals	5.0 p.p.m.
S.P.C.	10,000/gm.
Coliform	50/gm.
Insoluble material	0.5% in dilute Na_2CO_3 solution
Particle size	1% on 40-mesh screen
	10% on 60-mesh screen

[1]Not less than value shown.

whey is drained and the curd is washed twice with cold water, 40° to 45°F., facilitating whey removal.

The curd is then conveyed to a roller press for sizing and moisture reduction. Drying is accomplished by loading the curd onto trays in thin layers and transferring them to a tunnel drier. Temperature initially is controlled at 125° to 130°F. When drying is about two-thirds complete, the air temperature may be raised to 150°F. Good circulation of the air throughout the tunnel is important.

Now the dry casein is pulverized in a hammer mill in accordance with desired particle size. Next it is sifted, separating particle sizes and packaged in bags. The screen mesh on the sifter may be 30, 50, 70, or higher.

Buchanan (1957) described a casein plant with continuous hydrochloric acid precipitation having a capacity of 3000 gal. per hr. with only one person to operate it.

Small commercial operations can make casein with only a few items of equipment normally in a dairy plant. The skimmilk may be coagulated with lactic acid produced by inoculation and incubation with a culture at 70° to 90°F. When 0.64% acidity is attained (for 9.0% T.S.), the curd is stirred and heated 5°F. slowly. Whey is drained. Curd is washed with about one-half as much water as skimmilk (set) and held 10 to 15 min. with stirring. The whey-water is drained and the washing step is repeated again.

The pressing should not be less than 500 p.s.i. for 12 to 15 hr. after grinding, product is oven dried in trays at 125°F. for approximately 8 hr. The moisture should be lowered to 4%.

Sulfuric acid can be used instead of hydrochloric for coagulation.

Enzymes, e.g., rennet and salts, also can be used as the coagulating agent. Each coagulator, however, imparts different properties to the casein.

Composition and Quality Control

Each step in the manufacture of casein must be carefully controlled to obtain high quality, correct composition, and high yields. Two of the most important factors involved are: efficient separation of fat from the milk (0.05% or less) because fat lowers the quality of the casein; and thorough washing of the curd to reduce whey components, especially keeping lactose and minerals to a minimum. Lactose in the casein causes a brown discoloration to develop and lowers solubility.

Proper pH at the time coagulation is considered complete influences the particle size. The particles are too small if there is too much acid and too coarse if not enough acid is allowed to develop. Correct acidity when curd is cut also aids in obtaining a low mineral content of the casein as well as influencing odor and subsequent viscosity of the re-liquefied product. A high heat treatment (including drying temperature) will reduce dispersibility. Theophilus *et al.* (1935) observed that viscosity of casein solutions increased notably with higher precipitation temperatures.

Yields are affected by completeness of precipitation of the casein, loss of curd particles during whey and wash water drainage, completeness of washing, solids loss during pressing, and the final moisture content of the casein. Protein content of skimmilk, which normally ranges from 2.8 to 3.0 lb. per hundred wt., obviously affects the yields.

The composition of an acid coagulated casein may average within the following ranges:

	Per Cent
Protein (N × 6.38)	88.0 to 90.0
Fat	0.5 to 1.5
Ash	3.5 to 4.5
Moisture	4.0 to 12.0

Uses

Acid casein may be used in widely varying industries. It serves as an adhesive to bind clay pigments for the coating of paper. In this case freedom from insoluble particles and fat and the adhesive strength become important quality factors. Limited amounts of casein have also been used for glue, paint, and plastic manufacture. Food and medical products are other possible outlets for this product.

SODIUM CASEINATE AND CALCIUM CASEINATE

Processing Sodium Caseinate

The manufacture of dry sodium caseinate may be accomplished either by batch or the continuous operation. In either case, skimmilk is preheated to 115°F. Hydrochloric acid sufficient to bring the pH slightly below the isoelectric point is diluted with four parts of water and added to the skimmilk with thorough mixing. Precipitated casein and whey are separated by a sludge removing type of centrifuge. The curd is then milled to break up casein lumps. Water at 100°F. is used for washing the curd. One, two or more washings are required, depending upon desired purity for the final product. The water is removed by centrifugation or drainage. For a higher purity the curd is pressed to remove additional whey. The roller press is more suitable for large quantities.

A 0.5% solution of sodium hydroxide at 150°F. is used to disperse the curd and return the pH to the approximate neutral point of 7.0. Solids content of the slurry is usually controlled to 15 to 20% for spray drying. Precautions must be taken to have complete curd dispersion. For example, a centrifugal pump with fine mesh screen as its outlet or processing through a colloid mill will ensure a satisfactory dispersion.

The material is then spray dried using an outlet air drying temperature which reduces moisture to 3.5 to 4.5%. Special efforts to reduce the per cent of fines, such as low spray pressure and large orifice are recommended to prevent stack losses from becoming excessive. The final product then is sifted and packaged in dry milk containers. A normal yield is approximately 3 lb. per hundred wt. of skimmilk.

The composition of sodium and calcium caseinate is given in Table 22. The pH range in a good quality product will be 6.5 to 7.2. Sodium caseinate should be white or a very light cream color and free of off-flavors and -odors. Bulk density is 0.2 to 0.3 gm. per ml.

Utilization

Sodium caseinate functions as a binder, emulsifier, and whipping agent in food products. Because of its excellent nutritional value it is utilized as a source of protein in dry cereals or other starchy products and in special foods for infant, dietetic, and diabetic products. Sodium caseinate enhances whipping properties of ice milk or ice cream, toppings, and whipping cream products. It also improves body and texture and reduces shrinkage of these products. Usage generally is 2.0 to 4.0% of total solids of the product. It also functions as an important component in nondairy creaming and coffee coloring products, both liquid and dry. Sodium caseinate serves a useful purpose in sausage, other types of

TABLE 22

SODIUM CASEINATE AND CALCIUM CASEINATE COMPOSITION

Component	Sodium Caseinate	Calcium Caseinate
Protein	85.5%	86.2%
Ash	4.5	3.8
Lactose	4.0	3.5
Fat	1.5	1.5
Moisture	4.5	5.0

processed meat products and meat substitutes. It improves the protein quality of these products and binds the fat and water to inhibit shrinkage. The recommended amount usually is 1 to 2 % of the total meat product.

Manufacture of Dry Calcium Caseinate

In general, the equipment and procedure needed to manufacture dry calcium caseinate is similar to that used for manufacturing dry sodium caseinate. Hydrochloric acid may be used as the precipitating agent. A calcium hydroxide solution is used to neutralize the curd to a pH of 6.6 to 7.5 and at the same time affect its redispersion.

Calcium chloride and disodium phosphate also can be used to precipitate the casein in skimmilk. Control of temperature and pH to prevent the formation of a firm curd during precipitation is very important. Otherwise it is very difficult to obtain a satisfactory redispersion of curd particles prior to spray drying.

MALTED MILK POWDER

Malted milk was first developed in 1883 with commercial marketing beginning a few years later. Since then production volume has gradually increased. But today only a few companies are responsible for the entire production in the United States. Malted powders are used preferably for beverages in home and soda fountains. The confection industry also is a market.

Production Procedure

Barley malt may be obtained in dry form from a company specializing in its production, or it can be prepared by the company making malted milk powder. Details of manufacture vary among companies. Malt barley is ground (not finely); thoroughly moistened at ratio of 1:6 (barley to water by weight); and incubated at 90°F. to dissolve the enzymes. Then a mash is prepared by the addition of approximately ten parts of wheat flour. Starch is converted into maltose and dextrin with heating to

166°F. The barley husks are removed and whole milk or partially skimmed milk is added. Malt syrup may be added as a flavor supplement.

The mixture is then concentrated in a vacuum pan to 60% solids; and may be dehydrated in a special pan with an agitator by a batch operation, or in a spray or vacuum drum drier. In the pan method, moisture is removed until a porous, firm mass is obtained. The product is broken into chunks and manually removed. The chunks are milled and the malted milk powder is packaged. Vacuum drum drying is used for a substantial amount of the commercial production of malted milk powder.

Since malted milk is very hygroscopic, it is packaged in glass bottles with screw type metal lids or in other containers impervious to moisture vapors.

Composition

The milk fat content of malted milk powder must be not less than 7.5% and the moisture not more than 3.5%. Its composition may be:

Dextrins	45.0 to 49.0
Lactose	20.0 to 23.0
Protein	13.0 to 15.0
Milk fat	7.5 to 8.5
Ash	3.0 to 4.0
Maltose	1.0 to 2.0
Moisture	2.0 to 3.5

Cocoa, sugar, lecithin, salt and/or vanilla may be included.

Keeping Quality

If made from good quality products in a satisfactory manner, properly stored malted milk powder has very good keeping quality as compared to dry dairy products containing milk fat. Stale and oxidation off-flavors are the two common defects that develop over prolonged storage periods.

DRIED HIGH ACID PRODUCTS

Several dry high acid dairy products have been manufactured from skimmilk, whole milk, creams, whey, buttermilk (previously discussed), and various combinations of the above. Baking and the prepared dry mix industries have been the main users of a high acid product. Small amounts also have been used in the manufacture of speciality products.

Processing Procedures

The manufacture of a high acid skimmilk-whey powder was delineated by Barnes *et al.* (1961). Skimmilk, pasteurized at 170°F. and held for

3 to 10 min., was concentrated to 31% solids and inoculated with one to two per cent *Lactobacillus bulgaricus* at a temperature of 115°F. During incubation the product was agitated. Acidity developed within the range of 3.5 to 4.0%. High acid cottage cheese whey was concentrated to 31% solids resulting in an acidity of 2.4 to 2.8%. The whey and skimmilk concentrates were mixed, tempered to 110° to 115°F. and spray dried with caution taken to minimize the adverse effects of heat. An off-color and a high insolubility index became progressively more objectionable with increasing severity of heat exposure during drying.

An alternative method consisted of inoculating one to two per cent *L. bulgaricus* into the pasteurized skimmilk and developing the acidity to 1.75%. An equal amount of cottage cheese whey (about 0.9% acidity) on the solids basis was added and the mixture concentrated to 31% solids.

The procedure for drying high acid whole milk and cream is similar. If a lower acidity is acceptable *Streptococcus lactis* cultures may be used. The amount of solids concentration before culturing varies with the specific culture used for inoculation. *S. lactis* growth in condensed skim-

TABLE 23

ANALYSIS OF DRIED HIGH ACID SKIMMILK-WHEY PRODUCT

Trial	Acidity, %	Moisture, %	Scorched Particles, Mg.	Solubility Index, Ml.	Color	Flavor
1	5.6	3.06	15.0	1.4	White	Clean, acid
2	8.7	3.35	22.5	6.5	White	Clean, acid
3	11.1	1.38	22.5	5.0	V. sl. brownish	Clean, acid
4	11.6	2.37	22.5	3.8	White	Clean, acid
5	11.1	1.45	22.5	4.8	V. sl. brownish	Clean, acid
6	11.2	1.68	32.5	5.3	V. sl. brownish	Clean, acid
7	11.3	1.02	32.5	4.8	V. sl. brownish	Clean, acid
8	11.6	1.01	32.5	4.5	V. sl. brownish	Clean, acid
9	10.1	3.35	22.5	5.8	White	Clean, acid
10	10.5	3.42	22.5	5.8	White	Clean, acid

Source: Barnes, *et al.* (1961).

milk is increasingly retarded as the solids content increases above the 15 to 18% level. Table 23 shows the results of drying to various moisture contents.

Foam Spray Drying

Applying U.S. Dept. of Agriculture foam spray process to drying high acid products has several advantages. These include: a more readily dispersible product when recombined with water, less insoluble material, and less loss of desirable volatile substances. The density of the high acid dry product is reduced 50 to 100% as compared to regular spray dried product.

MISCELLANEOUS PRODUCTS

Dietetic Products

Among the many dry dairy products classified as dietetic the 900-calorie and complete meal preparations have received the most attention in recent years. They provide people limiting their calorie intake a nutritionally balanced food which is consumed after reliquefication. The general procedure in manufacture of such products is to fortify skimmilk with a protein such as sodium caseinate along with vitamins, minerals, and other essential growth factors. Chocolate, vanilla, and strawberry are but some of the common flavorings which may be added for taste appeal. But the most simple of these diet products contains flavored dry skimmilk plus fortification. When reconstituted according to directions, a quart contains 900 calories and the average daily requirements of minerals and vitamins.

The specific details of formulation and manufacture are restricted information of the producing company.

DRIED SKIMMILK-SHORTENING PRODUCTS

Nonfat dry milk-shortening products have been produced in small amounts for many years. They have been used chiefly by the baking industry. The prepared dry mix manufacturers also are a potential market for this product.

Skimmilk is concentrated to 20 to 30% solids, then vegetable oil of shortening type is added. This mixture is homogenized at 2000 to 3000 p.s.i. and 150° to 160°F. The mixture is spray dried under conditions similar to those for drying cream. The same precautions must be taken that are necessary in dry cream manufacture. Continuous removal of the product from drying chamber and cooling immediately to 90°F. or below are a necessity. Failure to properly cool results in rapid brown or black discoloration. Bags and drums have been most commonly used as containers.

The composition of the dried product is:

	Per Cent
Vegetable fat	40.0 to 60.0
Milk solids-not-fat	38.0 to 58.0
Moisture	0.5 to 2.0
Emulsifier	0.5 to 1.0

Butter Powder

Butter with modifications has been dried successfully by Australian scientists (Anon. 1963B). The fat content of the dried product is identical to regular butter, but other components differ. Consequently, the classification of this dry form as strictly a dry butter by regulatory officials is uncertain.

Following is an example of ingredients and composition:

132.0 lb. of 32% pasteurized cream	81.98 milk fat
6.0 lb. of dry acid casein	3.50 emulsifier
1.0 lb. of sodium citrate	6.70 nonfat milk solids
20.0 lb. of skimmilk (9% T.S.)	6.70 caseinate-citrate
7.0 lb. of 2% sodium hydroxide	0.50 free-flowing agent
3.5 lb. of 36% GMS	0.60 moisture
0.5 lb. of sodium aluminum silicate	0.02 antioxidant
10.9 gm. of BHA	

In production, pasteurized cream is separated to 62% fat. Sodium citrate is dissolved in the cream and the product is homogenized at 110°F. and 1000 to 500 p.s.i. High grade casein is dispersed in skimmilk at 150° to 180°F. and sodium hydroxide added slowly. Gycerol monostearate (GMS) is melted and the antioxidant, butylated-hydroxyanisole (BHA) added. The mixture is blended into the caseinate solution. This is added to the cream solids. After heating to 150°F. the product is spray dried, cooled immediately, and dry blended with a free flowing agent composed of 80% sodium aluminum silicate and 20% calcium phosphate.

Butter powder is useful for the preparation of baking goods, dry mixes, sauces, icings, fillings, and ice cream.

In the processing of butter powder for making cakes, complexes of fat protected by proteins should be avoided. It is desirable to keep fat in a form readily converted to free fat. This is accomplished by use of a relatively low homogenization pressure and temperature and by adding a portion of the caseinate and solids-not-fat to main ingredients of the formula after homogenization. The dispersed emulsifier also should be added after homogenization. These precautions increase the amount of free fat released during the cake baking.

Dry Milks and Cream Products in Blocks or Tablets

Numerous attempts have been made to compress dry milks and cream products into tablet or block form to improve the convenience and increase the density, but maintain good dispersibility. Commercial success of these products in the United States has been negligible.

Korolev *et al.* (1964) reported a procedure for making milk blocks. Milk is concentrated in an evaporator to one-third its volume. Seventy to seventy-seven per cent sugar solution is added to increase the total solids to 52 to 57%. After homogenization at 1500 to 2000 p.s.i. the moisture is removed by evaporation under vacuum in a vertical cylinder using a scraper unit until solids are 84 to 90%. The product, now a pasty consistency, is pumped into an aluminum or stainless steel mold for solidification into blocks at 68°F. The block of milk is removed from the mold and wrapped.

Another approach is to compress foam spray dry milks or agglomerated dry milks into blocks. These are disintegrated into the powder form before reconstituting. Hanrahan and Konston (1965) used 600 p.s.i. to compress foam spray nonfat dry milk into cakes with a bulk density of 0.45 gm. per ml. (Fig. 74). The cakes were relatively stable during handling

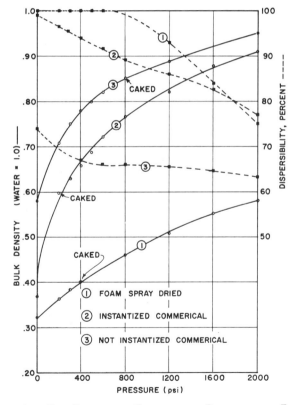

FIG. 74. THE EFFECT OF COMPRESSION PRESSURES ON BULK DENSITY AND DISPERSIBILITY OF NDM

yet "repowdered" easily. One hundred per cent dispersibility was retained at that pressure.

These investigators used 200 p.s.i. to compress agglomerated nonfat dry milk. The resulting density was 0.6 gm. per ml. After "repowdering" and reconstituting the sinkability was lost, but the dispersibility was only slightly impaired.

In the preparation of "cream" tablets, milk and cream are standardized in a ratio to obtain the desired fat content of the dried product (30 to 55%). The liquid product is preheated and homogenized thoroughly. After concentration in an evaporator, a stabilizing salt, e.g., sodium citrate, or disodium phosphate is added along with a substantial amount of lactose which is usually within 15 to 30%. An antioxidant (0.02%) may be included. The mixture is spray dried. More lactose (5 to 25%) may be blended into the dry mixture and the product is compressed into tablets using just enough pressure so the tablet maintains the pressed form with normal handling.

Hansen (1959) reported on a similar procedure for making cream tablets: lactose is added to 20% (fat) coffee cream, and the mixture spray dried. More lactose is blended into the dried product. It is agglomerated by using a fine spray of moisture. After sifting the product can be machine pressed into tablets at the rate of 200 to 400 per min. Product moisture is further reduced to 2.0 to 2.5% by drying the tablets. They are then machine wrapped in aluminum foil. These tablets contain approximately 55.2% lactose, 35.5% fat, 7.0% protein, and 2.3% moisture.

Meringue Powder

Buchanan (1965) outlined a procedure for processing a meringue base from skimmilk. It is a substitute for egg whites. Skimmilk is condensed to 25% T.S. and tempered to approximately 40°F. Six and one-half per cent (based on weight of total solids of skimmilk) calcium hydroxide is mixed into 10% cold water. The slurry is added rapidly to the condensed skimmilk with vigorous agitation. The mixture is held about 72 hr. at 45°F. or less. It is spray dried, cooled, sifted, and packaged.

Sponge Powder

Australian investigators developed a sponge powder for the baking industry. The procedure was reported by Buchanan (1965). Skimmilk is condensed to 25% t.s. and the temperature adjusted to 68°F. On the basis of total solids of the skimmilk, 3.5% sodium hexametaphosphate is dissolved in water and added. The mixture is heated to 140°F. within approximately 23 min. The pH is increased from 6.5 to 9.0 by the addition of 4.0% trisodium phosphate and 0.7% sodium hydroxide in aqueous solution. The product is immediately cooled to 86°F. or lower. It is spray dried, sifted, and packaged.

Milk-Potato Product

Insignificant amounts of milk-vegetable mixtures have been spray or drum dried. One of these consists of milk and potato solids. The potatoes are peeled and cooked by steam pressure. Condensed milk or skimmilk is standardized with comminuted potatoes at a solids ratio of 2 to 1. The mixture should have a solids content of approximately 30%. It is heated to 150°F., homogenized at 2500 p.s.i. and spray dried to 3.5% moisture.

Seasoning, if it does not contain particles that will clog the nozzle orifice, and salt may also be added to the mixture before homogenizing and drying. If 20% of the milk solids in the dry product is milk fat the flavor is improved compared to lesser amounts. However, the keeping quality is significantly reduced.

Stevens (1949) presented a procedure for drying a mixture of cheese whey and potatoes by the spray or drum process. Potatoes were peeled and cooked for 20 min. at 15 p.s.i. of steam. They were ground and combined with whey (32.5 lb. of potatoes and 100 lb. whey). The mixture was agitated, heated to 145°F., and homogenized at 1500 p.s.i. It was then spray dried with an 0.02-in. orifice at 2700 p.s.i. or drum dried. The acidity of the whey must not exceed 0.38% to produce a satisfactory quality by the drum process.

The author suggested dry whey-potato product may be used in soups, pancakes, and various bakery products.

REFERENCES

AMMON, R. 1956. The manufacture of dried cream. XIV Int. Dairy Cong. *1*, No. 2, 329–331.

ANON. 1949. Experiment concerning the manufacture of rennet and acid casein. Rept. 56, State Expt. Sta. Creamery, Hillerød, Denmark.

ANON. 1955. Casein manufacture. Aust. J. Dairy Tech. *10*, 140–142.

ANON. 1963A. Makes human-milk substitute. Food Eng. *35*, No. 8, 40.

ANON. 1963B. Butter powder gains acceptance. Food Eng. *35*, No. 8, 43.

ANON. 1963C. Dries heat sensitive cheese with infrared. Food Eng. *35*, No. 6, 93.

ANON. 1965. Handling dairy wastes. Manuf. Milk Prod. J. *56*, No. 5, 7–8.

BABCOCK, C. J. 1922. The whipping quality of cream. U.S. Dept. Agr. Bull. *1075*.

BARNES, J. C., STINE, C. M., and HEDRICK, T. I. 1961. Edible high-acid powder made from skimmilk and cottage cheese whey. Milk Prod. J. *52*, No. 4, 8, 9, 34.

BRADLEY, R. L. 1964. Spray-drying of natural cheese. Ph.D. Thesis, Michigan State University, East Lansing, Mich.

BRADLEY, R. L., and STINE, C. M. 1964. Foam spray drying of natural cheese. Manuf. Milk Prod. J. 55, No. 6, 8–11.

BUCHANAN, R A. 1957. Hydrochloric acid casein. The continuous precipitation process. Victorian Department of Agriculture J. Agr. 55, 527–529, 531, 533–534.

BUCHANAN, R. A. 1965. Special dried milk products. Aust. J. Dairy Technol. 20, 57–58.

BULLOCK, D. H., HAMILTON, M. O., and IRVINE, D. M. 1963. Manufacture of spray-dried Cheddar cheese. Food in Can. 23, No. 3, 26–30.

CAYEN, M. N., and BAKER, B. E. 1963. Some factors affecting the flavor of sodium caseinate. J. Agr. Food Chem. 11, No. 1, 12–14.

CHUCK, F. Y. 1935. Process for stabilizing milk, milk powder and similar colloidal products. U.S. Patent 2,016,592.

CHUCK, F. Y. 1939. Manufacture of lactose containing materials. U.S. Patent 2,174,734.

CORBETT, W. J. 1948. Powdered ice cream mix. Food Ind. 20, 534–537.

COULTER, S. T. 1946. The manufacture of dry ice cream mix. Milk Plant Mo. 35, No. 7, 84–85.

CZULAK, J., HAMMOND, L. A., and FORSS, D. A. 1961. The drying of Cheddar cheese. Aust. J. Dairy Technol. 16, 93–95.

ELDREDGE, E. E. 1933. Process of preparing whey concentrate. U.S. Patent 1,923,427.

FRITZBERG, E. L. 1944. Precipitation of proteins. Can. Patent 422,477.

HANRAHAN, F. P., and KONSTON, A. 1965. Properties of compressed nonfat dried milk. J. Dairy Sci. 48, 1533–1535.

HANRAHAN, F. P., and WEBB, B. H. 1961A. Spray drying cottage cheese whey. J. Dairy Sci. 44, 1171.

HANRAHAN, F. P., and WEBB, B. H. 1961B. U.S. Dept. Agr. develops foam-spray drying. Food Eng. 33, No. 8, 37–38.

HANSEN, R. 1959. Cream tablets—a new dairy product. Dairy Ind. 24, 895–896.

HARFORD, C. G. 1940. Recovery of solids from buttermilk. U.S. Patent 2,209,694.

HEDRICK, T. I. 1953. Spray dried buttermilk. Milk Plant No. 42, No. 9, 19–22, 72, 73.

HOLM, G. E. 1945. Dried milks. Milk Plant Mo. 34, No. 8, 23, 52, 54, 65–67, 69.

HUNZIKER, O. F. 1949. Condensed Milk and Milk Powder. Published by author, LaGrange, Ill.

KEMP, J. D., DUCKER, A. J., BALLANTYNE, R. M., and ACHESON, G. C. 1957. A study on the flexible packaging of dry cream and potato powders. Def. Research, Med. Lab Report 174-3, p. 16.

KOROLEV, A., FAVSTOVA, V., and SKOROMNIKOVA, V. 1964. Block milk. Mol. Prom. 25, No. 3, 32–34 (In Russian) Dairy Sci. Abs. 26, 369.

KRAFT, G. H. 1938. Drying whey. U.S. Patent 2,118,252.

LAVETT, C. O. 1939. Process for drying whey. U.S. Patent 2,172,393.

LAVETT, C. O. 1940. Process of drying lactose. U.S. Patent 2,197,804.

LAVETT, C. O. 1941. Process of drying whey. U.S. Patent 2,232,248.

LOWE, V. J., and BEAN, A. W. 1940. Production of cooked acid casein. U.S. Patent 2,225,387.

MANNING, P. D. V. 1938. The manufacture of casein and of by-products from whey. Proc. 11th Annual State Coll., Washington Institute of Dairying, Pullman, 38–47.

MULLER, L. L. 1959. Investigations in casein manufacture and quality. Aust. J. of Dairy Technol. 14, 81–88.

MULLER, L. L. 1960. New Zealand and Australian developments in casein manufacture. Aust. J. Dairy Technol. 15, 89–95.

MULLER, L. L., and HAYES, J. F. 1960. Improved equipment for the manufacture of casein. Aust. J. Dairy Technol. 15, 201–205.

MULLER, L. L., and HAYES, J. F. 1963. The manufacture of low-viscosity casein. Aust. J. Dairy Technol. 18, 184–188.

MUSSETT, A. T., and MARTIN, W. H. 1948. Manufacture, use and storage of dehydrated sweetened condensed skimmilk. Ice Cream Trade J. 44, No. 12, 54, 56, 81, 82, 84, 85.

OATMAN, W. F. 1945. Casein. U.S. Patent 2,388,991.

OBERG, E. B. 1940. Apparatus for preparing casein. U.S. Patent 2,190,136.

OBERG, E. B. 1941. Preparing casein. U.S. Patent 2,228,151.

OETIKER, N. 1960. The manufacture of lactic acid and its use as a precipitant for casein. Aust. J. Dairy Technol. 15, 69–76.

OTTING, H. E. 1941. Casein. U.S. Patent 2,225,506.

PAGEL, L. C. 1950. Casein. U.S. Patent 2,184,002.

PATTEN, J. L. 1955. The manufacture of casein. I. Lactic casein. II. Hydrochloric and rennet casein. Aust. J. Dairy Technol. 10, 160–164.

PEEBLES, D. D. 1938. Method for manufacture of stable powdered food products containing milk sugar. U.S. Patent 2,126,807.

PEEBLES, D. D. 1943. Process of manufacturing a concentrated milk product. U.S. Patent 2,336,634.

PEEBLES, D. D., and MANNING, P. V. 1933. Method for manufacture of lactose containing material. U.S. Patent 1,928,135.

PEEBLES, D. D., and MANNING, P. V. 1937. Manufacture of stable powdered products containing milk sugar. U.S. Patent 2,088,606.

PINCKNEY, M. W. 1960. Protein extraction from whey. Proceedings of the 8th Annual National Dairy Eng. Conf., Michigan State University, East Lansing, pp. 27–29.

PYENSON, H., and TRACY, P. H. 1948. Manufacture of powdered cream for whipping by aeration. J. Dairy Sci. 31, 539–550.

REYNIERS, J. A. 1951. Process for preparing casein. U.S. Patent 2,518,493.

SHARP, P. F. 1951. Casein manufacturing process. U.S. Patent 2,519,606.

SHARP, P. F., and SHIELDS, J. B. 1951. Casein. U.S. Patent 2,562,646.

SIMMONS, N. L. 1930. Process of making whey products. U.S. Patent 1,763,633.

SMART, P. H., and NELSON, C. E. 1957. Casein. U.S. Patent 2,807,608.

SPELLACY, J. R. 1939. Method of drying milk whey. U.S. Patent 2,163,331.

SPELLACY, J. R. 1942. Casein. U.S. Patent 2,304,429.

SRINIVASAN, M., MURTHY, P. N. A., SREENIVASAN, A., and SUBRAHMANYAN, V. 1963. Preparation of calcium caseinate with calcium hydroxide in sucrose solution. Food Technol. 17, No. 3, 112–113.

STEVENS, A. H. 1949. The preparation of dried whey-potato mixtures. Nat'l. Butter & Cheese J. 40, No. 4, 28, 29, 60, 62.

STRIBLEY, R. C. 1965. Ionselective membrane electrodialysis, a tool for the food industry. Proceedings of the 13th Annual National Dairy Eng. Conf., Michigan State University, East Lansing. pp. 34–40.

SUPPLEE, G. C. 1939. Treatment of whey. U.S. Patent 2,173,922.

THEOPHILUS, D. R., HANSEN, H. C., SNYDER, R. S., WOOD, R. E., and OLMSTEAD, R. L. 1935. Effect of various phases in the manufacture of casein by the natural sour method on its physical and chemical properties. Idaho Agr. Expt. Sta. Bull. 212. Moscow, Ida.

TRACY, P. H. 1948. The manufacture of powdered cream. Can. Dairy & Ice Cream J. 27, No. 11, 52–53.

TREXLER, P. C. 1952. Method of producing casein. U.S. Patent 2,618,629.

VON LOESECKE, H. W. 1955. Drying and Dehydration of Foods. 2nd Edition. Reinhold Publishing Corp., New York.

WEBB, B. H., and WHITTIER, E. O. 1948. The utilization of whey: a review. J. Dairy Sci. 31, 139–164.

WENDT, E. J. 1945. Precipitation of casein. U.S. Patent 2,369,095.

WHITTIER, E. O., and WEBB, B. W. 1950. Byproducts from Milk. Reinhold Publishing Corp., New York.

WILLIAMS, D. H., and POTTER, F. E. 1949. Dried ice-cream mixes. Bur. Dairy Ind. Mimeo. DBIM-Inf-74. U.S. Dept. Agr., Washington, D.C.

ZIEMER, K. A., AMUNDSON, C. H., WINDER, W. C., and SWANSON, A. M. 1962. Dried buttermilk and its properties in ice cream. J. Dairy Sci. 45, 659.

CHAPTER 9

Quality Control and Sanitation

QUALITY OF RAW MILK

Introduction

The important effect of raw milk quality on the quality of dry milks has been stressed for many years, but rarely overemphasized. Most dry milk products in the past have been processed from manufacturing grade milk or a mixture including surplus Grade A raw milk. The amounts from Grade A milk have been rapidly increasing in recent years and the trend may continue. Results of a survey in 1963 (Hedrick and Hall 1964) indicated 28% of the milk plants dried only surplus Grade A and an additional 41% dried some surplus Grade A milk.

The change from cans to farm bulk tank system on the farm in general has improved the quality of manufacturing grade milk, producing less developed acidity and flavor deterioration in the milk. Also the current trend toward fewer farm herds and more cows in the remaining herds may have a beneficial effect on quality of the milk.

Dairy Farm

A farm producing milk must meet minimum requirements for facilities and herd health (tuberculin and Brucellosis tested). Farm sanitation and practices involve: housing and premises, milk room, utensils, milking, and water supply. These requirements are presented in detail by the U.S. Department of Agriculture (1966) and American Dry Milk Institute (1961). The production facilities and sanitation practices should meet one of these farm surveys. An example of the USDA Dairy Farm Certificate Report and ADMI Dairy Farm Report is given in the Appendix. The farms should be inspected at least once and preferably twice per year to determine compliance.

Milk Quality

When milk is received at the plant it should be 45°F. or lower, the producer having cooled or delivered it within 2 hr. after milking. However, the regulatory requirement may be 50°F. The milk should be clean, sweet, free of off-flavors and -odors, and reasonably free of extraneous material. Contamination by antibiotics, pesticides, or other chemical residues and metals is highly undesirable. No abnormal milk should be

197

accepted, thereby excluding milk from cows with mastitis. Acid development is objectionable and indicates an excessive bacterial count. All milk should be checked at the plant by sensory tests and milk with developed acidity rejected. If milk is handled in cans, they should be clean, well-tinned, and free of open seams. Exposed iron in cans is an important source of metallic contamination of the milk.

Representative samples from each producer should be taken once or preferably twice a month on unannounced and variable days of the month. Results should comply with the limits as shown in Table 24 for sediment and bacterial standards. Figure 75 illustrates the sediment discs for the

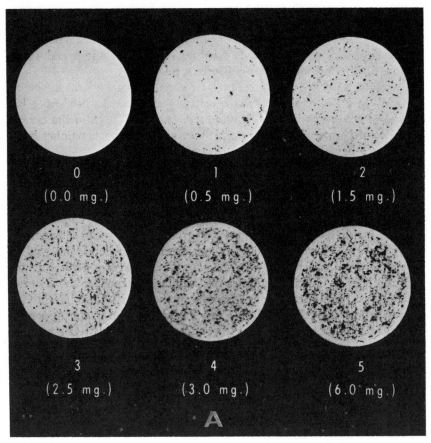

Fig. 75.   U.S. Department of Agriculture Sedi

A is the classification for the off-the-bottom method

various standards. The other tests also should comply with the minimum requirements of the USDA or ADMI. In addition, a test on the farm samples several times a year for added water has merit. Laboratory equipment and procedures for testing farm milk samples are described in the latest edition of "Standard Methods" by the American Public Health Association (1971).

Raw milk for Grade A dry milk must comply with the more strict standards of the U.S. Public Health Service. This includes the minimum facilities and sanitation procedures on the farm as well as the quantitative tests for bacteria.

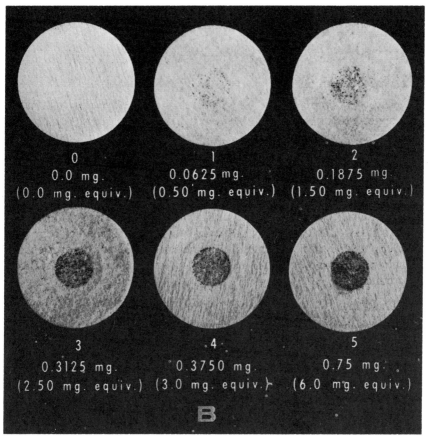

MENT STANDARDS FOR MILK AND MILK PRODUCTS

for cans and B is for 1 pt. of mixed bulk milk.

TABLE 24

U.S. DEPARTMENT OF AGRICULTURE PROPOSED SEDIMENT AND BACTERIAL ESTIMATE STANDARDS
ON MILK FOR MANUFACTURING PURPOSES

Classification	Milk in Cans Off-Bottom Test $1^1/_8$-in. Diam. Disc	Milk in Farm Tanks; Mixed Sample 0.4-in. Diam. Disc	Direct Microscopic Clump Count or S.P.C., Per Ml.	Methylene Blue Decolored in	Resazurin Reduction Time to Munsell Colors P7/4
No. 1 acceptable	Not more than 0.50 mg.	Not more than 0.0625 mg.	Not more than 500,000	Not less than 4.5 hr.	Not less than 2.25 hr.
No. 2 acceptable	Not more than 1.50 mg.	Not more than 0.1875 mg.	Not more than 3,000,000	Not less than 2.5 hr.	Not less than 1.50 hr.
No. 3 probational	Not more than 2.50 mg.[1]	Not more than 0.3125 mg.[1]	More than 3,000,000[2]	Less than 2.5 hr.[2]	Less than 1.50 hr.[2]
No. 4 reject	More than 2.50 mg.	More than 0.3125 mg.	. . .	. . .	. . .

Source: U.S. Dept. Agr. 1966.
[1] Probational not more than 10 days.
[2] Probational not more than 4 weeks.

CARE AND HANDLING MILK IN PLANT

Stainless Steel Equipment

The use of stainless steel equipment meeting 3-A standards will minimize opportunities for metallic contamination and assure ease of cleaning. All milk contact surfaces of equipment and the tubing for recycling the cleaning solutions, should be constructed of an approved stainless steel, glass, or plastic material. Even a short contact in limited areas such as with a white metal valve can cause a noticeable increase in the metallic content of milk. Cleaning solutions may pick up metallic ions and replace them on the stainless steel during cleaning-in-place. Later these ions are released into the milk during processing. Presence of copper is especially undesirable, but iron contamination should not be overlooked. These metallic ions catalyze the milk fat oxidation reducing shelf-life of the product.

Milk Handling

Quality of raw milk is not improved by storage and handling; it may easily deteriorate during these operations. Consequently, milk should be held for the minimum time on the farm, in receiving station and drying plant before being pasteurized and processed. During plant storage a holding temperature of 45°F. or a temperature preferably below this will minimize microbiological changes in the milk. Chemical and enzymatic changes are not prevented even at colder temperatures although the rate may be slower.

Recent trends in the industry are actually resulting in longer milk storage periods instead of shorter. There has been increased reliance on mechanical refrigeration to justify every-other-day farm milk pickup and longer plant storage. This has the general effect of encouraging growth of psychrophilic bacteria, rather than the lactic acid types. Some psychrophiles produce enzymes causing lipolytic and proteolytic deteriorations that are very objectionable.

Care of milk during processing directly influences the quality of dry milk. The forward flow during preheating, concentrating, reheating the condensed milk, and drying of the product should include a minimum of recycling. Detrimental effects on flavor have been demonstrated with 4 or 5 recycling points in the system although adverse effects of one or two recycles may not be detectable by sensory tests.

High and excessively long heat treatments of the milk beyond those needed to give baking qualities can increase the rate of off-flavor development during storage of the dry products. Excessive heat may cause more casein denaturation in condensed milk than in uncondensed milk.

Microorganisms

Growth of microorganisms may occur during storage and processing of the milk. Temperature influences the type and rate of organism growth. Bacterial acid producers, for example, *Streptococcus lactis*, grow well between 70° and 95°F. Thermophilic types multiply at temperatures of 115° to 145°F. or higher. Species of *Staphylococcus* bacteria can grow in the range of 90° to 115°F., but apparently not at 118°F. or above.

The growth of many types of microorganisms is detrimental to milk quality despite subsequent destruction of these organisms by the ensuing heat treatment. By-products and enzymes of the microbiological growth reduce keeping quality, increase intensity of off-flavor, and some organisms (*Staphylococci*) produce highly toxic products which are heat resistant. Another objectionable type is the heat resistant bacteria in dry milk which is detrimental to the fermentation process during bread production.

The prevention of bacterial growth in milk at the plant is likely to be more of a problem during hot weather, mainly because the milk arrives at the plant with a higher contamination; the refrigeration system may be taxed by high volume in relation to plant capacity. The chief areas of concern center around storage of raw milk and the processing steps: preheating for separation, concentrating, and holding of excess condensed milk. Even though the temperature of the milk may be too high for bacterial growth, the foam temperature can be low enough at some stages to permit their growth, particularly thermophilic organisms. Foam in hot wells, evaporators, and balance or surge tanks can support the development of excessive numbers of bacteria. Conversely, foam in storage tanks, etc., may be warm enough to allow organisms to grow even though the milk or concentrate under it are cold.

Continuous Processing Period

During the season of peak milk production, there is an inclination to maintain long continuous operations with fewer cleaning periods. When these processing periods are excessively long, product quality usually suffers. A maximum time consistent with good quality of dry product should be established and closely followed.

Cooling and Heat Effects

Dry milk products should be exposed to the least amount of heat once the desired moisture content is obtained. Complete and prompt continuous removal of the dry product from the drying chamber and collector system are important to prevent heat damage to the product. Dry milks packaged while still hot in 50- or 100-lb. bags or 200- to 250-lb. drums and stored compactly in a warm storage space may be several days dropping to the ambient temperature. The same is true of uncooled dry milks in large bins. Consequently, lumpiness due to "heat caking" can develop and in such cases the period before storage off-flavor becomes noticeable is decreased. Nonfat dry milk that is still above 125°F. after the second day of manufacture usually develops small lumps. These become larger with a continued delay in cooling. However, heat lumping is not a common problem in plants with a cooling system.

The dry milks with fat, dry whole milk and dry creams, have a greater need for cooling than the dry nonfat products. The longer the product is hot the higher the percentage of free fat. The rate of fat oxidation is accelerated by failure to cool fat products sufficiently and promptly. Uncooled products having fat under certain conditions such as holding in a drum will develop noticeable discoloration within a few hours.

Prompt cooling of the dry products after dehydration is accomplished by several types of cooling systems. The use of ambient air is least effective during the warm months and during this period dry product cooling is slowest in storage. Therefore, the use of refrigerated air is more effective for cooling, particularly during hot weather. However, there is the possibility of moisture condensation being absorbed by the product. Furthermore, some refrigeration evaporators are difficult to keep sanitary.

Satisfactory cooling is now being accomplished by spraying liquid carbon dioxide or nitrogen into dry milk as it comes from the drier. Condensation is not a problem, but the cooling must be controlled so the product temperature remains above the dew point.

SANITATION AND HOUSEKEEPING

General

Good sanitation includes the outside premises as well as the interior of the building and its facilities; all equipment and personnel habits. Good housekeeping contributes directly to efficiency. The esthetic values of good sanitation and housekeeping are also an important consideration.

The hazards of poor sanitation are equally pertinent in causing product contamination by microorganisms and extraneous matter along with lowering the sensory evaluation. Good quality products require more effort than simply meeting the legal minimum requirements established by regulatory agencies.

Building and Equipment

Good ambient conditions of the plant's location are essential for adequate sanitation. Satisfactory material and workmanship in the structure, particularly the floors, walls, and ceiling, are also necessary. Materials must be appropriate for wear, impervious, and sufficiently smooth for easy cleanability. For example, the floors under the evaporator, high pressure pump and around the drier should be constructed of creamery tile and grouting resistant to wear and acids. But floors in sifting rooms or storage rooms need only be durable to the wear of the traffic. All floors should have good drainage.

The lighting must be adequate for the work—not less than 30 to 50 ft. candles of intensity in the processing and analytical areas. Ventilation should be sufficient to maintain fresh, comfortable air without excessive humidity. If possible, the use of ventilation ducts in a dry milk plant should be avoided (see Insect Control in this chapter). In order to facilitate insect control separate ventilation systems for various areas of the plant are advantageous rather than one central system. The system should be planned to prevent the spread of milk dust as much as possible.

The plant provisions for rest rooms, hand washing facilities, storage closets, and rooms for equipment cleaning facilities contribute directly to sanitation and housekeeping. An ample and dependable supply of utilities is necessary. This includes safe water and no cross pipes between safe and unsafe water.

The receiving, holding, and processing equipment should meet 3-A standards if they have been developed for the particular items involved. Otherwise, the equipment including bulk tankers should be of sanitary design, durable, nontoxic, smooth, and easy to clean. Containers for the dry product also must be sanitary, nonreactive, impervious, and safe.

Employees

The employees should work in a dry milk plant only when able to pass a medical examination for all infectious diseases. Physical cleanliness of body and garments of employees requires continual attention. In fact, clean personal habits should be a prerequisite for working in a dry milk plant. Also the use of tobacco in processing areas should not be permitted.

Cleaning Evaporators

Introduction.—Cleaning evaporators and vacuum pans presents more difficulties than most other dairy equipment. The combination of relatively high temperature in heating medium, large surface area of equipment, long period of operation with ample opportunities for burn-on of milk solids is the principal cause of cleaning difficulties. CIP techniques generally have replaced the manual method. The manufacturers of cleaning compounds have been very helpful during the conversion to CIP. Equipment companies are now including CIP facilities with the fabrication of many items of equipment.

Spray, circulation, or a combination of both techniques should be adapted to type and design of the evaporator. Figure 76 shows different shapes and hole patterns of spray devices for evaporators. The spray

Courtesy of Wyandotte Chemicals Corporation

Fig. 76. Spray Devices with Different Hole Patterns to Clean Separators, Chests, Vapor Lines, and Tangential Inlets on Evaporators

method pumps cleaning solutions through spraying devices which are located to insure complete coverage of all surfaces. Proper solution temperature is maintained by direct steam or a preheater. Coils or tubes used for heating the product during condensing are less desirable for heating the cleaning solutions.

In the circulation system the cleaning solutions cover the same surfaces and maintain the same patterns as the product flow during processing. An outside heating source is preferable, but the coils or chest tubes also can be used in the cleaning process. Since not all surfaces are adequately contacted by cleaning solution in the circulation method, sprays are used to reach certain specific areas such as the separator, vapor tube, vapor dome, etc.

Details of Circulation Cleaning.—One cleaning procedure for the evaporator is: (a) immediately after shutdown of the condensing operation and product removal has been completed, rinse water at 115°F. is flushed through the system; (b) when rinsings are clear, an alkaline cleaning solution of one to four per cent is heated to 180°F. and circulated or sprayed for 45 to 60 min. throughout the system; (c) solution is drained and the equipment surface rinsed with warm water; (d) an acid solution of 0.3 to 0.5% at 160°F. is circulated for 20 to 30 min. to remove soils of mineral nature; (e) the acid solution is drained and the system rinsed with water at 140°F.; (f) the outside of the evaporator, deck, and floor are scrubbed with an appropriate cleaning solution, then are rinsed thoroughly; (g) just before using the evaporator is sanitized. The effective use of steam or water at 180°F. is difficult because of the large surface area of the evaporator.

An alternative sprays or circulates a chemical sanitizer at 75 p.p.m. of chlorine or its equivalent for 5 min. Fogging completely with 500 p.p.m. of chlorine solution is still another alternative. In these latter methods, the system should be drained soon after sanitizing to inhibit corrosion from chlorine compounds and product condensing initiated without delay.

Cleaning Driers and Related Items

Dry and Wet Methods.—Cleaning of the drier is more likely to be neglected in a plant than the evaporator, particularly if the spray nozzles functioned properly during the operation and no improperly dried solids have soiled the surface of the drier. The cleaning method must be adapted to the drier design. A practical policy for a flat bottom drier is to dry clean by a vacuum procedure each time the evaporator is cleaned and to wet wash the drier once a week. Wet washing should be a routine whenever the product adheres to the drier surface despite the vacuum treatment. Dry cleaning a cyclone drier is not practical. It should be wet washed after each day's operation.

Spray ball or similar devices and a high pressure gun are common methods. The spray devices must be located for adequate coverage. They usually are installed into the permanent lines just before cleaning and are removed after cleaning in order not to interfere with the air currents during the drying process. The wet wash should be preceded by removal of the dry product with vacuum cleaner in driers where practical. If augers are used for continuous removal of product, they should be removed from the drier, washed, rinsed, and replaced after drier is cleaned.

Procedure for CIP.—The steps for cleaning the drier and cyclones are: (a) rinse with water at 110°F. until clear and drain; (b) recycle alkali cleaning solution at a concentration of 0.5 to 1.0% and a temperature of 180°F. for 45 to 60 min.; drain; (c) spray rinse with water at 160°F., drain, and dry. If the drier is soiled excessively, a half hour circulation of a 0.4% acid cleaning solution becomes necessary and is followed by a water rinse at 160°F. before drying. Drier must be completely dry before it is used.

Soon after cleaning to avoid corrosion, the surfaces may be dried by air limited to 120°F. circulated through the system by the drier fans. The drier may be sanitized by hot inlet air until the surface is heated to not less than 180°F. for 10 min. For the manufacture of dry milks with very low recontamination of microorganisms sanitizing with a chemical may be more effective. A sanitary closet near the drier will provide storage for brushes and other cleaning tools plus uniforms for entering a drier with edible product in it.

Sifter.—Cleaning procedure for the sifter consists of removing the loose dry milk, disassembling, rinsing, washing all the parts, and thoroughly rinsing with water at 180°F. or above. After the parts are dry they are reassembled. An alternative is to dry clean the sifter with a vacuum cleaner.

Sanitation of Related Areas

The maintenance of clean floors and walls of the room housing the drier is an important part of a good plant sanitation program. Periodic scrubbing with a recommended floor cleaning compound is necessary. Clean uniforms and shoe covers should be used when entering the drier with edible product in it. Refuse must be removed daily and either burned or hauled away a sufficient distance so flies and rodents will not be conveyed back to the plant.

Copies of the Plant Inspection Form of the USDA and the Plant Sanitation Survey of the ADMI are presented in the Appendix.

INSECTS IN DRY MILK PLANTS

More than 750,000 species of insects are known. About 5000 are of concern to man and relatively few of these are a problem in a dry milk plant. However, any infestation is serious, thus control is essential. Insects which infect a dairy plant may be roughly classed into one of three groups—flies, cockroaches, and dermestid beetles.

Flies

The house fly is the most common of the flies. The female lays the eggs on moist, decaying organic material. Maggots hatch, feed, and in a few days complete their cycle, evolving into the adult. Flies enter a dry milk building through any opening of suitable size, but mainly through windows and doors. Screens or utilization of properly designed and mounted fans are major control measures. The premises must be kept free of breeding conditions to maintain effective control.

The exhaust air from the drier is certain to contain fine product particles, especially if a cyclone collector system is used. The immediate area around the exhaust vent must be washable. The roofing adjacent to the exhaust drying air vent is frequently overlooked, and if not cleaned regularly is a breeding place of insects.

Cockroaches

Cockroaches of several species breed in the habitat of a dry milk plant. Initial or continual entrance may be along with utilities, sewer, or other sub- or ground-surface openings. Incoming shipments of supplies and empty containers are frequently another source of cockroaches.

The American (reddish-brown), German (tan), and the Oriental (black) roaches are the more common. Eradication or control requires monthly or even weekly spraying of an effective mist throughout the complete plant. Special attention must be given to all possible hiding places such as cracks and crevices in walls, floors, and equipment, especially electrical switch boxes and controls.

Selection of the insecticide requires care in order to comply with the state and federal requirements. Chlorinated hydrocarbons are not generally recommended for use in food processing or storage areas. Where permitted malathion, chlorodane, and lindane are effective and have a residual effect.

Beetles of Dermestid Type

Infestation.—The two most common insects of the third group are the cabinet beetle, commonly called Trogoderma (*Inclusum glabrum*), and black carpet beetle (*Attagenus piceus*). The most likely areas of infestation are: cracks and crevices in floors, walls, and ceiling, particularly if these lead to space resulting from a false floor, wall, or ceiling; the top of beams, fuse boxes, shelving, etc., or any piece of equipment in the dry milk processing, handling, and storage rooms where dust or dry milk may accumulate. The warehouse in which bags or drums of dry milk are stored and the storeroom for used containers are other potential habitats for these insects.

Life Cycle of Trogoderma.—Trogoderma have four stages—adult, egg, larva, and pupa. Eggs can hatch in one week if temperature is warm. Larva grow to approximately $1/4$ in. long and shed their "skins" several times. These "skins" together with distinctive crawling trails in dust residue are frequently the first and most obvious evidence of Trogoderma presence. The larva stage can last from 3 months to more than a year, depending upon environmental factors. The pupa and adult beetle stages are each 2 to 4 weeks long. The adult beetle is $2/16$ to $3/16$ in. long and black with brown mottling. Several generations can be produced during a year. Mortality of the larva to insecticides is shown in Table 25.

TABLE 25

MORTALITY OF *Trogoderma Inclusum* AND BLACK CARPET BEETLE IN LARVAE CYCLE[1]

Insecticide	Trogoderma Inclusion % Mortality to Exposure		Black Carpet Beetle % Mortality to Exposure	
	1 Day	6 to 7 Days	1 Day	6 to 7 Days
Acetone	0	15	0	0
Chlorodane	0	68	. . .	. . .
DDT	0	33	. . .	. . .
Diazinon	100	100	88	100
Lindane	80	93	60	100
Malathion	43	100	85	100
Malathion[2]	50	100	83	100
Methoxychlor	3	30	. . .	. . .
Pyrethrins plus piperonyl butoxide[3]	28	35	. . .	. . .

Source: Marzke (1961).
[1] Continuous exposure to masonite panels treated with insecticides at rate of 50 mg. per sq. ft.
[2] Malathion at 100 mg. per sq. ft. (2%).
[3] Pyrethrins at 5 mg. and piperonyl butoxide at 50 mg. per sq. ft.

Stages of Black Carpet Beetle.—The black carpet beetle passes through the same four stages. Eggs are white to light grey, long oval, hatching in 1 to 2 weeks if the temperature is 75°F. or higher. The larva are dark brown, about $1/4$ to $1/2$ in. long, narrow, and tapered toward the rear. During this stage they must have food some time during the first 2 weeks or they die. The larva stage exists from 9 months to several years. The pupal stage is 2 to 3 weeks in duration. The adult beetle averages $^3/_{10}$ in. long and lives 2 to 4 weeks. This stage of the cycle usually emerges in the spring and early summer and is the period of most of the egg laying. These beetles are shiny black (U.S. Dept. Agr. 1962).

Insect Control

Maintenance of Facilities.—Remodeling or improving the construction of old buildings may be essential for control of insects. Repair of walls, ceiling, floors, doors, and windows to eliminate holes, cracks, and other unnecessary openings is essential. False walls, ceilings, and floors should be removed or sealed off completely to prevent means of access for the insects. However, effective sealing is very difficult. Proper screening and maintenance also contributes to adequate control of insects.

Eliminate Organic Material.—Milk dust should be confined to as few rooms as possible. Dust traps and filters should be used and doors should be kept closed and reasonably air-tight, maintaining a slight positive pressure in the room. All areas not closed off must also have regular cleaning. Careful attention must be given to the complete removal of milk dust and extraneous soil from all objects within each of these rooms. Areas sometimes neglected are loading docks, conveyors, fork lift trucks, chutes under platforms, conduits, switch boxes, beams near the ceiling, etc.

The surface not receiving a wet wash regularly must be dry cleaned. Even the more inaccessible surfaces which collect dust should be vacuum cleaned twice a month. This includes ventilation ducts and fans.

Soil from the vacuum cleaner and all other refuse must be promptly disposed of in a manner preventing insect recontamination. Prompt removal from the premises is advocated. Garbage containers should be wet cleaned each time soon after they are emptied.

Receiving Supplies and Shipping Dry Milks.—To be safe, all shipments of supplies received at the plant would have to be fumigated. If this is impractical each order should be inspected, with those having insect contamination in any form or even suspected of contamination fumigated before the shipment is taken into the plant. An added safeguard is to store the dry milk supplies in a separate area from the product storage. *Trogoderma inclusum* with food can penetrate a 2-mil polyethylene liner in 11 weeks and *T. glabrum* even without food, can penetrate polyethylene in 2 weeks. Black carpet beetle under some conditions requires 12 weeks (Marzke 1961).

Trucks and railroad cars become a source of possible insect infestation when shipping dry milks. These vehicles must be cleaned adequately and sprayed with an approved insecticide before loading the product.

Plant Inspection and Spraying.—Some plants spray routinely not less than once a month during the warm months. Otherwise control requires constant vigilance by plant inspections twice a month. At the first evidence of infestation thorough application of approved insecticides is important. Marzke (1961) suggested the following guide: (a) spray with a residual insecticide such as a two per cent malathion in areas in which insects may crawl (that are difficult to reach), (b) use an aerosol for emergency treatment only or in areas that cannot be sprayed; (c) blow an insecticide dust into cracks and crevices that are difficult to reach with a liquid spray and around electrical connections; (d) fumigate tightly enclosed spaces that have an infestation.

A good control program includes a complete file of records showing the dates and results of inspection, date of spraying areas treated, insecticide concentration, and related information.

GRADES FOR DRY MILK PRODUCTS

Grades for some dry milk products have been promulgated by the American Dry Milk Institute and U.S. Department of Agriculture. The test procedures used to establish some of the grading factors may differ between the two agencies. The terminology used by these two agencies is similar, but not identical. The U.S. Department of Agriculture grades have the "U.S." prefix. Federal Specifications have been prepared by U.S. Department of Agriculture on nonfat dry milk, dry whole milk, and malted milk. U.S. Armed Forces have issued military specifications for purchasing numerous dry milk products, which include: cultured, dry buttermilk solids; coffee type dry cream; anhydrous milk fat; cold-water dispersible, dry nonfat, cocca-flavored milk products; dry whole milk; and nonfat dry milk.

U.S. GRADES FOR NONFAT DRY MILK

Table 26 presents the laboratory requirements for U.S. grades of non-fat dry milk. The flavor and odor for U.S. Extra (instant) and U.S. Extra (regular) of spray process nonfat dry milk after reliquefying should be sweet and natural without off-flavors, but may possess slight chalky, cooked, feed, and flat; product shall be uniform, natural white to light cream and free of lumps, except those that break up with very slight pressure. Product shall be practically free from visible dark particles. Scorched particles shall not exceed 15 mg. per gm.

U.S. Standard Grade permits flavor and odor to a definite degree of flat, feed, cooked, and chalky taste. Slight storage, stale, bitter, oxidized, utensil, and scorched flavors are permitted in Standard Grade. The color may be slightly unnatural with a reasonable amount of dark particles allowed. Scorched particles shall not exceed 22.5 mg. for spray process. The product shall be free of lumps except those that break up with slight pressure.

TABLE 26

ANALYTICAL REQUIREMENTS FOR GRADES OF NONFAT DRY MILK

	Instant	Spray Process		Roller Process	
	U.S. Extra	U.S. Extra	U.S. Standard	U.S. Extra	U.S. Standard
Moisture (not more than), %	4.5	4.0	5.0	4.0	5.0
Milk fat (not more than), %	1.25	1.25	1.50	1.25	1.50
Titratable acidity (not more than), %	0.15	0.15	0.17	0.15	0.17
Scorched particles (not more than), mg.	15.0	15.0	22.5	22.5	32.5
Solubility index (not more than), ml.	1.0	1.2[1]	2.0[2]	15.0	15.0
Standard plate count (not more than), per gm.	35,000	50,000	100,000	50,000	100,000
Coliform count (not more than), per gm.	90	...	...	...	...
Dispersibility (not less than), gm.	44.0	...	...	...	...
Direct clump count (not more than), per gm.	75,000,000	150,000,000	150,000,000	150,000,000	150,000,000

Source: U. S. Dept. Agr. (1959, 1963).
[1] 2.0 ml. for high heat NFDM.
[2] 2.5 ml. for high heat NFDM.
General specifications—not more than 4 mg. of phenol per ml. of reconstituted nonfat dry milk, no addition of buttermilk, neutralizer, or other chemicals permitted.

Flavor and odor requirements for U.S. Extra and U.S. Standard roller process nonfat dry milk are the same as for spray process except that a slightly scorched flavor is acceptable for U.S. Extra while a definitely scorched flavor will pass for U.S. Standard. A higher scorched particle content is permitted in both grades than for the spray processed nonfat dry milk.

The U.S. Department of Agriculture will not issue a grade if the direct microscopic clump count is in excess of 150,000,000 per gm. of nonfat dry milk for regular nonfat dry milk or 75,000,000 in the case of instant nonfat dry milk.

U.S. GRADES FOR DRY WHOLE MILK

The microscopic count of raw milk must be 5,000,000 per ml. or less by weighted average for U.S. Premium Grade dry whole milk. When recombined with water dry whole milk must be sweet and free of off-flavors and -odors with the exception of a slight cooked flavor. The dry form shall be a natural light cream color. Product shall be free of lumps excluding those which break under slight pressure.

TABLE 27

ANALYTICAL REQUIREMENTS FOR DRY WHOLE MILK GRADES

	Spray Process			Roller Process	
	U.S. Premium	U.S. Extra	U.S. Standard	U.S. Extra	U.S. Standard
Moisture (not more than), %	2.25	2.5	3.0	3.0	4.0
Milk fat (not less than), %	26.0	26.0	26.0	26.0	26.0
Solubility index (not more than), ml.	0.5	0.5	1.0	15.0	15.0
Scorched particles (not more than), mg.	7.5	15.0	22.5	22.5	32.5
Titratable acidity (not more than), %	0.15	0.15	0.17	0.15	0.17
Standard plate count (not more than), per gm.	30,000	50,000	100,000	50,000	100,000
Coliform count (not more than), per gm.	90	...	...	...	...
Copper[1] (not more than), p.p.m.	1.5	1.5	...	1.5	...
Iron[1] (not more than), p.p.m.	10	10	...	10	...
Oxygen (not more than), %	2	3 (if gas packed)	...	3 (if gas packed)	...

Source: U.S. Dept. Agr. (1954C).
[1] This test is not required if equipment surfaces in contact with the milk are free of copper, iron, or copper alloys.

In the reliquefied form, U.S. Extra Grade dry whole milk (Table 27) may have definite cooked and other off-flavors and odors to a slight degree. Product shall have a light cream color and be free of lumps that do not break with moderate pressure.

U.S. Standard Grade product after recombining with water may have definite scorched and storage flavors, but no other objectionable flavors or odors. The product shall have a light cream color and no lumps which do not break under moderate pressure. A moderate amount of brown or black scorched particles is permitted.

U.S. GRADES FOR BUTTERMILK AND DRY WHEY

After recombining with water the U.S. Extra Grade dry buttermilk (DBM) (Table 28) shall be free of off-flavors and odors, cream to light-brown in color, free of lumps not breaking with a slight pressure, and practically free of black and brown scorched particles.

The U.S. Standard Grade DBM permits not more than slight unnatural flavors and odors and no offensive flavors and odors. The color may be cream to light brown. Product must be free of lumps that will not break with moderate pressure and not more than moderate amount of brown or black scorched particles are allowed.

The only grade currently established for dry whey is U.S. Extra. In addition to the requirements shown in Table 28 the product must be

TABLE 28

ANALYTICAL REQUIREMENTS FOR GRADES OF DRY BUTTERMILK AND DRY WHEY

	Spray Process DBM		Roller Process DBM		Dry Whey[1]
	U.S. Extra	U.S. Standard	U.S. Extra	U.S. Standard	U.S. Extra
Moisture (not more than), %	4.0	5.0	4.0	5.0	5.0
Milk fat (not less than), %	4.5	4.5	4.5	4.5	1.25% (not more than)
Solubility index (not more than), ml.	1.25	2.0	15.0	15.0	1.25
Scorched particles (not more than), mg.	15.0	22.5	22.5	32.5	15.0
Titratable acidity (not more than), %	0.18	0.20	0.18	0.18	0.16
(not less than), %	0.10	0.10	0.10	0.10	
Bacteria count (not more than), per gm.	50,000	200,000	50,000	200,000	50,000
Ash alkalinity (not more than), ml. of 0.1N HCl/100 gm.	125	125	125	125	225

Source: U.S. Dept. Agr. (1954A, 1954B).
[1] Not applicable to cottage cheese whey.

free of off-odors and flavors, have a uniform light color, no lumps except those that break with moderate pressure and be practically free from brown and black scorched particles.

QUALITY DEFECTS IN FRESH DRY MILKS

Moisture

High moisture is a common reason why nonfat dry milk would fail to meet the U.S. Extra Grade standards. A high moisture content in the fresh spray dried product usually is caused by too low a temperature of drying as indicated by exhaust temperature. Atmospheric changes such as a rapid increase in relative humidity on warm days necessitate higher air drying temperatures in order to obtain the desired moisture.

Spraying orifice too large for drying conditions and forcing too large a volume of liquid into the drying chamber may also cause high moisture in the product. If filters are allowed to collect air-borne particles to the extent the efficiency of airflow is impaired, the reduced air volume may not be sufficient for satisfactory moisture removal, assuming other conditions remain the same.

Scorched Particles

Scorched particles in dry milks are overheated or burnt particles which range from light brown to black in color. Figure 77 shows the standard discs for grading the scorched particles content. There are several areas

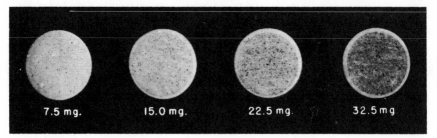

FIG. 77. U.S. DEPARTMENT OF AGRICULTURE SCORCHED PARTICLE STANDARDS FOR
DRY MILKS

where scorching of particles occurs. The most frequent is due to a defective spray which results in partially dried product adhering to the hot surface of the drying chamber. The heat will darken or char this material. Pieces break off and are mixed with the product. Improperly stopping the drying operation may cause a back draft which carries milk particles into the inlet air ducts. Particles deposit onto the hot surface of this duct and also on burners or steam coils. The particles become burned during the cooling period and again when the hot air is started into the drying chamber. Incidence of scorched particles increases when temperatures are increased to speed drying or obtain very low moisture content of powder.

If filters are not used or are inadequately maintained, the drying air may contribute discolored particles. This is most likely when inlet drying air is taken from a plant room having drying equipment. Some efficiency is gained because this air may be warmer than outside air. But it may contain milk dust which becomes scorched when passing over burners or heating coils. This is corrected by effectively filtering the inlet drying air. Likewise adequate filters should be at the inlet of the supplementary air supply to the burners if direct flame heating is used.

Dry milks allowed to remain in contact too long with hot air or metal surface during the drying operation will discolor or darken in varying degrees. Thin layers of product adhering to the hot surface in the drying chamber or collectors provide one opportunity for discoloration from excessive heat. In certain types of driers, small amounts of dried product may become inaccessible to the continuous removal system. The longer the operation the more severe the heat discoloration. Product collectors may add scorched particles if they are not properly maintained. In the drier with the bag system, a mechanical shaker helps free the powder allowing it to drop into the main stream. However, some fine particles tend to adhere to the cloth, darkening in direct relation to exposure time and temperature of the hot exhaust air. These particles gradually feed into the main dry product stream. Daily vacuum cleaning of bags is a preventive measure. Also the cyclone collector system can contribute scorched particles if not cleaned regularly.

The main causes of scorched particles in drum dried milks are: (a) failure to exclude the very first product dried during start-up; (b) neglected or unsatisfactory cleaning of the drum drier between operations; (c) pits in the drum surface; and (d) faulty operation of the drier, e.g., excessive heat on the drums, insufficient pressure on blade when removing the dried film, dull blades, etc. The corrections are obvious to an experienced operator.

Discoloration

Overheating in an attempt to dry product to relatively low moisture tends to darken the dry milk and also can cause scorched particles. High acid products, for example, buttermilk and skimmilk, easily brown during drying. Special care must be exercised in establishing the temperature and time of exposure based upon the acidity of the product. Barnes *et al.* (1961) observed a brown off-color when the moisture was reduced below 3.35% during spray drying. The acidity of the dried skimmilk-whey was 11.1 to 11.6%.

Decreased Solubility

A high solubility index refers to an above normal amount of "insoluble" components in the dried milk. Generally the casein is denatured. It, therefore, does not form a stable dispersion when recombined with water, but settles to the bottom of the vessel.

The following factors adversely affect the solubility index of spray dried product: (a) high accumulative heat treatment; (b) poor quality milk; (c) certain additives; and (d) drying conditions not properly adjusted for the drier design. Storage conditions such as high temperatures or long periods cause an increase in the solubility index.

Preheating of the milk for optimum stability decreases the protein denaturation during the ensuing dehydration. Effects of heating increase in magnitude as the total solids of the milk increases, especially when the concentration reaches "dough" consistency. Slow drying at this stage results in rapid denaturation of the casein. Spraying larger droplets to increase production above rated drier capacity also can increase the solubility index. Large droplets require more time to dry to a specific moisture content. One cause is enlargement of nozzle orifice from wear. Deterioration in quality of the milk and changes in the composition of the milk may influence the casein denaturation. As the acidity increases, a given heat treatment denatures more casein. The addition of salts has a similar effect.

Severe heating of the milk film inherent in the drum drying process denatures a substantial amount of the milk proteins. No means have yet been devised to prevent this change during atmospheric drum drying. For more detailed information see Reconstitutability, Chapter 10.

Lumpiness

The chief body and texture defect in dry milks is lumpiness. Such lumps vary in size and firmness. This problem may be induced by moisture or be caused by excessive heating and very slow cooling after drying and packaging.

Dry milks, particularly the nonfat and high lactose types, are quite hygroscopic. Therefore, lumping due to high moisture may be caused not only by insufficient drying, but also by subsequent moisture pickup. Consequently, immediately after drying the humidity of the atmosphere should be excluded in handling, packaging, and storage.

If dry milks are packaged while still hot and remain hot for several days, the lactose may soften and cause lumpiness. The problem is more acute in the warm months and if the uncooled packaged product is stacked too closely in the warehouse cooling is delayed. High stacks of bags or large bins place greater weight on the product in the lower tiers, intensifying lump formation. The correction is to cool the dry product to less than 100°F. before packaging in commercial bins, bags, or drums.

Off-Flavors

Flavor of the dry milks is of paramount importance for many uses. Evaluation of flavor characteristics should be made only on reconstituted samples; flavor determination on the dry product is of limited value. In the fresh product the greatest contributing factor is quality of the raw materials. Assuming that the milk is normal, then bacteriological, enzymatic, or chemical deterioration during processing and storage will contribute to the off-flavors. The processing procedure and drying method also influences the flavor, particularly the intensity of the "cooked" flavor. In high-heat nonfat dry milk the presence of cooked flavor is considered natural for the product. When readily detected in low-heat product it may be considered a defect.

The most common flavor defects observed in freshly manufactured nonfat dry milk besides cooked and occasionally high acid (sour) are feed or weed, scorched, utensil (includes unclean), metallic and foreign. To this list may be added rancid and oxidized off-flavor for dry whole milk, dry buttermilk, and dry cream.

Preventive measures are self-evident in the fresh dry milk products. The first precaution is to grade out all raw milk with off-flavors and odors. If the milk is held below 45°F., well protected and promptly processed by approved methods, the other opportunities for off-flavor development are minimized. Keeping quality defects will be discussed later in this chapter.

Microorganisms

The microbiological content of dry milks is of concern quantitatively and in respect to types of organisms. The lowest maximum limit among all grades of dry milk products is 30,000 per gm. by the standard plate count. Compliance generally is not difficult. Some industrial users specify a lower limit, however, and when the heat treatment must be confined to minimum pasteurization high quality milk and more care are necessary. The standard plate count is an index to the viable cells remaining after the heat treatments and includes the ensuing microbiological contamination.

The direct microscopic clump count (DMCC) is an indication of maximum bacterial content throughout the history of the milk. It is not likely to show a correlation with the standard plate count. Although it has been an important factor in failure of nonfat dry milk to grade U.S. Extra it has limitations. One of the chief drawbacks is that disintegrated cells will be missed in counting. The gram negative bacteria are not detected by the routine testing technique. This is unfortunate as some species are psychrophilic. Their growth in milk is particularly objectionable because they can cause off-flavors in the reconstituted products. A direct microscopic clump count above the customary limits of 75,000,000 per gm. (School Lunch Program and instant nonfat dry milk) or 200,-000,000 for U.S. grades of nonfat dry milk can be corrected without much difficulty by good sanitation, cooling on the farm, and following good established practices in the plant processing.

The incidence of *Staphylococci* in dry milks requires attention if not serious concern. George *et al.* (1959) has reported the existence and growth of *Staphylococci* which can and do occur in commercial processing plants. They observed growth of *Staphylococci* in condensed skimmilk with solids from 30 to 50%. The usual vacuum pressure in an evaporator would not prevent growth.

The presence of viable *Staphylococci* in dry milks, even in low numbers, constitutes a health hazard if the product is reconstituted and improperly held. If large numbers grow before heat destruction illness can result from enterotoxin produced by these bacteria. Enterotoxin is not destroyed by the heat commonly used in processing and drying milk.

In addition to good sanitation, care must be exercised to prevent growth of *Staphylococci* during processing. The vulnerable areas are the evaporator, balance tanks, and in the foam where the temperature may be in the *Staphylococci* growth range of approximately 90° to 118°F.

Low Spore Count

Industrial customers may have special microbiological specifications for their dry milks, particularly in nonfat dry milk. A low spore count, for example, is one of these. Since heat treatments during processing generally do not destroy the spores, milk must come from farms having a history of low spore count. The alternative is to preheat the fluid milk to a sterilization temperature, for example, 290°F. for 4 sec. Preheating, presumably in the range of 220° to 240°F., seems to activate the spores for rapid growth after the dry product is reconstituted. This seriously impairs the keeping quality of fluid milks fortified with dry milk having these organisms despite pasteurization after the dry product was added.

Zero Mold Count

Since molds are relatively heat sensitive, preheat treatments of pasteurization or above destroy them. But molds are easily air-borne so subsequent contamination is a problem. To manufacture dry milk that will pass with zero mold count, good sanitation of drying and packaging equipment, premises and storage room for the bags is necessary. Unfiltered air must be excluded from contacting the product from dehydration through completion of packaging.

Low Hemolytic Estimate

Dry milk for baby food purposes may have to meet a very low viable hemolytic requirement. If the original milk is adequately pasteurized and good sanitation is practiced no difficulties should be encountered in maintaining a low hemolytic count.

Low Thermophilic and Thermoduric Count

The precautions prescribed for maintaining a low spore count are applicable. Attention should be given also to the prevention of bacterial growth at approximately 115°F. The product or foam in the hot well, evaporator, hot milk balance tanks should be checked at hourly intervals for thermophilic growth during the entire operation.

KEEPING QUALITY OF DRY MILKS

General Factors

Keeping quality of dry milks has become increasingly important in recent years. Changing plant practices along with storage of surpluses (usually as nonfat dry milk) have increased the significance of good keeping quality. The trend toward shipping dry milk to under-developed countries of the world having widely ranging conditions and the lack of facilities also has increased the need for good storage quality.

Although average nonfat dry milk when properly packaged is relatively stable, additional improvement anticipating extreme conditions would be of value. Much remains to be accomplished to economically extend the keeping quality of dry whole milks and other milk fat containing products.

Deterioration occurring during storage periods involves principally flavor, color, and solubility index of the dried product. The rate of these changes is influenced by composition, quality of milk from the cow, care during producing, handling and processing to inhibit chemical and microbiological changes, processing conditions (especially heat treatment), moisture content, metallic contamination, packaging method, and material and conditions of storage (temperature, humidity, and light rays).

In commercial practice the three most important factors are fat content, moisture percentage, and storage temperature. Since milk fat deteriorates rapidly, the content should be restricted in nonfat dry milk to 1.25% or less. Most of the product will be below 0.8% with adequate separation of the milk. As the percentage increases, especially above 1.5%, the keeping quality is rapidly shortened, but nonfat dry milk with more than 1.5% fat fails to meet the Standards of Identity.

Moisture likewise adversely affects storage life, as the content increases above the Extra Grade limit of 4.0%. To prevent absorption of moisture during storage the package must be relatively impervious to moisture vapor. The American Dry Milk Institute recommendation for dry milk packaging materials is that vapor permeability shall not exceed 0.35 gm. per 100 sq. in. at 100°F. and 90% R.H. Other recommendations include: the package must not transmit volatile odors or give off objectionable odors; must seal satisfactorily; and must be elastic from —20° to 120°F. In commercial practice, the polyethylene film is widely used. It serves adequately, but allows the passage of volatile odors.

High storage temperatures are detrimental to keeping quality of dry milks. Consequently, temperatures of less than 75°F. are desirable. Refrigerated storage should be used for long storage in warm climates.

Nonfat Dry Milk Storage

Table 29 illustrates the influence of moisture, temperature, and nitrogen packing on keeping quality of nonfat dry milk. If processed properly from good quality raw milk and packaged in vapor-proof containers, nonfat dry milk should have good keeping quality at 70°F. No flavor changes other than loss of freshness should be detectable within 1 to $1\frac{1}{2}$ years or longer at this temperature. The most common flavor change when one does develop is storage or stale. If this off-flavor exists relatively soon, the direct cause usually is high moisture in the dry product or too high a storage temperature.

TABLE 29

THE EFFECT OF MOISTURE CONTENT OF NONFAT DRY MILK UPON THE NUMBER OF DAYS
REQUIRED TO RENDER IT UNPALATABLE

Storage Temperature		3.0% Moisture		5.0% Moisture		7.6% Moisture	
°C.	°F.	Air-pack, Days	N-pack, Days	Air-pack, Days	N-pack, Days	Air-pack, Days	N-pack, Days
20.0	68.0	500	>700	410	>700	115	340
28.5	83.3	370	650	290	590	12	51
37.0	98.6	240	450	145	260	2	18

Source: Henry *et al.* (1948).

Dry Whole Milk Storage

Oxidation.—Dry whole milk and other dry milks having substantial percentages of milk fat undergo oxidative deterioration (also called tallowy) in storage. There is a utilization of oxygen and the formation of peroxides, volatile aldehydes, and other compounds.

Assuming average conditions of milk quality and processing, oxidation off-flavor becomes noticeable by sensory tests in roughly three months when dry whole milk product is held at 70° to 75°F. (air packed). Many factors influence the rate of the milk fat oxidation. The extent of the metallic contamination from cooper and iron is one of the most important. These metallic ions are believed to act as a catalyst. To maintain the recommended low levels of not more than 1.5 p.p.m. of copper and 10.0 p.p.m. of iron, milk should contact only stainless steel surfaces during production, handling, and processing. The trials of Lea *et al.* (1943) show that at room temperature dry whole milk becomes tallowy in 9 to 10 months with 0.7 to 1.0 p.p.m. of copper; 5 to 7 months with 3 to 7 p.p.m.; and less than 50 days with 10 to 15 p.p.m. Sunlight and ultraviolet irradiation will accelerate fat oxidation in dry milk. Presumably the rate is associated with the total energy of light to which the product is subjected. Under normal commercial conditions of packaging and storage, light would not be expected to have a significant effect on the development of a tallowy off-flavor.

Lecithin is a contributing factor to the keeping quality of dry whole milk and dry creams. Dahle and Josephson (1943) prepared fat containing dry milk products with 0.75, 0.48, and 0.37 gm. of lecithin per 100 gm. of fat. The rate of oxidation was substantially decreased with the decrease in lecithin content. However, lecithin associated with the fat may also act as an antioxidant.

Proper preheating of the milk will delay oxidative changes during storage of the dry whole milk and cream products. The temperature of preheating should be above 170°F. Studies of Mattick et al. (1945) indicated that preheating the milk up to 190°F. increased its resistance to tallowy off-flavors. Crossley (1945) also reported that preheating to 190°F. improved the keeping quality. Harland et al. (1952) obtained a better keeping quality of the dry product with preheats of 194° to 205°F. for 60 to 80 sec. when compared to 171° to 205°F. for 30 min.

Improvements in keeping quality caused by preheating have been attributed principally to the effect of releasing sulphydryl compounds which react as antioxidants. There may be other beneficial effects of the action of heat on the proteins. Another proposed possibility is the more complete destruction of oxidizing enzymes occurring naturally in the milk or released by the growth of microorganisms (Mattick et al. 1945).

Antioxidants.—Adding antioxidants in small amounts to the fluid product before dehydration may have some effect on extending the keeping quality. The specific amount of influence may vary with the compound and storage condition, but none seem to be as effective as the removal of oxygen. Usually the amount necessary to be effective produces an objectionable flavor. Furthermore, antioxidants are not included in the Standards of Identity for dry whole milk.

A few of the antioxidants that have been used experimentally are: ascorbic acid, ascorbic acid plus sodium citrate, oat flour, tocopherols, gallic acid, ethyl gallate, nordihydroguaiaretic acid, gum guaiac, butylated hydroxyanisole, and butylated hydroxy toluene. Different results among investigators probably vary because of the varying conditions of product quality, storage temperature, moisture in the dry milk, etc.

Findlay et al. (1945) observed an improvement in keeping quality with 0.07% ethyl gallate added to the fluid milk. Gallic acid was less effective than the ethyl gallate and tended to give a hint of a bitterness. The gallates of a higher series caused bitter off-flavors. Oat flour was slightly effective at 0.25% (based on milk). Above this level an oat flavor was imparted to the dry milk.

The addition of 0.05% (based on fluid milk) of sorbitan monostearate, polyoxyethylene sorbitan monostearate, and glycerol monostearate materially improved the keeping quality of dry whole milk stored at 85°F. (Hibbs and Ashworth 1951). These products are primarily emulsifying agents. Presumably they enhanced the protective effect rather than reacted with the oxygen.

Oxygen Limitation.—The most practical commercial procedure for extending keeping quality has been to lower the oxygen content of the dry product. By reducing the oxygen to two per cent or less in the headspace gas, the keeping quality is roughly doubled or more. The product may be vacuum packaged with this low oxygen level. However, replacing the partial vacuum with nitrogen is more common in the dry milk industry.

Lea *et al.* (1943) reported that dry whole milk with 0.01 ml. of oxygen per gm. of product stored well at room and at high temperatures. A good commercial operation should keep the oxygen content below 0.02 ml. per gm. of product.

In recent years, oxygen scavengers have been utilized to reduce the oxygen in products packaged in impervious containers (Scott and Hammer 1961). The glucose oxidase-catalase packet will reduce oxygen, but is fragile and slow. It is also costly. Another effective procedure is the use of 10% hydrogen with the nitrogen for gas packaging, and addition of a packet with a palladium pellet to the container.

Rancidity.—Hydrolytic rancidity may cause off-flavor development in the milk fat of dry milks. Lipase enzymes are responsible for this deterioration by producing free fatty acids. The defect occurs very infrequently since adequate pasteurization and the customary high preheat treatment is sufficient to destroy the lipases. However, if lipolysis occurs in the milk before pasteurization the free fatty acids will be present to accelerate oxidation in the dry product.

Staleness.—Dry milks may develop a stale off-flavor during storage. Numerous investigators associate stale flavor with the milk fat. Others attribute it to protein-lactose reactions. A third group believes that staleness is a conglomeration of off-flavor compounds occurring simultaneously in the lipids and the protein-lactose complexes.

Tarassuk and Jack (1946) considered oxygen necessary for staleness to form believing the milk fat in dry whole milk or dry ice cream mix was the precursor. They stated preheating to 190° to 195°F. delayed staleness compared to 165°F. in the dry products held at room temperature. Whitney and Tracy (1949, 1950) also associated stale flavor with the milk lipids. Milk fat from stale dry whole milk contained a stale flavor component.

Stale flavor may be delayed in storage by reducing the temperature. Air packed dry whole milk at 45°F. had a slightly superior flavor at the end of one year than when it was freshly processed as well as the gas packed and held at 85°F. for one year (Christensen et. al. 1951). Excessive heating of the fluid milk (for example 200°F. for 30 min.) or the exposure to heat after drying is complete, hastens the appearance of the stale flavor. Dry milk products with moistures above normal show a direct relationship to the appearance and intensity of staleness.

Insolubility.—The rate of solubility index (denaturation) increase in dry milk during storage is directly correlated to the time, temperature of storage, and moisture content of the dry milk. The amount of denaturation of a good quality product with three per cent moisture and storage at room temperature will be slight after one year. If either moisture or storage temperature is high, a definite denaturation will take place within a few months.

Color Change.—Browning during commercial storage usually is not a problem. Discoloration will progress slowly during prolonged storage or rapidly if either the moisture content of the powder or the storage temperature is high. Delayed cooling of the dry whole milk after drying is contributing factor. Concomitant changes in flavor deterioration are usually evident before a color change is serious. Browning during storage is attributed to the reaction between free amino groups and lactose, a reducing sugar. Low moisture content delays this reaction. The conditions of manufacture and storage that delay staling are effective for delaying browning.

Effect of Agglomeration on Keeping Quality of Nonfat Dry Milk

Various investigators have reported no adverse effects of agglomeration and redrying on the keeping quality of dry milk, and in particular nonfat dry milk. Undoubtedly, some change takes place. Probably it is too small to be detected under average conditions of agglomeration processing. Severe redrying conditions or residual moisture after redrying above 4.0 to 4.5% will be detrimental on prolonged storage.

LABORATORY TESTING OF DRY MILKS

General

Close cooperation between laboratory and production personnel and coordination of both activities is necessary for effective composition and quality control. Prompt laboratory reports on the product during manufacture will serve as a useful guide. Testing after the production is of less value.

Accurate testing of the fluid and dry milk is essential for quality, composition, and efficiency control throughout processing. Furthermore, the test selected although acceptable may affect the results. This fact can have important implications when large volumes of product are involved or minor differences are enough to influence the standard of a borderline factor. Common examples are the difference in results between direct microscopic and resazurin test on raw milk or the moisture test on dry milk by the toluol, Cenco, or Fischer methods.

Despite the original intention of Dr. Babcock to obtain identical results, fat tests by the Babcock procedure may differ from an ether extraction procedure (Mojonnier) (Fig. 78). Tests among different laboratories

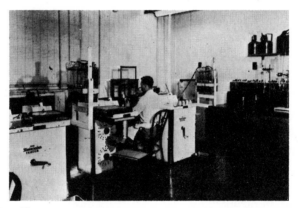

FIG. 78. MILK FAT TESTING BY THE ETHER EXTRACTION
METHOD

on duplicate samples have varied much more than the experimental error of the method. Consequently, much attention should be given to technique, particularly for the routine analyses. Regular periodical comparative tests should be made with other laboratories to check validity of results. In addition the laboratory equipment, reagents, and procedures should be inspected at regular intervals for adequacy. Too often discrepancies develop so gradually that error in the results is unnoticed. Results of an unofficial test procedure should be checked with an official procedure occasionally. This is particularly true of the moisture determination.

The inherent weakness of sensory testing increases the need for verification by a checking system with another laboratory. A taster may become accustomed to a slight off-flavor through repeated exposure until it no longer seems objectionable.

Sampling

The importance of obtaining representative samples must be emphasized otherwise the significance of test results is decreased. Obtaining a representative fluid sample from a batch operation usually is not difficult since milk can be adequately mixed by agitation. Dry materials in large batches requires a systematic approach for representative sampling.

Representative sampling of a dry product during continuous manufacture requires both experience and a system. The frequency of tests depends upon several factors such as the variability of the product and the limits of the standards that have been established. Using a statistically selected procedure of sampling may prove to be the most efficient. A good quality control program requires standard size lots of dry milks and consecutive numbering of the bags or larger containers. U.S. Department of Agriculture restricts a lot of nonfat dry milk to 4000 lb. and dry whole milk to 2000 lb. Automatic sampling devices for dry milk during production have now been developed and are in limited use (see Fig. 79).

Test Procedures

Routine Grading Tests.—Table 30 presents references for the common test procedures used currently by the dry milk industry. These apply

Courtesy of Automatic Sampling Systems

FIG. 79.   AUTOMATIC SAMPLING DEVICE FOR DRY MILKS

TABLE 30

REFERENCES FOR TESTING MILK AND DRY MILK PRODUCTS

Test	Source[1]				
	ADMI (1965)	APHA (1971)	AOAC (1970)	MIF (1967)	U.S. Dept. Agr. (1961)
Standard plate count	+	+	−	+	+
Methylene blue	+	+	−	+	−
Resazurin	+	+	−	+	−
Freezing point	−	+	+	+	−
Microscopic count	+	+	−	+	−
Coliform	+	+	−	+	+
Yeast and mold count	+	+	−	+	+
Sediment	+	+	+	+	+
Milk fat	+	+	+	+	+
Titratable acidity	+	+	+	+	+
Direct microscopic clump count (dry milk)	−	+	−	−	+
Moisture	+	−	+	+	+
Solubility index	+	−	−	+	+
Scorched particles	+	−	−	−	+
Alkalinity of ash	+	−	+	+	+
Oxygen	+	−	−	−	+
Copper	+	−	+	+	+
Iron	+	−	+	+	+
Flavor and odor	+	−	−	+	+
Dispersibility	+	−	−	−	+
Whey protein nitrogen	+	−	−	+	+
Protein	−	−	+	−	+

[1] See complete reference at end of Chapter 9.

to raw milk as well as the dry milk products. Copies of each publication have been available by contacting the organization or agency.

Routine tests on the raw milk supply primarily consist of a bacterial estimate, sediment and flavor, and odor. Special tests that should be conducted a few times a year are freezing point for adulteration with water and inhibitory substances.

Procedures for Special Tests.—The procedures for the common tests on dry milks are referenced in Table 30. The U.S. Department of Agriculture procedures can be used to compare results with the analyses for official grading factors. A brief procedure of special purpose tests is outlined in the following paragraphs of this chapter. These tests are presented for processing control purposes and are not necessarily appropriate for research purposes.

Moisture-Cenco Procedure.—This test requires a Cenco Moisture Tester supplied by Central Scientific. The Cenco method consists of turning the unit "on" for 60 sec. to stabilize (dry) the clean pan. Tare the empty pan by adjusting knob to obtain a reading of 100. Add sample by uniformly covering bottom of pan until the indicator is at 0; this will be 5 gm. Dry sample by turning on the heat lamp until a constant weight is obtained without scorching the sample. Occasionally the knob is used to bring the indicator back to zero to compensate for loss of moisture weight. The per cent moisture is read directly from the scale. The test takes approximately five minutes.

Moisture Absorption-Farinograph Method.—A Brabender Farinograph (Fig. 80), balance, spatula, and laboratory accessories are needed. Weigh accurately 300 gm. of spray dried milk tempered to 86°F. and transfer to bowl of Farinograph. Depress starting switch for "slow" to turn mixing blades over several times obtaining an even distribution of

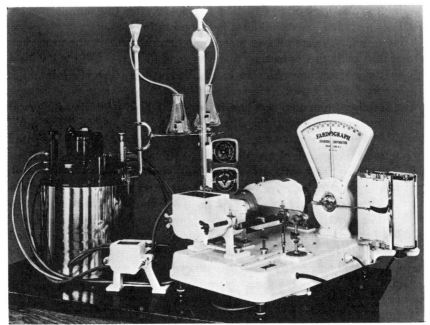

Courtesy of Brabender Corporation

Fig. 80.   Apparatus for Testing Dry Milks Moisture Absorption and Development Time

powder in the bowl. Drain 170 ml. of distilled water at 86°F. into bowl containing the nonfat dry milk. Immediately start the mixing blades on slow speed and allow to run for $1^1/_2$ min. Then stop the blades and scrape down the bowl and mixing blades taking 30 sec. Again, start on slow speed and allow to run 30 sec. and repeat the scraping procedure. Start blades on fast speed and continue mixing for 20 min. Record time for indicating pen to first reach 500 units (development time) and total units attained as well as units at end of the 20-min. mixing period. An excellent development time is within 5 min. and excellent total absorption is 800 units or above. Hoffman *et al.* (1948) reported a different Farinograph procedure using flour of known absorption to test drum dried nonfat dry milk for absorption.

Heat Stability Test.—An oil bath thermostatically controlled within 0.2°F., balance and screw cap test tubes are used. The caps must seal adequately.

Reconstitute nonfat dry milk with distilled water, 10 gm. to 100 ml. or 12.5 gm. of dry whole milk and allow to hydrate for 2 hr. Pipet 10 ml. of sample into a test tube and heat rapidly to 250°F. The time required for the first particles of coagulation to appear after the sample reaches 250°F. is recorded. Sixty minutes is a good rating.

To determine the stability to a sodium chloride solution the procedure is repeated with the exception that the sample is reconstituted with a 2% solution of sodium chloride in distilled water. Twenty minutes is satisfactory for most uses.

The stability in an acid medium is ascertained by acidifying the water (before reconstitution with the dry milk) with lactic acid to 6.0 or a pH that simulates utilization of the dry milk and then proceeding as indicated for a regular reconstituted sample.

Bulk Density Test.—Apparatus for bulk density consists of 100 ml. graduated cylinder, balance, funnel with large diameter stem, and ringstand. The tare weight is obtained on the cylinder. The funnel stem is placed at mouth of cylinder and the powder is allowed to flow freely to 100 ml. mark. The net weight is obtained and the results expressed as gm. per ml. (loose density). To determine the "packed" bulk density the cylinder of powder from above procedure is tamped on a rubber mat until the volume is reasonably constant. The volume of powder is read in ml. and density recorded as gm. per ml.

Accuracy may be increased by using a larger graduated cylinder and sample. Another alternative is to place the cylinder with the sample on a vibrator for a specific time, such as 5 min., instead of tamping manually to ascertain the "packed" density.

Simple Dispersbility Test.—The apparatus used is a balance, beakers, thermometer, and a teaspoon.

Into a 250-ml. beaker is added 100 ml. of water. It is tempered to 60°F. and 10 gm. of nonfat dry milk is gently placed upon the water. Sinkability is immediately observed for 10 sec. The content is stirred with the teaspoon for 10 sec. The dry milk should be well dispersed with no lumps on the bottom and no lumps or particles floating on the surface.

The same steps are followed to test dry whole milk except that 12.5 gm. are placed on water adjusted to 115°F. Stirring time is 25 sec.

The dispersion of dry coffee creaming product may be tested by heating 200 ml. of distilled water to 190°F. in a 400-ml. beaker. Two grams of instant coffee are dispersed into the water. Then 4 gm. of product are added and stirred with a teaspoon using a circular motion ten turns clockwise and ten in the reverse direction taking 25 sec. After stirring no particles should be floating, and not more than a trace of melted fat should be visible.

Free Fat Test.—The major equipment is a balance, pipets, Erlenmeyer flasks, and stoppers.

Add 100 ml. petroleum ether to 10 gm. of sample in 250-ml. Erlenmeyer flask. Shake ten times with a vertical motion. Allow to settle 15 min. Filter petroleum ether layer through No. 42 Whatman filter paper catching the solvent in a tared Mojonnier fat dish. Repeat with a second extraction by the same procedure. Evaporate the ether (Mojonnier method for fat), weigh sample, and calculate results as percentage of free fat in the 10 gm. sample.

Flowability Test.—A funnel with a stem having an inside diameter of $1/2$ in., a ringstand, balance and beakers are needed for this test. A pint sample is poured into the funnel and is then allowed to free-flow out of the funnel. Good flowability is indicated by a sample that rapidly drains from the funnel. If the sample tends to bridge to some extent, or is slow, but drains from the funnel unassisted, it is rated fair. A poor sample does not drain completely from the funnel without aid.

Test for Inhibitory Substances.—The chief equipment required is a constant temperature incubator, balance, thermometer, graduated cylinder, glass bottles, and Mafis acidity tester with N/10 sodium hydroxide.

Reconstitute 10 gm. of nonfat dry milk (or 12.5 gm. of dry whole milk) per 100 ml. of distilled water. The equipment and water should be sterile and precautions taken to prevent contamination. Add 1% active culture and mix thoroughly. Adjust temperature to 70°F. and incubate 15 hr. Titrate the acidity and also test the sample for flavor and odor. For a 4 hr. incubation adjust temperature to 90°F. and add 5% culture.

REFERENCES

American Dry Milk Institute. 1961. Recommended sanitary/quality standard code for the dry milk industry. Bull. *915* (Revised), Chicago.

American Dry Milk Institute. 1965. Standards for grades for the dry milk industry including methods of analysis. Bull. *916*, Chicago.

American Public Health Association. 1971. Standard Methods for the Examination of Dairy Products. New York.

Association of Official Agricultural Chemists. 1970. Official Methods of Analysis of the Association of Official Agricultural Chemists. Washington 4, D.C.

BARNES, J. C., STINE, C. M., and HEDRICK, T. I. 1961. Edible high-acid powder. Milk Prod. J. 52, No. 4, 8, 9, 34.

CHRISTENSEN, L. J., DECKER, C. W., and ASHWORTH, U. S. 1951. The keeping quality of whole milk powder. I. The effect of preheated temperature of the milk on the development of rancid, oxidized and stale flavors with different storage conditions. J. Dairy Sci. 34, 404–410.

CROSSLEY, E. L. 1945. Spray-dried milk powder. Commercial observations over two years of the effect of high temperature preheating. J. Dairy Res. 14, 160–164.

DAHLE, C. D., and JOSEPHSON, D. V. 1943. Improving the keeping quality of dry whole milk. Milk Plant Monthly 32, No. 10, 28–29.

FINDLAY, I. D., SMITH, J. A. B., and LEA, C. H. 1945. Experiments on the use of antioxidant in spray-dried whole-milk powder. J. Dairy Res. 14, 165–175.

GEORGE, E., JR., OLSON, J. C., JR., JEZESKI, J. J., and COULTER, S. T. 1959. The growth of Staphylococci in condensed skimmilk. J. Dairy Sci. 42, 816–823.

HARLAND, H. A., COULTER, S. T., and JENNESS, R. 1952. The interrelationship of processing treatments and oxidation-reduction systems as factors affecting the keeping quality of dry whole milk. J. Dairy Sci. 35, 643–654.

HEDRICK, T. I., and HALL, C. W. 1964. The dry milk industry: status and practices. Manuf. Milk Prod. J. 55, No. 12, 5–6.

HENRY, K. M., KON, C. H., LEA, C. H., and WHITE, J. C. D. 1948. Deterioration on storage of dried skimmilk. J. Dairy Res. 15, 292–363.

HIBBS, R. A., and ASHWORTH, U. S. 1951. The solubility of whole milk powder as affected by protein stabilizers and emulsifiers. J. Dairy Sci. 34, 1084–1091.

HOFFMAN, C., SCHWEITZER, E. N., and DALBY, G. 1948. Evaluation of the baking properties of roller process nonfat dry milk solids by Farinograph procedure. Cereal Chem. 25, 385–390.

HOLLANDER, H. A., and TRACY, P. H. 1942. The relation of the use of certain antioxidants and methods of processing to the keeping quality of powdered whole milk. J. Dairy Sci. 25, 249–274.

LEA, C. H., MORAN, T., and SMITH, J. A. B. 1943. The gas-packing and storage of milk powder. J. Dairy Res. 13, 162–215.

MARZKE, F. O. 1961. Entomological studies as related to plant sanitation and operations. Copy of talk given at Am. Dry Milk Inst. Annual Meeting, Chicago, Ill. April 12.

MATTICK, A. T. R., HISCOX, E. R., CROSSLEY, E. L., LEA, C. H., FINDLAY, J. D., SMITH, J. A. B., THOMPSON, S. Y., and KON, S. K. 1945. The effect of temperature of pre-heating, of clarification and of bacteriological quality of the raw milk on the keeping properties of whole milk powder dried by the Kestner spray process. J. Dairy Res. 14, 116–159.

MILK INDUSTRY FOUNDATION. 1967. Laboratory Manual Methods of Analysis of Milk and Its Products. 3rd Edition. Washington 6, D.C.

SCOTT, D., and HAMMER, F. 1961. Oxygen-scavenging packet for in-package deoxygenation. Food Technol. 15, No. 2, 99–104.

TARASSUK, N. P., and JACK, E. L. 1946. Biochemical changes in whole milk and ice cream powders during storage. J. Dairy Sci. 29, 486–488.

U.S. DEPT. AGR. 1954A. United States standards for dry whey. Federal Register 19, 3349–3351 (June 7), Washington 25, D.C.

U.S. DEPT. AGR. 1954B. United States standards for grades of dry buttermilk. Federal Register 19, 3955–3957 (June 30), Washington 25, D.C.

U.S. DEPT. AGR. 1954C. United States standards for grades of dry whole milk. Federal Register *19*, 4899–4902 (August 5), Washington 25, D.C.

U.S. DEPT. AGR. 1959. United States standards for grades of nonfat dry milk, spray process and roller process. Federal Register *24*, 1363–1365. Washington 25, D.C.

U.S. DEPT. AGR. 1961. Methods of laboratory analyses for dry whole milk, nonfat dry milk, dry buttermilk and dry whey. Agricultural Marketing Ser., Dairy Div., Inspection and Grading Branch Laboratory, Chicago, Ill.

U.S. DEPT AGR. 1962. Insect prevention and control in plants processing dry milk. Agricultural Marketing Serv., Market Quality Res. Div., *AMS-302* (Revised) Washington, D.C.

U.S. DEPT. AGR. 1963A. United States standards for instant nonfat dry milk. Federal Register *28*, 2956–2957 (March 26), Washington, D.C.

U.S. DEPT. AGR. 1966. Minimum standards for milk for manufacturing purposes and its production and processing recommended for adoption by state regulatory agencies. Federal Register *31*, 12354–12386. Washington 25, D.C.

WHITNEY, R. McL., and TRACY, P. H. 1949. Stale flavor components in dried whole milk. I. The distribution of stale flavor between fractions of reconstituted stale whole milk powder. J. Dairy Sci. *32*, 383–390.

WHITNEY, R. McL., and TRACY, P. H. 1950. Stale-flavor components in dried whole milk. II. The extraction of stale butter oil from stale dried whole milk by organic solvents. J. Dairy Sci. *33*, 50–59.

CHAPTER 10

Properties of Dry Milk

PHYSICAL, BIOCHEMICAL, AND SENSORY

The properties of dry milks are usually divided into the three categories of physical, biochemical, and sensory. Much interrelationship exists. Consequently, the dry product characteristics and their changes to a large extent are the result of these three types and their interactions.

Principal physical properties of dry milks that are germane embrace particle size, surface, and shape; density; flowability; dustiness; physical state and solids composition of particles and the sorbed gases and liquids. The biochemical characteristics of major importance include heat induced changes of the various constituents, especially proteins during manufacture and the reactions that occur during storage or utilization. The dry milk sensory characteristics of chief importance include flavor and appearance plus tactual sensations in the dry form, reconstituted form and when utilized either reconstituted or dry by incorporation into food products.

FACTORS AFFECTING PROPERTIES

Practically every item from the milk supply to the equipment, processing, packaging, and storage will influence the properties of dry milk products. Figure 81 illustrates the multitude of factors that affect the total properties of a dry milk product. Obviously the magnitude and complexity of the effects on a single characteristic are tremendous.

Raw Milk

Well known is the fact that raw milk varies in composition not only among localities, but within the same supply. Some components vary from day to day and others seasonably. Since more than 100 substances are known in milk, and the chances are large that additional ones will be identified, the effect of composition on properties can be large.

Although many of the compounds or elements in milk are present in minute quantities their influence may be pronounced. An example is the addition of a few parts per million of citrate. or phosphate or the removal of a minuscule proportion of the calcium can increase heat stability of heat labile milk; in other cases the reverse may be true.

Heinemann (1948) reported minor constituents that varied in a Missouri milk supply embraced calcium, phosphorus, acid soluble phosphate,

Composition of Raw Materials
  Skimmilk
  Buttermilk
  Whole milk
  Cream
  Whey
  Nondairy ingredients
    sugar
    stabilizer
    buffering salts
    flavorings
Processing Operations
  Separation-efficiency (nonfat dry milk)
  Preheat-temperature and time for each step
  Homogenization-efficiency-temperature and pressure
  Evaporation
    atmospheric
    vacuum
      single-effect evaporator
      multiple-effect evaporator
      regular-heat evaporator
      low-heat evaporator
Dehydration
  Drum drying
  Spray drying
    nondairy compounds
    gas—oxygen, nitrogen, and carbon dioxide
    no heat (tower) drying
    foam spray drying
    pressure spraying
    centrifugal spraying
    venturi spraying
  Foam mat drying
    atmospheric drying
    vacuum drying
  Freeze drying
  Wetting and redrying (instantizing)

Storage Condition of Dry Product
  Moisture content
  Storage temperature
  Storage time
  Packaging material
  Oxygen content
    vacuum packaged
    gas-nitrogen packaged
    oxygen scavenger
Physical-Chemical Properties of the Dry Product
  Structure compounds
    lactose
    proteins—soluble, dispersible, denatured
    fats
    moisture
  Primary structure
    dispersed components
    continuous components
    physical interrelation
  Secondary structure
    milk particles
      size
      shape
      gas content
    air
  Practical characteristics
    keeping quality
    flavor and odor
    appearance
    reconstitutability
      dispersibility
      solubility
      sinkability
      wettability
    density
      bulk
      particle
    flowability

Fig. 81.   Raw Materials and Processing Factors Interrelated to the Physical-Chemical Properties of Dry Milk Products

magnesium, and citric acid. Numerous other investigators have shown substantial variations in inorganic and minor organic components as well as major nutrients (lactose, protein, and fat).

The quality of milk as influenced by microbiological deterioration and the kind and extent of adulteration by chemicals can affect the properties of dry milk.

Processing Operation

Each step in the processing operation and the packaging can contribute to the immediate and eventual properties of the dry product. One of the most important is the total heat treatment—the time and temperature for each step in processing. The method of heating as well as the total heat influences protein denaturation, keeping quality, flavor, product dryness, and many other characteristics.

The type of the processing equipment, for example, drum, spray or freeze drier, materially affects the properties mentioned above. In spray drying, the atomizing device influences size, shape, and density, etc., of the particles. The type of collector controls the recovery of fine particles—a factor in reconstitutability. Secondary heaters and cooling equipment have a bearing on the moisture and keeping quality. The packaging material determines the permeability of moisture vapors, odors, and gases which can be absorbed by the product.

The magnitude of the interactions of processing and greater details on specific properties will be discussed subsequently under the individual characteristic.

Size and Shape of Particles

Drum Dried.—Size of particles of drum dried milk depends primarily upon the film thickness and the pulverizing performed by the hammer mill. Increasing the total solids of the milk will increase the particle thickness of drum dried particle. The thickness is roughly 8 to 20μ. The

TABLE 31

PARTICLE SIZE DISTRIBUTION OF VARIOUS SPRAY DRIED WHOLE MILKS

Sample	Type of Drier	No. of Particles Measured	Percentage of Particles Having Diameters of				
			$0-$ 60μ	$60-$ 120μ	$120-$ 180μ	$180-$ 240μ	Over 240μ
A-1	Pressure spray 1	481	2.9	41.4	29.1	16.8	9.8
A-2	" " "	448	12.0	45.8	21.9	15.4	4.9
B-1	Pressure spray 1	705	12.1	44.8	24.1	12.8	6.7
B-2	" " "	454	8.8	44.5	30.6	11.5	4.6
C-1	Pressure spray 2	1165	59.7	29.7	7.3	2.3	1.0
D-1	" " "	1008	66.4	24.6	5.4	2.2	1.5
E-1	Univ. of Minn.	1023	71.9	23.6	3.4	1.0	0.1
F-1	Air whipped into condensed milk	1050	73.7	22.6	3.1	0.4	0.0
D-2	Centrifugal spray	150	2.0	31.2	24.0	18.6	24.0

Source: Coulter *et al.* (1951).

shape of the particle is irregular, angular with rough edges and wrinkled surfaces. Normally air cells are not present inside the drum dried particles.

Spray Dried.—Spray dried particles are usually spherical shaped, but some may be elongated and generally range in diameter from 10 to 250μ. Table 31 shows the percentage of five size ranges for several types of driers and spray systems. Figure 82 consists of microphotographs of three different samples of nonfat dry milk.

The surface of the spray dried milk particles is usually smooth. The tendency to show roughness or wrinkles increases with the larger temperature gradient between the hot air and milk particles during drying (Hayashi 1962).

Particle size is influenced by the milk characteristics, drying system, and operating conditions. Centrifugal spray gives a larger particle in

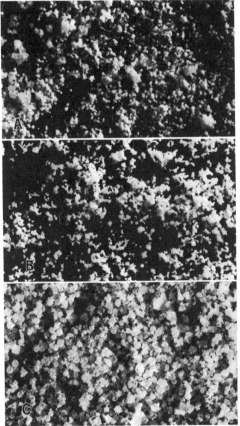

Courtesy of Niro Atomizer

FIG. 82.   MICROPHOTOGRAPHS OF NONFAT DRY MILK FROM
THREE DIFFERENT PROCESSES

A. Regular nonfat dry milk particles.
B. Particles after pneumatically recycling fines back into drying chamber.
C. Aggregates after agglomeration.

general than the pressure nozzle. Higher total solids in the condensed milk causes a larger particle. Foam spray drying results in relatively large sphere-shaped particles (63 to 104μ) that tend to form aggregates (141 to 429μ) (Hanrahan *et al.* 1962). Very fine air cells were distributed through the particle. A few but larger bubbles were noticed in the interior of the particles.

STRUCTURE

The physical structure of dry milks is the way the chemical components are distributed and connected together. These relationships influence the particle characteristics. King (1965) believes dry milks have a dual structure—primary and secondary. He stated that the dry milk particle has a primary physical structure comprised of the milk solids in which is dispersed the moisture and air cells. The physical mass of the particle (nonfat dry milk and dry whole milk) is dominated by lactose presumably in which the protein, fat, and minerals are more or less dispersed. In the case of dry cream the predominance of fat undoubtedly changes this theory.

The bulk of particles surrounded by air (where particle surfaces are not in contact) constitute the secondary structure. The size and shape of the particles and degree of uniformity of these characteristics affect the secondary structure.

LACTOSE

Forms of Lactose

The lactose content may vary from approximately 75% in dry whey to 12% in dry cream. Lactose in dry milk products may be in the amorphous or glass (noncrystalline) form and as α monohydrate crystals or β anhydride crystals. Immediately after drying milk, the lactose exists in the glass form (Troy and Sharp 1930) and is distributed through the particle.

Sorption

In the amorphous state lactose is considered a supersaturated syrup with a low vapor pressure and a very high osmotic pressure. It is very hygroscopic. Sorption begins when the relative humidity of the air is roughly 50% or higher.

As sorption occurs, the lactose becomes sticky and this initiates the adherence of the milk particles to each other. Then solidification, commonly called caking, occurs. With sufficient moisture and time a solid mass is formed due to continued crystallization of the lactose.

Crystallization

Lactose crystals in dry milks are mainly α hydrate. Any initially formed β changes into the α form at a temperature below 200°F. (Fig. 83). Optical rotation of the α lactose is higher than the β. No crystallization takes place in dry milk that conforms to the legal Standards of Identity of 5.0% or less moisture. Crystallization into the α and β forms starts at roughly 7% moisture in the dry milk. Supplee (1926) observed moisture being absorbed and then some released during the crystallization in dry milk. Troy and Sharp (1930) attributed this phenomenon to the decrease in osmotic pressure and vapor tension increase.

With the moisture sorption the concentrated lactose (glass) is diluted permitting the lactose molecules sufficient mobility to form into crystals, usually α lactose monohydrate (Troy and Sharp 1930). But the lactose glass form is highly supersaturated with respect to the α and β. At room temperature and suitable conditions (sufficient moisture for crystallization), the β anhydride crystals were found in dry milk as well as the α hydrate crystals (Sharp 1938). He found no β crystals in fresh dry milk

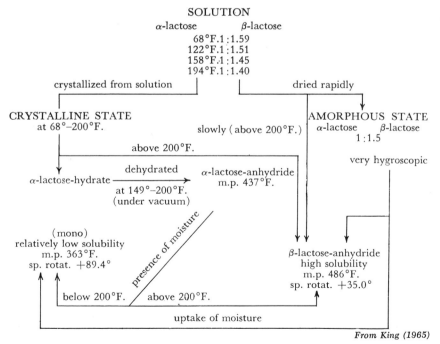

From King (1965)

FIG. 83. PHYSICO-CHEMICAL RELATIONSHIPS BETWEEN DIFFERENT FORMS OF LAC-
TOSE

or samples stored under good conditions for ten years, if the moisture content remained low. At 7.6% moisture crystallization in the dry milk occurred in 100 days at 68°F., 10 days at 83°F., and 1 day at 99°F. (King 1954).

Effects of Crystallization on Casein and Fat

Crystallization of lactose in dry milk causes fine cracks adjacent to the crystal surfaces. This network of cracks in the particles makes them much more permeable to gases and fat solvents (King 1965). During crystallization nonlactose components are expelled into the spaces of the network of the particle. Conditions for casein coagulation are enhanced because of loose packing of micelles and reaction of the concentrated milk salts. The piercing action of the sharp lactose crystals during formation may rupture the membrane of fat globules thus increasing the free fat.

Lactose Crystals in Dry Milks

Lactose crystals may be incorporated in freshly dried milks by altering the manufacturing procedure. These crystals develop before drying by high concentration of total solids in the condensed milks and holding the product cold. A common practice when the drier has a smaller capacity than the evaporator, is to cool and hold the surplus condensed milk until end of the operation. Some lactose crystals may or may not develop during agglomeration depending upon the procedure. If any crystallization occurs the methods must have product wetting above seven per cent for a sufficient period. Peebles and Clary (1955) claim crystallization of lactose takes place in their process of agglomeration.

PROTEIN

The protein ranges from approximately 38% in nonfat dry milk to 8% in dry cream. It is readily subject to changes during processing and to a lesser extent during storage under normal conditions. The changes adversely affect the dispersibility and may influence the sensory qualities as well as the baking qualities of the product.

Action of Heat

Of paramount consideration is the action of heat on the milk proteins which are usually classed into two groups, the casein and serum proteins. The heat treatment (temperature and time) destabilizes by inducing first stage denaturation and then irreversible denaturation (coagulation) as it becomes more severe. Wright (1933) differentiated protein destabilization by: (a) that due to the application of heat to milk in the liquid state and the resulting coagulation of milk proteins particularly the casein, and (b) destabilization of the protein in dry milks by heat which he associated with the release of bound water from casein. A destabilization also takes place slowly during prolonged storage. The rate is directly related to the storage temperature. It is considered to be a protein-lactose deteriorative reaction.

Destabilized serum proteins in dry milk are dispersible upon reconstitution unless the treatment has been extreme. Increasing the acidity or salt concentration has the same effect as elevating the heat. Heat application within the range of regular plant operations on the condensed milk does not denature the serum protein nearly as drastically as it does the casein. In fact, the caseinate micelles become nondispersible at an increasingly rapid rate with a given heat as the concentration of milk solids increases (Table 32). The rate is extremely rapid while the milk

TABLE 32

MILK SOLIDS CONCENTRATION AND TIME REQUIRED TO PRODUCE
50% INSOLUBILITY AT 212° F.

Total Solids, %	Heating Time
8	7.5 hr.
18	2.0 hr.
28	33.0 min.
38	8.0 min.
48	2.0 min.
58	31.0 sec.
68	8.0 sec.
78	2.0 sec.
88	0.4 sec.

Source: Wright (1933).

solids are in the "dough" stage of moisture removal. The effects of heat on the biochemical-physical properties will be discussed further in the consideration of dry milk reconstitutability.

Casein Micelles—Size and Structure

Casein micelles in dry whole and nonfat milks are between 20 and 300 mμ in diameter with an average of approximately 100 mμ (Roelofsen and Salomé 1961). These investigators reported no visible (by electron microscope) inner structure of the casein particle, but noticed separation of casein spheres from each other by a granular substance. King (1962) suggested that during drying the casein micelles contact each other with the formation of hydrogen bonds. Since the bonds are stronger in drum than spray dried milk, he thinks a second continuous phase is indicated probably by interconnected casein micelles.

MILK FAT

General

Milk fat content may be as low as 0.5% in dry whey or as much as 75% in dry cream with 26 to 28% in dry whole and 1% in nonfat dry milk. The fat may exist in dry milks as globules surrounded by a membrane covering of protein nature or as free fat. In the free state it is considered demulsified and is no longer protected by the membrane. Free fat contributes greasiness to the dry product and an oily film to reconstituted milk.

Phospholipids

In addition to the true fats or glycerides, fat solvents such as ethyl ether also dissolve phospholipids that are fat-like substances with both aqueous and fat affinity. Dry natural buttermilk has the highest phospholipid content. The phospholipids are associated with the fat globule membrane, thus are concentrated with the cream. Churning separates the phospholipids from the fat globules and they remain in the serum (buttermilk).

Fat in Drum Dried Milk

In drum dried whole milk the fat is distributed through the particle as either small globules or a "pool" of free fat. Most of the fat is free and much of it is on the surface of the particles. At room temperature the milk fat will be in both the liquid and solidified state.

The free fat content of drum dried whole milk may be 85 to 95% of the total fat. This high percentage may be caused by rupturing the globule membrane by the hot drum and scraping action of the knife in removing the film. The free fat is also increased by increasing the fat content of the milk, exposure of the dry milk to the pulverizing in hammer mill and holding dry product in a hot condition. Choi *et al.* (1951) suggested that coagulation of the protein membrane contributed to freeing the fat.

Fat in Spray Dried Milk

In homogenized spray dried whole milk the small globules are scattered uniformly through the particle (King 1965). He suggested that deviations from a good drying operation or improper storage increase the free fat. This fat tended to surround the air bubbles in the particle or to form blotches on the outside surface of the particle. Size of fat globules with the membrane in the particle depends upon the efficiency of homogenization, but usually is 0.2 to 3μ in diameter (Roelofsen and Salomé 1961). Müller (1964) reported the thickness of the membrane was 8 to 10 mμ in unhomogenized, dry milk and half that thickness in homogenized, dry milk.

Analysis of free fat in spray dried whole milk by Lindquist and Brunner (1962) indicated that it contained slightly more neutral glycerides, saturated fats of C_{10} to C_{18} and less mono- or diglycerides, phospholipids, and unsaturated fat (C_{18}) than the total milk fat.

Studies by Reinke *et al.* (1960) indicated that free fat in spray dried whole milk could be reduced by: (a) lowering the preheating temperature and/or time; (b) using whole milk rather than recombining skimmilk and butter oil; (c) decreasing the heat treatment of the condensed; (d) using a large nozzle orifice at a low pressure (1000 p.s.i.); and (e) homogenizing the condensed milk thoroughly.

Treatment of the milk with a pancreatic enzyme increased the free fat. Excessive action of an agitator or pump on the warm milk or condensed milk will result in more free fat.

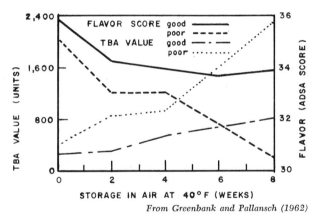

FLAVOR SCORE good ——————
poor — — — —
TBA VALUE good — · —
poor ············

STORAGE IN AIR AT 40°F (WEEKS)

From Greenbank and Pallansch (1962)

FIG. 84. STORAGE CHANGE IN THIOBARBITURIC ACID VALUE AND FLAVOR SCORE OF DRY WHOLE MILK WITH DIFFERENT AMOUNTS OF FREE FAT

Free fat promotes oxidation, especially free fat on the particle surface. It is more rapidly oxidized—2 to 4 weeks according to the investigation of Greenbank and Pallansch (1962) (Fig. 84). Oxidation is reported in Chapter 9 and the effect of free fat on dispersibility will be discussed later in this chapter.

GASES IN DRY MILK

The gases of concern in dry milk are nitrogen, oxygen, and carbon dioxide. Normally air occupies the interstices among the particles and is occluded as cells (bubbles) inside the particle. It may be absorbed by the particles or dissolved in the moisture or other particle components, e.g., milk fat. Dry milk with a high air content may have particles that are large air bubbles covered with a thin shell of solids.

Only a portion of the particles contains air bubbles. Coulter *et al.* (1951) observed with different types of spray systems a range of 25 to 89% of the dry whole milk particles containing air bubbles. Variations occurred between two samples from the same drier. Also, the number of air bubbles per particle was variable. Milks dried by the centrifugal system have a higher percentage of particles with air bubbles. Air incorporated in the condensed milk directly influences the air content of the dry milk. As the total solids increases in the condensed milk the air content decreases in the dried particles.

Milks immediately after drying have less air than later. Coulter and Jenness (1945) attributed this fact to the vacuum created in the entrapped bubbles during cooling which causes the air to be drawn into these bubbles. Also, the pressure of water vapor in the air cells within the particle is decreased by cooling and milk solids sorption which increases the vacuum.

The amount of air may be ascertained by:

$$\text{volume of particles} - \frac{1}{\text{density of particles}} = \text{volume of trapped air}$$

To obtain the volume of particles petroleum ether is added to a known amount of dry milk or Hetrick and Tracy (1948) suggested the following equation:

$$\left(\frac{22,400 \; P_e}{W \; R \; T}\right) \times \left(V_c - \frac{W}{1.31}\right) = \text{ml. of gas/gm.}$$

where:

P_e = mm. of Hg of equilibrium pressure
W = gm. of dry milk
T = absolute temperature
R = 62,400 ml. mm. gas constant
V_c = ml. of container
1.31 = gm./ml. density of powder without air.

The residual gas (that remaining after removal of air from interstices) contains more oxygen than regular air. Analysis by several investigators showed the oxygen of this air was 33 to 39% and averaged approximately 35%. The selectivity of oxygen out of air may be caused by differential solubility in the fat or adsorption.

Dry milk absorbs a substantial amount of carbon dioxide if ambient conditions are favorable. In gas packaging with carbon dioxide Coulter and Jenness (1945) observed a sorption of 48 ml. of carbon dioxide per 100 gm. of dry whole milk.

The oxygen content of dry whole milk and related products with a substantial per cent of milk fat is important in oxidative deterioration. Further discussion on oxidation is included in Chapter 9.

MILK ENZYME SYSTEMS

Enzymes in Raw Milk

Numerous enzymes have been found in raw milk (Whitney 1958). The more common ones include alkaline phosphatase, lipase and esterase, lactoperoxidase, catalase, xanthine oxidase and protease. Each has individual characteristics of destruction by heat treatment. In general, these enzymes in dry milk appear to have little if any activity. However, in the reconstituted products enzyme induced changes are of more significance.

Alkaline Phosphatase.—This enzyme is inactivated by pasteurization. Its presence in dry milk is objectionable principally because of the association with unsatisfactory pasteurization and thus the possibility of pathogenic organisms in the product.

Lipase.—It is associated with the lipids of milk and thus is more closely related to the dry whole milk and dry cream products. Lipase consists of several enzymes that vary in properties and hydrolyze many types of fat. They have not all been clearly defined. Lipase associated with hydrolytic rancidity in raw milk is inactivated by a heat treatment below pasteurization—131°F. for 30 min. Esterases catalyze the hydrolysis of esters. Therefore, they are closely allied with lipase. Much more knowledge is necessary to establish the number of enzymes in this group and their individual characteristics.

Lactoperoxidase.—It requires a high temperature for inactivation (above pasteurization) so many undergo only partially reduction in processing of spray dried milk. The severe heat in drum drying inactivates it. This enzyme catalyzes hydrogen peroxide breakdown. Storage of dry milk with a high moisture reduces the lactoperoxidase activity.

Catalase.—Catalase also catalyzes the hydrogen peroxide decomposition in milk. It is inactivated by heating the milk to 158°F. for 30 min. so normally would not be present in dry whole milk.

Xanthine Oxidase.—This enzyme is not inactivated by temperatures up to 176°F. It catalyzes oxidation or reduction reactions. Undoubtedly, it is present in low-heat spray dried milk.

Protease.—Hydrolysis of proteins is catalyzed by protease. Warner and Polis (1945) observed that almost all proteolytic activity in milk was precipitated with the casein by acidifying the milk. They were able to increase the activity 150 times by concentration. Some enzyme activity was obtained in both the α and β fractions after the first precipitation, but did not appear to be present in the purified β fraction. Hipp *et al.* (1952) also reported purified α casein had a small proteolytic activity.

Protease appears to be inactivated at 176°F. for 10 min. At a pH of 4.5 the enzyme is slightly soluble in the presence of salt or hydrolysis products of casein.

MOISTURE

Moisture content of dry milks probably receives more attention during the drying operation than the other factors. Customary range is 3 to 4 and 2 to 3% for nonfat dry milk and dry whole milk, respectively. In addition to its effect on lactose (discussed previously in this chapter) the rate of several deteriorative changes are influenced by the moisture content of the dry milk. The sensory changes include stale (burnt feathers) and browning. Chemical changes from high moisture are: increased protein "insolubility," acidity, carbon dioxide, acid ferricyanide values, and fluorescent substances. Low moisture (below 2.0%) in dry whole milk may slightly accelerate oxidative changes compared to three per cent.

Among the constituents of milk solids, lactose and protein have the greatest effect on the moisture equilibrium. Salts in solution can decrease the vapor pressure by an osmotic effect (Coulter *et al.* 1951). These authors reported that at a relative vapor pressure of 0.1, dry whole milk with 3.0% moisture had 0.9 gm. in the 27 gm. of caseinate, 0.35 gm. in the 10 gm. of serum protein, and 0.5 gm. in the 49 gm. of lactose per 100 gm. sample. The remainder of 1.25 gm. of moisture was attributed to the salts. Milk fat is presumed to have no direct influence on the moisture retention of dry milk.

The hygroscopicity of the amorphous lactose probably has the greatest practical effect on the moisture equilibrium of dry milk after its manufacture. Should crystallization occur the α hydrate retains more moisture than β anhydride. Proteins, mainly casein, hold moisture by sorption. Heat treatments during the various activities associated with drying influence the moisture retaining capacity of the dry milk through its effect on casein.

The importance of moisture is further elucidated in the discussion on keeping quality (Chapter 9).

DENSITY

Practical Effects

The density of dry milk products has several practical implications. The principal one is its pronounced influence on packaging costs. Recovery of milk particles in cyclone collector is affected and density directly contributes to the amount of particles that become air-borne in the drying, sifting, and handling areas. Dry milk dust in the atmosphere is more of a nuisance problem than one of economics.

Classification

Densities are classed into three groups—bulk (apparent) density, particle density and density of the dry milk solids. Bulk density is regarded as weight per unit volume and frequently is expressed as gm. per ml. It is divided into packed and loose densities. Packed bulk density is determined after the sample has been tapped or vibrated until the volume is relatively constant. Loose bulk density is ascertained prior to the vibrating or tapping.

Factors Affecting Bulk Density

The packed bulk density of nonfat dry milk has a wide range of 0.18 to 1.25 gm. per ml., but regular spray dried generally is 0.50 to 0.60 gm. per ml. and drum dried nonfat dry milk is 0.30 to 0.50 gm. Table 33 pre-

TABLE 33

DENSITY OF NONFAT DRY MILK SAMPLES PREPARED BY THREE DIFFERENT PROCESSES

	Density of Powders Determined by Two Methods					
	Die Method				Cylinder Method	
Type of NDM	1	2	3	Ave. of 1–3	Untapped	Tapped
U.S. Dept. Agr. foam spray	0.316	0.319	0.320	0.318	0.30	0.34
Commercial instant	0.260	0.264	0.269	0.264	0.24	0.29
Commercial regular	0.580	0.593	0.600	0.591	0.50	0.67

Source: Hanrahan and Konston (1965).

sents the bulk density of nonfat dry milk processed by three methods and determined by two methods. Dry whole milk is slightly less in bulk density. Foam spray dried milk has the least, drum is next, and spray the highest. The degree of pulverizing drum dried milk in the hammer mill affects bulk density. Agglomeration reduces it roughly 40 to 60%. Dry milk from centrifugal sprays has a trifle lower density than the pressure spray product.

The manufacturing procedure and conditions greatly influence bulk densities, principally by the effect on air content. Obviously steps to reduce the occluded air will increase the density. One means is to minimize the air content of the concentrate before drying. Increasing the total solids of the concentrate, reducing the spray pressure, and using a large orifice increases the bulk density. Increasing the inlet air up to 430°F.

decreases the density slightly at a uniform rate and more rapidly above 430°F. The outlet air temperature seems to have no direct correlation. Less uniformity in particle size distribution can result in closer packing and thus a higher bulk density. The shape as well as size of the particle will also affect the closeness that particles pack together.

Particle and True Density

Particle density is influenced principally by the amount of entrapped air. One of the main processing factors that contributes is viscosity and air incorporation into the concentrate ahead of drying. The type of spray atomization has an effect on the air retention. Centrifugal spray dry milks have more entrapped air than pressure spray products.

True density refers to the air free solids and may be calculated from the formula:

$$\text{density} = \frac{110}{\dfrac{\% \text{ fat}}{0.93} + \dfrac{\% \text{ SNF}}{1.6} + \dfrac{\% \text{ water}}{1.0}}$$

The true density of nonfat dry milk is 1.44 to 1.48 gm. per ml. and of dry whole milk 1.26 to 1.32. The moisture content and the ratio of solids-not-fat to fat are the two chief variables affecting true density. In a dry high fat product an increase in moisture reduces density and the opposite is true for nonfat dry milk. A decrease in fat will increase true density.

Increasing Density

With the occluded air plus the space in the interstices, dry milk offers an opportunity for compression. The application of 100 to 150 p.s.i. to dry whole milk can eliminate one-half or more of the space. Densities of drum or spray dried products were increased to 1.1 or above. Hanrahan and Konston (1965) obtained a bulk density of 0.45 by compressing foam spray nonfat dry milk at 500 to 600 p.s.i. The addition of nondairy ingredients, e.g., sucrose, to milk before drying will affect the bulk density of the dry product. Certain additives including sodium aluminum silicate and calcium silicate will increase the bulk density of dry whole milk.

FLOWABILITY

Characteristics of Dry Milks

Many dry materials must have good flowability for satisfactory use. This includes dry milks for several purposes. Flowability of dry milks utilized in vending is sufficiently important to necessitate standards for this characteristic. The movement of the product in vertical driers, cyclones, and storage tanks is affected by flow characteristics. The use of bulk tanks instead of drums or bags enhances the need for adequate flowability.

There are two types of flow. One is the independent movement of a particle and other is group flow. Nonfat dry milk may exhibit mainly particle flow while dry whole milk has a greater amount of group flow.

Some characteristics of dry milks that determine its flowability are particle size, shape, density, and electrostatic charge. Large particles tend to flow more easily than the fines. Consequently, agglomeration is beneficial. Uniformity of size is advantageous. A wide variation in particle size permits the fines to occupy the interstices among the large particles which results in closer packing with the greater friction reducing flowability.

A spherical shape is superior to the irregular shapes that are present in drum dried milks. High density particles tend to possess a more desirable flowability. A smooth surface of the particle reduces friction and thus enhances the flow property. Milk fat tends to inhibit flow with a greater effect at temperatures that melt the fat than 45°F. or lower.

Increased dryness is directly correlated with flowability of dry milks. Particularly detrimental is moisture adsorption, hence benefits are obtained by the addition of a free flowing agent. However, Taneya (1963) reported nonfat dry milk increased in fluidity when it absorbed seven per cent moisture. This was attributed to development of lactose crystals on the surface of the particles.

Free Flowing Agents

The addition of free flowing agents or moisture absorbing compounds to dry milk adds to its flowability. Calcium silicate at the rate of 1.5% before or after drying has been claimed to be beneficial (Cipolla *et al.* 1961). The use of 0.5 to 1.0% of sodium aluminum silicate (Zeolex and Zerofree) or 1.0% of pyrogenic silica (Cab-O-Sil) are effective (Linton-Smith 1961). Sodium aluminum silicate (0.5%) renders about eight per cent of the calcium in the dry milk unavailable.

Other compounds that have been suggested are amorphous silicon dioxide, tricalcium phosphate, aluminum oxide, and magnesium oxide, plus mixtures such as 4 parts of sodium aluminum silicate and 1 part of calcium phosphate.

MILK DUSTINESS

Small light dry milk particles have the objectionable property of becoming air-borne during movement subsequent to drying. These particles also increase the entrainment losses, are a health hazard to people allergic to protein and create sanitation problems, particularly for insect control. Small orifice, high pressure, low solids in the condensed milks and low moisture increase air-borne milk dust. An increase in fat content of the product slightly reduces the problem. Agglomeration also decreases the per cent of milk particles becoming air-borne.

RECONSTITUTABILITY

Functions in Recombining Dry Milks and Water

An important characteristic of dry milks is their reconstitutability. Other terms—solubility or dispersibility, are used to indicate the total phenomenom of recombining dry milk with water. But in a strict sense solubility, dispersibility, sinkability, and wettability are limited, but dis-. tinct in function. Principally lactose, undenatured serum protein and some of the salts are soluble. The casein is dispersible. Sinkability refers to the particle's ability to penetrate the aqueous surface tension and wettability is the penetration of water into the particles. Reconstitutability includes all these functions taking place in the process of recombining dry milks with water.

The ideal is a dry product that rapidly recombines with water without agitation and reacquires the same characteristics of regular milk. Ease of reconstitution influences many uses—beverages, recombined milk, milk solids fortification of liquid milk products, etc. Common defects are incomplete or slow dispersion, stratification (higher solids at top or bottom) and scum residue on container surface.

Contributing Factors

Factors that contribute to reconstitutability may be the physical properties: size, shape, surface, density, uniformity, air content, composition, especially ratio of solids-not-fat to fat of the particles, and the presence of additives; the chemical properties which include the amount of protein denaturation (heat or storage) and the degree of proteolysis; and the conditions for recombining: the water and dry milk temperature, the nature and amount of water hardness, and time and nature of the agitation.

The property of dry milk reconstitutability is affected by the equipment and processing conditions: the drier and system of atomization, heat stability of the milk, preheat treatment of milk and the concentrate, total solids of the concentrate, outlet air temperature and contact time, recycling of fines, selectivity of screening, agglomeration, and storage temperature and time.

Dispersibility

For good dispersibility an important consideration is the total heat treatment on the casein during processing. Increased heat application with increasing total solids causes larger amounts of irreversible denaturation. Under normal reconstitution procedures this denatured casein does not form a stable dispersion. The American Dry Milk Institute solubility index is a measure of the denatured casein.

Drum drying exposes the milk film to severe heat with the resulting high per cent of denaturation. Spray dried milk normally has very little denatured casein. But the percentage of denaturation will be increased by heating the condensed milk too high and long or subjecting the milk particles to above normal air temperatures in the drier. Denaturation from this cause is more prevalent when attempting to obtain production above the normal capacity of the drier.

Large particles of dry milks are generally recognized as necessary for good dispersibility. Apparently there are exceptions. Konston et al. (1965) noticed that very large particles of foam spray dried whole milk have less dispersibility than the small ones. Particles above 710 μ in average diameter showed a linear relationship with dispersibility. Ten per cent of the large foam spray particles reduced dispersibility 4%.

Wettability

The tendency of dry milk to form lumps upon addition to water indicates a lack of wettability. Small particle size and symmetrical shape enhance close packing of particles and thus inhibit penetration of the water. In general, larger particles more irregular in shape provide more space in the interstices for wetting. The increased size by formation of aggregates and larger spaces among them as a result of agglomeration also favor wetting (Brockian et al. 1957). An orientation of soluble solids toward the particle surface was thought to occur and thus improves wetting. More uniformity of large particles results in less fines to fill in adjacent spaces which restrains wetting.

The amount and dispersion of fat affects wettability (Coulter et al. 1951). These authors reported no difference in wettability of dry milk with fat between 18 and 32%. Since fat is hydrophobic it probably inhibits wetting in relation to the amount present up to approximately 18% although the test may not be sensitive enough to detect the difference in small increments.

Fractionation of the fat indicates that the milk dried with low melting point fats have better wettability and dispersibility. As little as five per cent of the high melting point fats decreased sinkability (Nelson and

Winder 1961). Coulter *et al.* (1951). were not successful in attempts to improve wettability of dry milk by the addition of various ionic wetting agents to the milk before drying, to the dry milk or to the water before reconstitution.

Additives

Hibbs and Ashworth (1951) added 0.05% (reconstituted basis) emulsifiers-sorbitan monostearate, polyoxyethylene sorbitan monostearate, and glycerol monostearate to the milk before drying. These emulsifying agents caused the fat to churn during reconstitution. Some success in control of the churning is possible with a combination of 0.05% sorbitan monostearate and 0.05% polyoxyethylene sorbitan monostearate.

However, the addition of up to 25% sugars, e.g., sucrose, corn syrup solids and castor sugar to the milk before drying or by dry blending the sugar in the granulated form improves wettability and dispersion. The granulation is significant since very fine (powdered) sugars are less effective.

Storage Change

The reconstitutability of dry milk is diminished during storage. The same factors that enhance keeping quality apply to reconstitutability chiefly through the affect on destabilizing the protein. High moisture equilibrium of the dry milk and high storage temperature are directly proportional to "solubility" degradation.

Film or Scum

The objectionable film or scum residue on some container surfaces of reconstituted whole milk has been attributed to denatured protein and to the fat. The scum always has a higher percentage of fat and protein and less lactose and ash than the "soluble" portion (Litman and Ashworth 1957) (Table 34). King (1960) stated that the scum was clusters of casein micelles and fat globules. Some of the fat also was in the free state with casein and air bubbles. The fresh dry whole milk had only a few undissolved particles in the serum, but there was a noticeable increase from an old product. These particles were in the 10 to 15 μ range in diameter. Aging may destabilize casein and increase fat-casein complex. Wilster *et al.* (1946) mentioned that thorough homogenization (4000 p.s.i.) of the condensed milk and spray drying small particles as well as not allowing overheating of the product reduced the scum. The addition of 0.2% sodium citrate was helpful. Disodium phosphate may replace the sodium citrate, but both of these salts are of limited value.

TABLE 34

ANALYSES OF SCUM AND NORMAL FRACTIONS OF DRY WHOLE MILK
(average of 10 samples stored at room temperature)

Component	Scum Fraction Solids	Normal Fraction Solids
Total protein, %	30.0	26.2
Fat, %	48.2	27.4
Iodine No.	30.4	32.6
Melting point, °F.	95.0	85.1
Ash, %	4.2	5.8
Calcium, % of ash	42.9	19.0
Phosphorus, % of ash	16.6	12.0
Carbohydrate, % by diff.	14.6	40.6

Source: Litman and Ashworth (1957).

Sinkability

Sinkability is closely related to wettability. The amount of occluded air in the particle has a pronounced influence on sinkability. Hence, foam spray dried milks have very poor sinkability. Agglomeration of regular spray dry milks improves sinkability chiefly by increasing the aggregate weight. The effect of agglomeration on sinkability is greater in nonfat dry milk than dry whole milk. The relation may be indirect since fat has a negative influence on agglomeration. Harper *et al.* (1963) claim that bulk density of dry milk for self-dispersion should not be above 0.4 gm. per ml. despite the improved sinkability. Protein is destabilized by the high localized concentration of lactose and salts during rehydration.

Equipment and Procedure

In reconstitution by mechanical means the vigorousness of agitation directly affects the dispersion efficiency, including dry milks with denatured casein. Time of agitation also is a factor. But excessive agitation in recombining dry whole milk and water may cause churning of the fat. The water can be too cold or hot for best results. A range of 80° to 90°F. for nonfat dry milk and 110° to 115°F. for dry milk is optimum. High calcium hardness of the water may adversely affect reconstitutability. The addition of sodium citrate or phosphate will negate the calcium effect.

HEAT TREATMENT

Numerous heat labile constituents of dry milk are affected by the normal processing procedures. The extent depends primarily upon the severity of the temperature, exposure time and the total solids content. The effect of heat on protein denaturation is discussed earlier in this chapter and its effect on keeping quality is elucidated in Chapter 9.

Cooked Flavor

The heat treatment necessary to ensure good baking qualities of nonfat dry milk and the production of antioxidants in dry whole milk for improved keeping quality results in the liberation of sulfhydryl compounds from the serum protein probably β-lactoglobulin. These volatile sulfides are associated with cooked or caramel flavor.

Gould and Sommer (1939) established the fact that the appearance of cooked flavor in milk was a function of temperature and time. They detected the liberation of sulfides using a lead acetate test when milk was heated to 169° to 172°F. momentarily; 165° to 169°F. for 3 min.; and 158° to 162°F. for 30 min. A decrease in pH and increase in fat content and aeration during heating correlated directly with the amount of cooked flavor. Milk from various sources may vary in sulfide release by a given heat treatment. Supposedly the intensity of cooked flavor directly corresponds. Browning also was correlated with temperature and time (Table 35).

TABLE 35

INFLUENCE OF HEATING MILK FOR PROLONGED PERIODS AT 203° F. UPON THE LIBERATION OF SULFIDES, THE NITROPRUSSIDE REACTION, THE COOKED FLAVOR AND THE DISCOLORATION

Period of Exposure to 203° F., Min.	Volatile Sulfur, Mg./l.	Nitro-prusside Test[1]	Flavor	Browning[1]
0	0.158	4	Cooked	0
5	0.165	4	"	0
15	0.192	4	"	0
30	0.203	4	"	0
60	0.208	3	Caramel	0
90	0.091	2	"	1
120	0.060	1	"	1
150	0.053	1	"	2
180	0.012	?	"	3
210	0.000	?	"	4

Source: Townley and Gould (1943).
[1] Intensity indicated numerically: 0 = no coloration, 4 = strong coloration.

Reducing Complexes

The values of reducing complexes of dry milks vary according to the test used. These include acid ferricyanide, 2.6 dichlorobenzenone indophenol and thianmin disulfide, sodium nitroprusside, and lead acetate.

Harland et al. (1940) observed no significant change in acid ferricyanide reducing complexes in dry milk due to preheat treatment of the milk. But preheat was a pertinent factor for thiamin disulfide reducing values. Table 35 shows a decline in sulfide release after a maximum was attained by continued heating. The same trend is indicated by thiamin disulfide reducing complexes. During storage reducing complexes increase in dry milk as shown by the acid ferricyanide and thiamin disulfide values.

Loaf Volume Depressant

Adequate heat treatment of the milk before spray drying is essential for good bake test qualities. Insufficient heating depresses loaf volume. Figure 85 shows the effects of various preheat treatments of nonfat dry milk on the bread score from under, normal and overmixed dough. Ultra high temperature of 300°F. for 30 sec. was not as effective as 185°F. for 20 min.

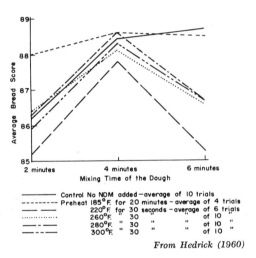

FIG. 85. AVERAGE BREAD SCORE OF DOUGHS WITH NDM FROM SKIMMILK WITH VARIOUS PREHEAT TREATMENTS

——————— Control No NDM added—average of 10 trials
- - - - - - Preheat 185°F. for 20 minutes – average of 4 trials
— — — 220°F. for 30 seconds – average of 6 trials
.... 260°F. " 30 " " of 10 "
—··—··— 280°F. " 30 " " of 10 "
— ·· — ·· — 300°F. " 30 " " of 10 "

From Hedrick (1960)

The complete identity of the depressing substance is not yet known. But Ashworth and Harland (1942) suggested the compound reacted with the gluten matrix causing it to soften. Later studies indicated the factor was associated with the serum protein. Stamberg and Bailey (1942) believed that cystine + cysteine were involved. These are important amino acids in milk with sulfur groups and the sulfhydryls are altered by the heat treatment, thus cystine + cysteine probably are responsible. A number of other investigators have come to the same conclusion.

Cheese Coagulum

Preheat treatments at atmospheric pressure that are sufficient to denature a substantial amount of serum protein may not measurably increase the casein denaturation. Yet the firmness of the casein coagulation is decreased. Consequently, reconstituted milks from high-heat dry milks may not be satisfactory for the manufacture of cheese by regular procedures. The specific reactions induced by heat that contribute to the coagulum formation need further elucidation. Changes have been attributed to precipitation of calcium probably as a complex phosphate.

REFERENCES

ASHWORTH, U. S., and HARLAND, H. A. 1942. Chemical factors affecting the baking quality of dry milk solids. II. The effect of milk on gluten fractionation. Cereal Chem. *19*, 830–837.

BROCKIAN, A. H., STEWART, G. F., and TAPPEL, A. L. 1957. Factors affecting the dispersibility of "instant dissolving" dry milks. Food Research *22*, 69–75.

CHOI, R. P., TATTER, C. W. and O'MALLEY, C. M. 1951. Lactose crystallization in dry products of milk. II. The effects of moisture and alcohol. J. Dairy Sci. *34*, 850–854.

CIPOLLA, R. H., DAVIS, D. W., and VANDER LINDER, C. R. 1961. Free flowing dried dairy products. U. S. Patent 2,995,447.

COULTER, S. T., and JENNESS, R. 1945. Packing dry whole milk in inert gas. Minn. Agr. Expt. Sta. Tech. Bull. *167*.

COULTER, S. T., JENNESS, R., and GEDDES, W. F. 1951. Physical and chemical aspects of the production, storage and utility of dry milk products. Advances in Food Research, *3*, 45–118. Academic Press, New York.

GOULD, I. A., and SOMMER, H. H. 1939. Effect of heat on milk, with especial reference to the cooked flavor. Mich. Agr. Sta. Tech. Bull. *164*.

GREENBANK, G. R., and PALLANSCH, M. J. 1962. The progress of oxidation in the milk powder granule. XVI Intern. Dairy Cong. B 1002–1008.

HANRAHAN, F. P., and KONSTON, A. 1965. Properties of compressed nonfat dried milk. J. Dairy Sci. *48*, 1533–1535.

HANRAHAN, F. P., TAMSMA, A., FOX, K. K., and PALLANSCH, M. J. 1962. Production and properties of spray-dried whole milk foam. J. Dairy Sci. *45*, 27–31.

HARLAND, H. A., COULTER, S. T., and JENNESS, R. 1949. Some factors influencing the reducing systems in dry whole milk. J. Dairy Sci. *32*, 334–344.

HARPER, M. K., HOLSINGER, V., FOX, K. K., and PALLANSCH, M. J. 1963. Factors influencing the instant solubility of milk powders. J. Dairy Sci. *46*, 1192–1195.

HAYASHI, H. 1962. Studies on spray drying mechanism of milk powders. Rept. Res. Lab. Snow Brand Milk Prod. *66*, 42–46.

HEINEMANN, B. 1948. Importance of milk composition in manufacturing dry milk solids. Nat'l Butter and Cheese J. *39*, No. 5, 49, 50, 52, 54, 56, 58.

HETRICK, J. H., and TRACY, P. H. 1948. Manometric measurement of the gas desorbed from vacuumized whole milk powder. J. Dairy Sci. *31*, 831–838.

HIBBS, R. A., and ASHWORTH, U. S. 1951. The solubility of whole milk powder as affected by protein stabilizers and emulsifiers. J. Dairy Sci. *34*, 1084–1091.

HIPP, N. J., GROVES, M. L., CUSTER, J. H., and McMEEKIN, T. L. 1952. Separation of α-, β- and γ-casein. J. Dairy Sci. *35*, 272–281.

KING, N. 1954. The physical structure of milk powder. Dairy Ind. *19*, 39–43.

KING, N. 1960. The microstructure of some non-dispersible particles in milk powder. Aust. J. Dairy Technol. *15*, 77–79.

KING, N. 1962. Microscopy of dispersion phenomena in milk powders. XVI Int. Dairy Cong. B 977–980.

KING, N. 1965. The physical structure of dried milk. Dairy Sci. Abs. 27, No. 3, 91–104.

KONTSON, A., TAMSMA, A., and PALLANSCH, M. J. 1965. Effect of particle size distribution on the dispersibility of foam spray dried milk. J. Dairy Sci. 48, 777–778.

LINDQUIST, K., and BRUNNER, J. R. 1962. Composition of the free fat of spray-dried whole milk. J. Dairy Sci. 45, 661.

LINTON-SMITH, L. 1961. Free flowing milk powder. Aust. J. Dairy Technol. 16, 22–24.

LITMAN, I. I., and ASHWORTH, U. S. 1957. Insoluble scum-like materials on reconstituted whole milk powders. J. Dairy Sci. 40, 403–409.

MÜLLER, H. R. 1964. Electron microscopical examination of milk and milk products. I. Examination of the structure of dried milks. (In German) Milchwissenschaft 19, No. 7, 345–356.

NELSON, W. E., and WINDER, W. C. 1961. Effect of solid-fat/liquid-fat ratio upon the sinkability of dried whole milk. Paper presented at 56th Annual Meeting, American Dairy Science Association, Madison, Wis.

PEEBLES, D. D., and CLARY, P. D. 1955. Milk treatment process. U.S. Patent 2,710,808.

REINKE, E., BRUNNER, J. R., and TROUT, G. M. 1960. Effect of variations in processing on free-fat, self-dispersion and flavor of whole milk powder. Milk Prod. J. 51, No. 9, 6, 7, 24, 25.

ROELOFSEN, P. A., and SALOMÉ, M. M. 1961. Submicroscopic structure of milk powder. Neth. Milk Dairy J. 15, 392–394.

SHARP, P. F. 1938. Seeding test for crystalline beta lactose. J. Dairy Sci. 21, 445–449.

STAMBERG, O. E., and BAILEY, C. H. 1942. The effect of heat treatment of milk in relation to baking quality as shown by polarograph and farinograph studies. Cereal Chem. 19, 507–517.

SUPPLEE, G. C. 1926. Humidity equilibria of milk powders. J. Dairy Sci. 9, 50–61.

TANEYA, S. 1963. Surface structure of dried skim-milk powder particle by electron microscopic observation. Jap. J. Applied Physics 2, 637–640.

TOWNLEY, R. C., and GOULD, I. A. 1943. A quantitative study of heat labile sulfides of milk. I. Method of determination and the influence of temperature and time. J. Dairy Sci. 26, 689–703.

TROY, H. C., and SHARP, P. T. 1930. α and β lactose in some milk products. J. Dairy Sci. 13, 140–145.

WARNER, R. C., and POLIS, E. 1945. On the presence of a proteolytic enzyme in casein. J. Am. Chem. Soc. 67, 529–532.

WHITNEY, R. McL. 1958. The minor proteins of bovine milk. J. Dairy Sci. 41, 1303–1323.

WILSTER, G. H., SCHREITER, O. M., and TRACY, P. H. 1946. Physico-Chemical factors affecting the reconstitutability of dry whole milk. J. Dairy Sci. 29, 490–491.

WRIGHT, N. C. 1933. Factors affecting the solubility of milk powders. J. Dairy Res. 4, 122–144.

CHAPTER 11

Markets and Uses

GENERAL ADVANTAGES

All dry milk products have several obvious advantages and some of these are becoming increasingly important with the greater emphasis on efficiency. Dry milks have a much longer keeping quality and can be held in unrefrigerated storage areas. Much less storage space is required per unit of solids. This fact plus the longer shelf life permit extensive stocking of dried nonfat products during peak production for a steady available supply throughout the year.

Costs of shipping dry products are much less because of the substantial weight reduction, and unrefrigerated trucks or railroad cars can be used. Distribution is possible to underdeveloped countries, particularly those with unfavorable conditions causing the perishable dairy products to be impractical. Consequently, dry milks have superiority both in economy and convenience.

Dry milks can be pretested and selected for specialized and critical uses to insure favorable results. Its advantage as a concentrated source of many essential nutrients should not be overlooked.

NUTRITIONAL VALUE

Milk has been recognized for many years as the most valuable source of nutrients available to man. Fortunately modern drying methods preserve these nutrients during processing so little or no losses occur. Table 36 lists major and minor nutrients present in the four most important dry dairy products.

The biological value of the dry milk proteins is 89; milk protein is highly digestible. All of the essential amino acids are contained in dry milks. Analysis of spray processed dry milk indicated a slight destruction of lysine and no difference in its availability compared to fresh milk (Mauron et al. 1955). The severe heat treatment of roller drying destroyed more lysine. There was no destruction of two amino acids, methionine and tryptophan, by either the spray or roller drying process.

Dry whole milk is a good source of vitamin A. Dry milks are rich in calcium and phosphorus. In fact, many nutritionists believe that milk in most forms is more important than any other food in providing a diet for humans adequate in calcium and phosphorus. Children and adults have a very high utilization of these two minerals from dry milk. Also dry

TABLE 36

CONSTITUENTS OF DRY MILK PRODUCTS
(Per 100 Gm.)

Nutrient	Nonfat Dry Milk	Dry Whole Milk	Dry Buttermilk	Dry Whey
Carbohydrate (lactose), gm.	52.3	38.2	50.0	73.5
Protein, gm.	35.9	26.4	34.3	12.9
Milk fat, gm.	0.8	27.5	5.3	1.1
Minerals, gm.	8.0	5.9	7.6	8.0
Moisture, %	3.0	2.0	2.8	4.5
Vitamin A, I.U.	30.0	1130.0	220.0	50.0
Calcium, mg.	1308.0	909.0	1248.0	646.0
Phosphorous, mg.	1016.0	708.0	970.0	589.0
Pantothenic acid,[1] mg.	3.3	2.9	3.1	. . .
Riboflavin (B$_2$ or G), mg.	1.8	1.5	1.7	2.5
Thiamin (B$_1$), mg.	0.4	0.3	0.3	0.5
Niacin, mg.	0.9	0.7	0.9	0.8
Potassium, mg.	1745.0	1330.0	1606.0	. . .
Sodium, mg.	532.0	405.0	507.0	. . .
Iron, mg.	0.6	0.5	0.6	1.4
Calories	363.0	502.0	387.0	349.0

Sources: Watt and Merrill (1963).
[1] American Dry Milk Institute (1962).

milks are a valuable source of riboflavin. Other vitamins or growth factors that are present include vitamins C and E and B$_{12}$, D and K with the fat soluble vitamins being proportionally related to the fat content of the dry milks.

OUTLETS FOR NONFAT DRY MILK

Exports

The domestic market price as a result of the government support price for surplus nonfat dry milk usually has been above the world market price in recent years. Consequently, any export sales usually are made with the assistance of a government subsidy; during 1968 the government supported exports of nonfat dry milk were 10,600,000 lb.

The total U.S. Government purchase in 1969 were roughly one-third the total production of nonfat dry milk. The major U.S. Government distributions were to domestic agencies and to underdeveloped foreign countries for relief food purposes.

Domestic Utilization

Table 37 presents the domestic consumption data of nonfat dry milk since 1935. Since domestic consumption was 70% of production the merchandising and market development practices should be continuously improved to utilize a much larger share of the production for commercial

TABLE 37

CONSUMPTION OF NONFAT DRY MILK AND DRY WHOLE MILK IN THE UNITED STATES
(Add 000,000)

Year	Nonfat Dry Milk		Dry Whole Milk	
	Total	Lb. Per Capita	Total	Lb. Per Capita
1935	200	1.6	17	0.13
1936	226	1.8	18	0.14
1937	244	1.9	14	0.11
1938	275	2.1	15	0.11
1939	285	2.2	17	0.13
1940	295	2.2	19	0.14
1941	325	2.5	21	0.16
1942	335	2.5	26	0.20
1943	273	2.1	50	0.39
1944	193	1.5	44	0.34
1945	248	1.9	48	0.37
1946	451	3.2	71	0.51
1947	417	2.9	65	0.45
1948	485	3.3	42	0.29
1949	481	3.3	37	0.25
1950	549	3.7	42	0.28
1951	637	4.2	41	0.27
1952	711	4.6	70	0.46
1953	649	4.2	37	0.19
1954	716	4.5	29	0.18
1955	889	5.5	42	0.25
1956	864	5.2	50	0.30
1957	895	5.3	37	0.22
1958	953	5.6	48	0.28
1959	1074	6.2	49	0.28
1960	1106	6.2	55	0.31
1961	1128	6.2	50	0.28
1962	1119	6.1	61	0.33
1963	1087	5.8	48	0.26
1964	1131	6.0	61	0.32
1965	1069	5.6	59	0.31
1966	1132	5.8	66	0.33
1967	1081	5.5	51	0.26
1968	1034	5.8	65	0.30
1969[1]	1017	5.7	50	0.26

Sources: American Dry Milk Institute (1951–1970). U. S. Dept. Agr. (1965).
[1] Preliminary.

food purposes. Three possibilities for improvement have been proposed by Miller (1965) for dry milk cooperatives, but they apply to all dry milk manufacturers. He specifically suggested that salesmen receive a more thorough training in the technical aspects of all commercial uses. More coordination would be beneficial between sales and the technical service needed by the user for obtaining optimum results with the product, and last, he reported the desirability of more direct contact between manufacturer and user.

More research is needed in the marketing of dry milk products. Also critical is the dearth of product development by the dry milk industry, particularly to meet specific needs of other food industries. Expanded studies on new uses and applications of both new and regular dry milks should aid in obtaining greater sales.

The three principal commercial outlets for nonfat dry milk currently are for bakery, home (retail), and dairy (Table 38). These three ac-

TABLE 38

MAJOR USES OF NONFAT DRY MILK DURING 1961 TO 1969
(Add 000,000 Lb.)

	1961	%	1963	%	1965	%	1967	%	1969	%
Bakery	267.0	31.4	261.4	30.0	241.3	26.3	249.8	26.5	187.0	18.4
Packaged for home use	218.5	25.7	248.3	28.5	231.2	25.2	231.9	24.6	324.3	31.9
Dairy	193.8	22.8	195.2	22.4	288.6	26.7	240.4	25.5	282.6	27.8
Meat processing	47.6	5.6	54.9	6.3	78.9	7.3	75.4	8.0	44.7	4.4
Prepared dry mixes	55.3	6.5	40.1	4.6	89.7	8.3	68.8	7.3	101.7	10.0
Confectionery	19.5	2.3	20.9	2.4	22.7	2.1	17.9	1.9	19.3	1.9
Institutions	13.6	1.6	9.6	1.1	3.7	0.4	3.8	0.4	4.1	0.4
Soft drink bottlers	3.4	0.4	4.4	0.5	5.5	0.6	3.8	0.4	6.1	0.6
Soup manufacturers	1.7	0.2	4.4	0.5	0.9	0.1	7.5	0.8	4.1	0.4
Chemicals; Pharmaceuticals	1.7	0.2	1.7	0.2	0.9	0.1	0.9	0.1	1.0	0.1
Animal feed	8.5	1.0	9.6	1.1	52.3	5.7	13.2	1.4	9.1	0.9
All other uses	19.6	2.3	20.9	1.1	23.9	2.6	29.2	3.1	32.5	3.2
Total domestic nongovernment use	850.2	100.0	871.4	100.0	917.6	100.0	942.6	100.0	1016.5	100.0

Source: American Dry Milk Institute (1965–1970).

count for approximately 80% of the domestic sales. Minor outlets in terms of amount used include production of processed meats, prepared dry mixes, confections, institutional foods, soft drinks, soups, chemicals and pharmaceuticals, and miscellaneous purposes.

BAKERY PRODUCTS

Bakery use of nonfat dry milk began in the early 1900's. Sales to this industry increased until 1960 when slightly more than 312,000,000 lb. were used.

Bread and Rolls

Advantages.—The principal reason bakers use nonfat dry milk in bread and rolls is the improvement of flavor. Other distinct benefits resulting from its use in bakery products include: longer keeping quality, more golden appearance of crust, a more tender, soft texture and uniform grain, and a reduction of crust and loaf shape defects. The toasting characteristics are superior.

Many experts in nutrition have reported on the improved nutrient value of bread made with nonfat dry milk. The protein quality of the bread and all other bakery goods is improved through a more adequate balance of its amino acid contents. Among the important amino acids is lysine which is deficient in bread products without nonfat dry milk. Calcium content is substantially increased.

Amount and Kind of Nonfat Dry Milk Used in Bread.—The recommended usage of nonfat dry milk in bread is at least 6% (range from 2 to 12%) based on the weight of flour. Rolls and buns warrant a similar percentage use of dry milk. Only the high-heat type of nonfat milk should be used in a bread formula. Low-heat nonfat dry milk contains a depressing factor on the loaf volume. The specific identity of this depressant has not yet been completely identified.

A rough correlation exists between the amount of serum protein denaturation and the depressing effect. Consequently, a maximum of 1.5 mg. of undenatured serum protein per gram of nonfat dry milk usually assures satisfactory baking qualities. However, the relationship is general so the safest procedure is to conduct a bake test on dry milk products if the undenatured serum protein content indicates the results are in the borderline range.

The baker also is interested in the moisture absorption rate and capacity. The absorption rate should not be slow. The total absorption capacity should be good and relatively uniform among various lots for the same bread formula.

Continuous Dough.—The conversion from the conventional to the continuous mix process in bread manufacture introduced new problems in the use of nonfat dry milk at the recommended level of six per cent. Based on commercial trials, Swortfiguer (1962) developed adjustments in the formulation for nonfat dry milk usage up to six per cent of the flour weight. The recommendations are presented in Table 39.

He reported the pH of the fermentation must be lowered because of the buffering effect of nonfat dry milk. The adjustment in pH consisted of an

TABLE 39

FERMENT ADJUSTMENTS FOR INCREASED LEVELS OF NONFAT DRY MILK IN DOUGH
(100 Lb. Flour Basis)

Ingredient	Control 2%	3%	4%	5%	6%
Water	63.00	64.00	65.00	66.00	67.00
Sugar	8.00	8.00	8.00	8.00	8.00
Salt	2.25	2.25	2.25	2.25	2.25
Nonfat dry milk	2.00	3.00	4.00	5.00	6.00
Yeast food (bromate type)	0.60	0.60	0.60	0.60	0.60
Calcium acid phosphate	0.30	0.3750	0.3750	0.3750	0.3750
Lactic acid—85%	0.00	0.0312	0.0312	0.0468	0.0625
Inhibitor (variable)	0.13	0.13	0.13	0.13	0.13
No. of enrichment tablets	1	1	1	1	1
Yeast	3.20	3.20	3.30	3.30	3.30
Ammonium chloride[1]	0.0625	0.0703	0.0781	0.0781	0.0833
(combined amount)	(1 oz.)	(1¹/₈ oz.)	(1¹/₄ oz.)	(1¹/₄ oz.)	(1¹/₃ oz.)

Source: Swortfiguer (1962).
[1] Total ammonium chloride, combining that in yeast food plus the amount added.

addition of lactic acid and calcium acid phosphate. Doughs sensitive to proofing and handling will benefit by increasing the addition of ammonium chloride. The addition of water is directly related to the increase in non-fat dry milk in the formula.

The specific gravity of the dough increased as per cent of nonfat dry milk was larger. About 0.5 oz. of more dough was delivered at the panner per pound loaf if no change was made in the weight adjuster.

The greater yield of dough may compensate for the additional cost of the nonfat dry milk. Bakery use of nonfat dry milk has been very sensitive to a price increase during recent years. Glabau (1964) reported only a very slight increase in cost of bread as the nonfat dry milk varied from 0 to 12% in the dough. He suggested a cost of $0.0557 per lb. of bread with 0% nonfat dry milk and $0.0575 with 12%.

Cakes

The baker improves cakes by using nonfat dry milk. At least 15% nonfat dry milk should be included in the formulas. The range is 5 to 25% of the flour, depending upon the kind of cake. A few of the common kinds which can be benefited are: white, devil's food, chocolate, sponge, angel food, chiffon, etc.

Nonfat dry milk enhances the flavor of cakes. It causes the batter to be more stable. Cakes are improved in color, tenderness, texture, and grain, besides having a longer keeping quality and increased nutritive value at a lower cost.

Doughnuts

Doughnuts are a bakery product that particularly benefits from the inclusion of nonfat dry milk, but the moisture absorption capacity of the dry milk must be good and the absorption rate reasonably rapid. Otherwise, the quality of the doughnut suffers beause of the increased tendency to absorb an excessive amount of fat. Both the cake and the yeast raised doughnut formulas should include at least six per cent nonfat dry milk. The amount may range from 2 to 25% of the flour.

Cookies and Wafers

Various types of cookies, wafers, and graham crackers have a more pleasing flavor, color, and texture if nonfat dry milk is an ingredient. At least six per cent should be used for most formulas, but the range is 2 to 25%.

Frostings

Nonfat dry milk is especially effective in icings, fillings, coatings, and glazes. It reduces the greasiness of these products and contributes to a better flavor by reducing excessive sweetness. Ordinarily these products also become smoother and easier to apply. The nonfat dry milk percentage may vary from 5 to 50% of the total frosting ingredients with 15% being common.

Other Baked Goods

The benefits of increased flavor appeal, improved nutritive value, and longer keeping quality also apply to the bakery products of lower production volume. Consequently, sweet dough, Danish pastry, biscuits, pretzels, gingerbread, pie fillings and crusts, etc., should contain nonfat dry milk. The amount included in the formulas depends upon the specific bakery product, but usually is within the 3 to 25% range.

PREPARED DRY MIXES

The prepared dry mix industry obtains the same benefits as bakers when nonfat dry milk is added to biscuit, cake, pancake, waffle, and frosting mixes in the dry form. The suggested amount for these prepared mixes is: biscuits, 6 to 8%; cakes, 10 to 20%; pancakes, 7 to 12%; waffles, 10 to 15%; and frostings, 15 to 25%.

PROCESSED MEATS

Usage Level of Good Sorption Nonfat Dry Milk

Data in Table 38 reveal the meat industry has been utilizing roughly six to eight per cent of the domestic market for nonfat dry milk. Federal regulations prohibit the addition of more than 3.5% of nonmeat ingredients other than seasonings in smoked and cooked sausage, frankfurters, and bologna. However, state laws vary and some permit a much higher percentage. There is no limitation on the amount of nonfat dry milk in meat loaf. The nonfat dry milk used in processed meats should have good total moisture absorption characteristics.

Improvements

Nonfat dry milk enhances several qualities of smoked and cooked sausage and of processed meat products, such as meat loaves. Despite the fact that meat contains the essential amino acids of protein, addition of nonfat dry milk does enhance the protein quality as well as supplying other nutrients, principally minerals. In general, bologna, sausage, frankfurters, meat loaves, and other processed meat products with three to six per cent nonfat dry milk have less shrinkage and leaking, better binding and slicing qualities (American Dry Milk Institute 1953).

Cook and Day (1947) stated that eight to ten per cent nonfat dry milk in sausage improved the flavor, color, binding, and yield. Nonfat dry milk can retain 1 to 1.7 times its own weight in moisture. Research by Saffle (1964) demonstrated that flavor acceptability of sausage increased as the nonfat dry milk was increased to six per cent of the total ingredients. Moisture loss in the smoke house was reduced from an initial 16.8% with zero per cent nonfat dry milk down to 11.3% with six per cent nonfat dry milk included. He also noticed that nonfat dry milk decreases the emulsion breakdown.

Other studies proved that nonfat dry milk in comparison with soy proteinate and potassium caseinate had the greatest emulsifying capacity in the pH range of meat (5.4) regardless of ionic strength (Pearson *et al.* 1965).

The inclusion of 3.5% nonfat dry milk in bologna significantly increases the yield. The tensile strength increased and texture-tenderness scored higher with increases in nonfat dry milk (Rongey 1961). However, no change in shrinkage or percentage of moisture and protein from the addition of nonfat dry milk was observed.

CONFECTIONS

Nonfat dry milk can be incorporated with relative ease into the mix of various kinds of candy. For most types of candy it can be blended with the sugar (dry). Nonfat dry milk requires a precaution in the candy manufacture. Care should be taken to be sure the scraper in the cooking kettles covers the heating surface adequately as a means of preventing the possibility of scorching when nonfat dry milk is used in the formula.

The leading reason for using nonfat dry milk in confections such as fondants, creams, caramels, and fudges is the more pleasing flavor. Furthermore, it enhances the nutritional value, uniformity and increases the yield. Less cooking time is required. Some candies have a smoother texture when nonfat dry milk is used. Nonfat dry milk is highly beneficial as a moisture retaining ingredient and thus extends the shelf life of candies. Hall and Fahs (1949) reported the amount of moisture retention was influenced by the type and amount of dry milk product used in the formula. Table 40 provides data on the moisture loss from fondant during storage (U.S. Department of Agriculture and National Confectioners Association 1948).

The customary range of nonfat dry milk usage is 3 to 20% of the total ingredients depending upon the kind of candy. Miller (1965) indicated that 5 to 18% nonfat dry milk was used in certain chocolates. A low moisture content in the nonfat dry milk is particularly desirable for use in chocolate coatings.

TABLE 40

MOISTURE LOSS FROM FONDANT CONTAINING 5 OR 10% OF VARIOUS DRY MILK PRODUCTS

Product	5% Added Dry Milk Product				10% Added Dry Milk Product			
	Initial Product Moisture	% Moisture Loss (Days)			Initial Product Moisture	% Moisture Loss (Days)		
		10	20	30		10	20	30
Control	12.14	4.82	6.59	7.65	12.14	4.82	6.59	7.65
Dry whole milk	11.76	4.31	5.85	6.73	10.90	3.34	4.56	5.42
Nonfat dry milk	11.63	3.82	5.61	6.51	12.00	3.68	5.21	6.19
Whey solids	11.91	3.98	5.49	6.53	11.52	3.98	5.38	6.18
Buttermilk solids	11.33	3.76	5.30	6.19	11.51	3.46	4.97	5.80
Lactalbumen	12.03	4.56	6.41	7.43	11.51	3.81	5.46	6.44

Source: U.S. Dept. Agr. and National Confectioners Association (1948).

There are limitations in the utilization of nonfat dry milk in confections. One example is nonfat dry milk in certain kinds of caramels increased coarseness.

DAIRY

Utilization

The utilization of nonfat dry milk in dairy products is approximately 28% of the total domestic market. Ready availability and a reliable source of concentrated milk solids are important reasons for the high utilization. Another is the trend by the industry to manufacture nonfat dry milk with optimum characteristics for specific dairy manufacturing purposes. The quality of the nonfat dry milk and its characteristics resulting from the processing procedure frequently are pertinent for a specific use in the dairy. Flavor is among the important considerations for most purposes. Heat treatment during manufacture and absence of inhibitory substances mainly antibiotics are others.

Reconstitution

Dairy plant water used for reconstitution can contribute to the end results. The type and amount of microbiological contamination in the water is highly important unless post pasteurization is required. The kind and extent of the chemical impurities are of concern for the manufacture of cultured reconstituted milk products. Chemicals constituting the water hardness can be a detriment to the flavor of all reconstituted products.

The dispersibility of nonfat dry milk and the amount of agitation supplied most dairy vats often are less than completely satisfactory. Best results in reconstitution are obtained with equipment adapted for this purpose, using an adequate procedure. One satisfactory arrangement

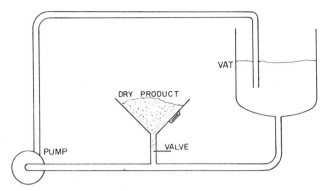

FIG. 86. BATCH RECONSTITUTION OF DRY MILK

(Fig. 86) includes a funnel shaped hopper, tubing ($1\frac{1}{2}$ in. diameter or larger), and a centrifugal pump (American Dry Milk Institute 1951).

The funnel is connected to the line between outlet valve of the vat and the pump. The tubing from the pump outlet extends back into the vat below the liquid level to reduce foaming. The water or milk in the vat is heated to 90°F. and the recycling started. The nonfat dry milk is placed into the hopper to feed into the liquid. When all the nonfat dry milk has been dispersed adequately and appears to be in "solution" the recirculating is terminated.

The recombining of dry milks and water also may be accomplished by extending the impeller of a high speed agitator into the vat contents. The assembly is mounted on a portable frame or onto the side of the vat which is less desirable.

FROZEN DESSERTS

Frozen dessert manufacturers use large amounts of nonfat dry milk. It can be a supplement or the main source of milk solids-not-fat in the ice cream mixes (1 to 12%), sherbets (1 to 3%), ice milk (1 to 14%), soft serves (1 to 14%), and mellorine products (1 to 12%).

There are conflicting reports in the choice of heat treatment of the nonfat dry milk for best results in ice cream. Apparently the quality associated with the nonfat dry milk is more important than the heat treatment although high heat would be expected to have more antioxidants to reduce oxidation if the frozen dessert has a tendency to undergo this chemical change.

Hedrick *et al.* (1964) reported ice cream and ice milk mixes made entirely with low-heat nonfat dry milk furnishing the milk solids-not-fat, had slightly less cooked flavor than those with high-heat nonfat dry milk.

But no consistent differences were detectable in flavor, body and texture, meltdown, and resistance to shrinkage in the samples of ice cream and ice milks stored at —15°F. Shrinkage of ice cream and ice milk stored at 10° to 15°F. was slightly more in those prepared with low-heat nonfat dry milk.

CULTURED PRODUCTS

High Quality Low-Heat Nonfat Dry Milk and Good Quality Water

Low-heat nonfat dry milk should be used for cultured products. In fact, not more than ten per cent of the serum proteins should be denatured, especially if the nonfat dry milk is intended for the manufacture of cottage cheese. The nonfat dry milk should have a standard plate count of not more than 50,000 per gm. or preferably below. The presence of psychrophilic bacteria is undesirable because the reconstituted nonfat milk may not be repasteurized. The nonfat dry milk should be free of all inhibitory substances such as antibiotics. Randolph and Kristoffersen (1961) noticed much difference in the cottage cheese making properties of reconstituted nonfat milk from different dry milk plants, but noted relative uniformity from the same plant source.

The water for reconstitution should be free of pathogens, low in count of all other types of microorganisms, free of sediment, and have no off-flavors or odors. Freedom from sulfur, sodium carbonate and bicarbonate, sodium chloride, phenols, and iron compounds is also desirable.

Cottage Cheese

In addition to providing a source of solids-not-fat during low milk production, low-heat nonfat dry milk can improve the cottage cheese operation. The yield of cottage cheese per pound of solids is increased if the total solids of the fluid skimmilk is increased above that of regular skimmilk. The vat production capacity can be substantially increased. Since regular milk varies in composition, nonfat dry milk can be used to obtain the same solids content for each operation, promoting greater uniformity in the procedure and the resulting cottage cheese.

The solids percentage in recombined skimmilk for cottage cheese ranges from 9 to 15%. Best results are obtained by recombining to the same solids content for each vat. Two hours are a desirable period for hydration of the proteins before setting, if the short-set method is used. The overnight method provides ample time after setting for protein hydration. The coagulation may be firmer if 2 ml. of 25% calcium chloride (diluted in water) is added to each 1000 lb. of recombined nonfat milk at the time of setting. Otherwise the same procedure is used as when making cottage cheese with fresh skimmilk.

The solids-not-fat in the creaming mixture can be provided by nonfat dry milk. By recombining it with milk fat and water plus salt, a good cottage cheese dressing can be prepared.

Bakers cheese is similar to cottage cheese, but has small, irregular curd particles. Excellent Bakers cheese can be manufactured from recombined nonfat milk using the same procedure used for fresh skimmilk. A common proportion is 11 lb. of nonfat dry milk to 89 lb. of water. The range is 9 to 15% solids-not-fat in the recombined nonfat milk.

Cultured Buttermilk

Reconstituted nonfat milk can replace 100% of the fresh skimmilk in making a cultured buttermilk which is excellent in flavor, body, and keeping quality. Customarily it is recombined at 9 to 11% solids-not-fat. Body and viscosity can be controlled to within desired customer standards by establishment of the per cent of solids that gives the preferred characteristics. Once the optimum composition of the reconstituted milk has been developed, close adherence gives greater uniformity of the cultured buttermilk. After reconstitution of the nonfat dry milk, the pasteurization, setting, incubation, cooling, and packaging procedures are the same as used for fresh skimmilk.

Cultures and Starters

Nonfat dry milk recombined with water to produce nine per cent solids is an excellent medium for the propagation of cultures and the preparation of cheese or buttermilk starters. Water with a low hardness and the selection of a good quality nonfat dry milk are important for best selection of a good quality nonfat dry milk are important for best results. A definite correlation exists between the undernatured serum protein and starter activity (Greene and Jezeski 1955). The reconstituted milks with extremely high- or low-heat treatment are less satisfactory than those with 2 to 5 mg. of undenatured serum protein per gm. of powder (indicating a medium-heat treatment). Reconstituted nonfat milk does not require any changes in the procedure for progagating cultures or making starters from that used for fresh skimmilk.

Cultured Cream

Low-heat nonfat dry milk may be recombined with water and anhydrous milk fat to give 18.5% fat and nine or ten per cent solids-not-fat for the preparation of cultured (sour) cream. The procedure and precautions are the same as required for the manufacture of cultured cream from fresh cream.

Cream and Neufchatel Cheese

Numerous cheeses (other than cottage and Bakers) can be made successfully with reconstituted nonfat dry milk and cream, unsalted butter, or milk fat. Two of these are cream and Neufchatel cheese. One of numerous formulas for the former is 18.00 lb. of low-heat nonfat dry milk, 100.00 lb. of 20% cream, 0.35 lb. gelatin, and 1.00 to 5.00 lb. of starter. Milk for Neufchatel cheese can be reconstituted from low-heat, nonfat dry milk, anhydrous milk fat and water to obtain 9.5% solids-not-fat and 4% fat.

After reconstitution the recombined cream or milk is homogenized and the regular procedure for cream cheese and Neufchatel cheese is followed. Both of these cheeses have numerous flavoring possibilities—pimento, olive, Blue cheese, pickles, relishes, nuts, or even honey.

Cured Cheeses

Milk for making Cheddar cheese can be prepared from reconstituted nonfat milk and cream or milk fat. If milk fat or unsalted butter is used the fat must be emulsified into the nonfat milk. The low-heat nonfat dry milk, fat and water are well mixed, heated to 130°F. and homogenized with just enough pressure to obtain a fat emulsion at least as stable as the normal fat globules in regular whole milk. This requires approximately 500 p.s.i., but should not be more than 1000 p.s.i.

Increasing the homogenization pressure adversely affects the matting, whey expulsion, and body and texture of the ripened cheese. But homogenization reduces the fat loss in the whey. Peters and Williams (1961) recommened 4 oz. of rennet per 1000 lb. of reconstituted milk for best flavor and body characteristics. Peters (1960) obtained an improvement in body and texture of reconstituted milk cheese by the addition of *Lactobacillus casei* to the milk at setting time. He added this culture at the rate of 20 ml. per 100 lb. of milk. In general, the Cheddar cheese manufacturing procedure for fresh milk can be followed.

Highly refined vegetable oils can be used as the source of fat in the preparation of a reconstituted filled-milk for cheese. The oil should have a melting point of 90° to 100°F. The filled-milk preparation and subsequent cheese manufacture are the same as for cheese made with reconstituted milk. The legal implication should be checked with appropriate regulatory officials.

Processed Cheese, Cheese Food, and Cheese Spreads

Nonfat dry milk may be blended into various processed cheese products. Their uses in processed cheese are limited primarily by the federal or state standards for the product.

However, larger amounts are utilized in the formulas of processed cheese foods and processed cheese flavored spreads. The amount of nonfat dry milk varies from 3 to 40% of the total ingredients depending upon the specific cheese product and the desired characteristics.

FORTIFICATION OF MILK AND MILK PRODUCTS

The fortification of fluid beverage milk products with additional milk solids-not-fat is common. Skimmilk, low-fat milk, and half-and-half are fortified in certain markets. The practice is also applied to chocolate milk drinks. Numerous investigators have demonstrated the physical and organoleptic benefits (Table 41). Obviously nutritional qualities are im-

TABLE 41

FORTIFICATION OF FLUID DAIRY PRODUCTS WITH ADDITIONAL MILK SOLIDS-NOT-FAT

Product	% Added (Optimum)	Final % solids-not-fat	Benefits	Reference
Milk, whole (1)	1	...	Highly significant increase in consumer acceptance	Stull and Hillman (1960)
(2)	1	10	Improved flavor	Custer et al. (1958)
Milk, low-fat (1)	1	...	Highly significant increase in consumer acceptance	Stull and Hillman (1960)
(2)	...	Up to 11	Increased acceptability	ADMI (1951)
(3)	2 to 3	11 to 12	More acceptability in flavor	Wahid-Ul-Hamid and Manus (1960)
(4)	...	10	Most acceptable	Custer et al. (1958)
Chocolate drink (1)	1 to 2	11	Richer flavor, slightly increased viscosity	Janzen (1962)
Half-and-half (1)	...	Up to 11	More desirable body, increased color-coloring, reduced feathering	ADMI (1951)
Buttermilk (1)	1 to 2	11	Fuller flavor, control body for uniformity by varying NDM added	Janzen (1962)
Buttermilk (2)	...	Up to 11	Aids desirable growth, produces full flavored beverage	ADMI (1951)
Skimmilk (beverage) (1)	1 to 1½	11	Adds flavor, body not as thin	Janzen (1962)
(2)	1	...	Highly significant increase in consumer acceptance	Stull and Hillman (1960)
(3)	...	Up to 11	Increased acceptability	ADMI (1951)
(4)	2	...	Usually best, increase in viscosity	Hollander and Winder (1951
(5)	1 to 3	...	Flavor acceptability improved	Weckel (1952)
Cultured sour cream (1)	1 to 3	...	Increases viscosity and improves the body	ADMI (1961)

proved. Where legal restrictions exist steps should be taken for their removal and the market milk industry encouraged to fortify with additional milk solids-not-fat for the benefit of the consumers.

Many people can detect and prefer the addition of one per cent solids-not-fat to fluid whole, low-fat, and skimmilk (Stull and Hillman 1960). With a processing procedure that prevented cooked flavor and removed volatiles, Pangborn and Dunkley (1964) observed that 67% of the time the addition of 0.5% milk solids-not-fat was favorably detected in homogenized milk. While the specific amount varies among individual consumers in general, the addition of approximately two or possible three per cent appears to be the upper limit. Larger amounts tend to cause a decline in flavor acceptability. In order to obtain the maximum benefits of the milk solids-not-fat addition to milk beverages, the nonfat dry milk must be of high quality. The flavor of the nonfat dry milk is an especially critical factor for the nonflavored milk beverages.

RECOMBINED MILK AND CREAM PRODUCTS

The need for reconstituted milks and creams is of little importance in the United States. In the underdeveloped countries and those with low milk production, reconstituted milk beverages can be an important source of nutrients. In fact, recombined, manufactured dairy products also have a good potential for improving the inadequate nutrition of the people of many underdeveloped countries.

In the preparation of recombined beverage milks and recombined creams, the nonfat dry milk of low- or medium-heat should be Extra Grade and have a minimum of "cooked" flavor. The reconstituted beverages can be processed to have a composition the same as whole milk, low-fat milk (two per cent), skimmilk, and flavored milk drinks. By approximating the standard composition, reconstituted cream products similar to half-and-half and coffee cream may be prepared from nonfat dry milk and milk fat.

A satisfactory procedure is to heat the water to 90°F. and completely disperse the nonfat dry milk in it. Heating is continued to 115°F. and the melted fat is added. Agitation must be adequate to keep melted fat in suspension until homogenization. The reconstituted whole and low-fat milks are homogenized at 2500 p.s.i. and not less than 145°F. Homogenization pressure should be 500 to 800 p.s.i. at 130° to 140°F. for half-and-half or coffee cream. Pasteurization and other processing steps are the same as for fresh milks and creams.

RECOMBINED STERILIZED MILK EVAPORATED MILK
AND SWEETENED CONDENSED MILK

The production of these recombined products provides an important source of milk nutrients in a number of countries or areas (see Fig. 87). Millions of cases of recombined milk products, and in particular recombined sweetened condensed milk, are produced annually in a few of these countries. Also production has been increasing. The low milk production in Asia and Africa and the lack of adequate refrigeration facilities present a large potential market for the sterilized milk products and sweetened condensed milk.

Courtesy of Beatrice Foods

FIG. 87. PLANT IN ASIA FOR PRODUCING RECOMBINED DAIRY PRODUCTS

Recombined milk products intended for subjection to high-heat treatment, e.g., sterilization, require special care in the selection of the dry milk product and water. Usually medium-heat nonfat dry milk having good heat stability is preferred.

The required water should have a low calcium and magnesium content. A high ion concentration of these elements in the water will decrease the casein stability of the reconstituted milk (Sargent *et al.* 1959). The effects of calcium and magnesium on the reconstituted products can be reduced by sequesting the water with sodium hexametaphosphate. Removal by a commercial water softening method is also an alternative.

Sterilized Recombined Milk

Low- or medium-heat nonfat dry milk is recombined with melted milk fat and water at 115°F. to obtain either nine per cent solids-not-fat and 3.5% fat or another desired composition. The mixture is heated to 160°F. and homogenized at 3000 p.s.i. or higher if the product is intended for extended storage. The reconstituted milk is filled into metal cans with a food lacquer lining or glass bottles and sealed.

The cans are sterilized at 240°F. for 15 min. or a minimum time and temperature for commercial sterilization. A batch or continuous operation procedure may be used. Glass bottles are usually sterilized by retort (batch) with care to heat and cool as rapidly as possible yet prevent breakage.

Recombined Evaporated Milk

For the most satisfactory results nonfat dry milk intended for recombined evaporated milk should be processed for good heat stability. This requires giving attention to optimum preheating before the original condensing and drying. The specific time and temperature varies with the milk supply according to feeds (season), breeds of cows, etc., but is approximately 200°F. for 10 min. or 240°F. for 1 min. A simple heat stability test on the nonfat dry milk will serve as a check.

A suggested composition of recombined evaporated milk is fat, 8%; and solids-not-fat, 16%. Medium-heat nonfat dry milk is dispersed into water at 90°F. and is warmed to 115°F. The melted milk fat is added. With sufficient agitation to insure uniformity, the mixture is heated to 160°F. and homogenized at 2500 p.s.i. In a continuous operation the recombined evaporated milk may be packaged and sealed in cans and processed through the sterilizer. In the batch system, the product should be cooled to 40°F. The cans are filled, sealed, and sterilized. The procedures are similar to those used with regular evaporated milk (given in Chapter 6).

Sweetened Condensed Milk

The formula and procedure per 100 lb. of product recommended by Pont (1960) consists of: (a) 11 lb. of sugar and 22 lb. of medium-heat nonfat dry milk are dry blended; (b) the mixture is dispersed in 25 lb. of water at 85°F. and heated to 140°F.; (c) 33 lb. of sugar and 9 lb. of milk fat are added; (d) the recombined product is homogenized at 150 to 300 p.s.i. and 122°F.; (e) it is cooled to 86°F., seeded with lactose or fine particles of nonfat dry milk; (f) the product is held an hour for crystallization with ample agitation; (g) it is then pumped through a deaeration chamber to remove air; (h) the recombined sweetened condensed milk is filled into cans, sealed, labeled, and cased.

Two to five per cent of glycerol monostearate on basis of the fat content improves the fat emulsion in the serum. It should be added to the milk fat before the fat is combined with the other ingredients.

The sequence of combining the ingredients affects their ease of dispersion and one of the product characteristics; mealiness results if sugar is dissolved in water before adding nonfat dry milk. Apparently the protein does not rehydrate satisfactorily if the sugar is dissolved prior to the dispersion of the nonfat dry milk.

A high homogenization pressure increases the viscosity and age thickening. Heat treatment of the nonfat dry milk influences the viscosity. Low-heat nonfat dry milk causes too low a viscosity and too much viscosity results if a high-heat product is used. The undenatured serum protein of dry milks is only a general indication of the heat treatment. Muller and Kiesker (1965) observed that despite rather severe preheating during late fall and early winter the serum protein test on the resulting nonfat dry milk indicated the product could be classed as low-heat. As spring advanced lower preheats gave lower serum protein results. Then as summer progressed the preheat was increased, but serum protein decreased. Consequently the serum protein index *per se* is not a completely reliable indication for predicting the viscosity of sweetened condensed milk. In controlling viscosity of sweetened condensed milk, the correct weight of each ingredient is important also.

Burgwald and Strobel (1959) suggested manufacturing sweetened condensed milk by thoroughly mixing 10 lb. of low-heat nonfat dry milk in 27 lb. of water, then adding 9 lb. of dry milk fat. The mixture is heated to 160°F. and homogenized at 2000 to 2500 p.s.i. The heating is continued to 180°F. After dry blending 11 lb. of nonfat dry milk with 43 lb. of sugar, the mixture is added slowly to the other product with vigorous agitation while holding the heat at 180°F. The product then is cooled rapidly to 86°F. and 2.0 lb. of the previous sweetened condensed milk or

2 oz. of powdered lactose is added. After one hour with vigorous agitation the product is cooled rapidly to 65°F. and is packaged in sterile containers.

RETAIL USES OF NONFAT DRY MILK

Nonfat dry milk for retail sale and home use has been steadily increasing since World War II. These sales in 1948 were only 2,300,000 lb. or 0.5% of the total domestic consumption of nonfat dry milk. During the last few years retail sales have jumped to 25-32% of domestic sales or to nearly 324,300,000 lb. in 1968.

Several reasons have been given for the upward shift since World War II. These include: (a) calorie consciousness (since at least 30% of the adult population believes it has a problem in controlling overweight); (b) the improved dispersibility and flavor; (c) the greater protection and convenience of the package; and (d) economical value in relation to the rising costs of other foods.

Home water supply in some areas is not always entirely satisfactory for recombining with nonfat dry milk. Hard water may impart an unnatural off-flavor to the reconstituted product when compared to the fresh product. Furthermore, water containing psychrophilic bacteria subjects the reconstituted product to the possibility of developing various fermented flavors during storage in the home refrigerator. This latter problem can be prevented by boiling and cooling the water before reconstitution.

Utilization of nonfat dry milk in the home is generally limited only by the imagination and resourcefulness of the user. Recombined with water it is served as a beverage directly or with numerous added flavorings—chocolate, eggnog, fruit juices or extracts, mint, and instant coffee plus sugar to the desired sweetness.

Nonfat dry milk is used in the dry or reconstituted form for cooking. A few of the hundreds of possibilities include the recipes for baked products, desserts (e.g., puddings, custards, cakes, pie fillings), gravies, sauces, soups, vegetable dishes and main dishes such as ground meat dishes (Swickard 1949).

INSTITUTIONAL UTILIZATION

The sales of nonfat dry milk in the preparation of meals for feeding public and private groups have been one to two per cent of the domestic consumption. The total amount used is larger due to government distribution of surplus nonfat dry milk to some of the public institutions. Nevertheless, this outlet offers a good opportunity for promoting much greater utilization in many foods not usually prepared with milk.

Advantages of nonfat dry milk such as improved nutrition and more pleasing flavored foods, are worthy of exploiting. Nonfat dry milk for the institution has the advantage of ready availability and a substantial economy of 40 to 70% (Aldrich and Miller 1958).

DRY WHOLE MILK UTILIZATION

Sales Outlets

The exports of dry whole milk have been roughly one-third or less of production. The consumption of dry whole milk currently is very small in comparison to nonfat dry milk (Table 37, p. 258). The confection industry buys approximately 70% of the domestic sales of dry whole milk (Table 42). The baking, dairy, and baby food industries have a combined

TABLE 42

DOMESTIC SALES OUTLETS OF DRY WHOLE MILK
(Add 000,000 Lb.)

	1963	1964	1965	1966	1967	1969[1]
Candy, chocolate-coating manufacturers	31.1	44.6	42.4	49.7	36.2	37.3
Bakery	4.1	4.3	9.0	5.5	4.6	3.9
Baby food manufacturers	2.9	3.3	2.3	2.8	3.1	2.7
Packaged for home use	1.2	0.3	0.1	0.1	0.7	0.1
Institutions	1.1	2.5	0.9	2.2	2.6	3.0
Dairy	0.9	3.4	1.2	1.1	1.0	0.8
Soup manufacturers	0.2	0.2	0.1	0.1	0.7	0.1
All other	1.0	1.0	2.5	4.4	2.3	2.3
Total domestic sales	48.5	60.1	58.5	65.9	51.2	50.2

Source: American Dry Milk Institute (1965–1970).
[1] Preliminary.

usage that varied from 7 to 12,000,000 lb. per year in recent years. Institutional utilization has been less than 3,000,000 lb. per year. Uses in the home, by soup manufacturer, and miscellaneous (all others) account for the remainder of the minor domestic sales.

Limitations

Dry whole milk utilization has been hampered by defects in the fresh as well as the stored product. Reconstituted milk from dry whole milk has a cooked or chalky flavor that limits its use for beverage purposes in many areas. There were other shortcomings listed by a group of housewives that were surveyed for likes and dislikes concerning the use of dry whole milk (Table 43).

The storage life of atmospheric packaged dry whole milk for commercial food manufacture is relatively short. The product develops an oxidized off-flavor even under favorable storage conditions. Therefore, dry whole milk is less likely to be manufactured during periods of peak milk production for supplementing needs during the months of low milk pro-

TABLE 43

RESPONSE OF HOUSEWIVES REGARDING LIKES AND DISLIKES OF DRY WHOLE MILK

Dislikes of 202 Housewives		Likes of 202 Housewives	
Quality	Per Cent[1]	Quality	Per Cent[1]
Poor taste	39	No qualities liked	37
No dislikes mentioned	36	Taste good	15
Preparation difficulties	22	Convenient (on hand)	14
Qualities not rich enough	3	Quality rich	12
Quality too rich	1	Keeps well	11
Peculiar odor	1	Ease to use	6
Keeping quality poor	1	Economical and save space	7
Is a substitute	1	Miscellaneous	7
No response	6	No response	8

Source: Cook and Day (1947).
[1] Respondents could have more than one answer.

duction. Although the keeping quality can be roughly doubled by commercial nitrogen packaging techniques, the extra cost of the equipment, processing and container are incurred.

The dispersibility of regular dry whole milk in water leaves much to be desired. This is not serious with satisfactory equipment for recombining. Unfortunately agglomeration does not improve dispersibility as much as it does for nonfat dry milk.

Undoubtedly, economics has been a major negative factor in dry whole milk purchases for industrial prepared foods. The much cheaper price of vegetable fats and certain types of animal fats has drastically limited the sales of dry whole milk despite its superior flavor characteristics.

Confections

Dry whole milk used by the confection industry in candy and chocolate may be the drum or spray type. Much of it is processed normally. But some users for certain confection products prefer a dry whole milk with the maximum amout of free fat, which is accomplished by drum drying unhomogenized milk.

One of the principal uses of dry whole milk is in milk chocolate. The range in amount is from 12 to 24% of the total ingredients (Gott and Van Houten 1958). Dry milk should give the chocolate a distinct, rich butter-like flavor for best results. This desirable "milk" flavor is correlated with the free fatty acid content (Hollander 1948). Controlled lipolytic reactions on the fats by lipase enzymes develop the desired flavor for use in milk chocolate.

Dry whole milk is used in fondants, fudges, and caramels. Hall and Fahs (1949) have reported that dry whole milk in fondants will reduce moisture losses and thus extend the freshness of the product. Table 40 (p. 264) presents data showing the reduction of moisture loss from fondants made with five or ten per cent dry whole milk.

Much of the specific details on the usage of dry whole milk in candies is restricted confection company information. The dry milk industry could profit by exploiting and developing the benefits of dry whole milk and disseminating the information.

Bakery, Dairy and Other Uses

The addition of dry whole milk to the formulae of bakery products (pastries, cakes, and frostings), substantially increases the desirability of the flavor. Other benefits include the improvement of the nutritive value, body and texture, and keeping quality. The principal restricting factor has been the increase in cost of the baked product.

During periods of shortage, dairy plants may utilize small amounts of dry whole milk to supplement the milk solids requirements. Frozen dessert mixes, especially those intended for chocolate and other strongly flavored products, may include up to 50% of the dry whole milk as a source of milk solids without a noticeable adverse effect on the flavor. Dry whole milk is feasible for the making of processed cheese spreads. Limited quantities of dry whole milk have provided the milk solids for the manufacture of recombined sweetened condensed milk in underdeveloped countries.

Institutions use dry whole milk mainly for cooking purposes either in the dry or reconstituted forms. The many possibilities include main dishes, vegetables, numerous kinds of dessert, sauces, and oven products.

Much of the dry whole milk in baby foods has been modified. These products are usually sold under a private brand. The modifications may involve the protein, lactose, fat, ash, or a combination of two or more of these components. The specific details generally have not been released by manufacturers for competitive purposes.

USAGE OF DRY BUTTERMILK

Domestic Sales

The sales of dry buttermilk for domestic human use have gradually increased with yearly variations. The total in 1969 was 66,000,000 lb. Baking industry purchased almost 33% of this amount or 21,400,000 lb. Sales to prepared dry mix, dairy and candy industries were 22,900,000, 15,800,000, and 3,000,000 lb. respectively (American Dry Milk Institute 1970).

The sales potential for dry sweet cream buttermilk intended for human food purposes is large. However, the quality must be good and its superior properties for specific purposes more fully appreciated and exploited. Dry buttermilk has a much higher phospholipid content than other dry milk products. Consequently, it possesses a natural emulsification property which is useful in the reconstitution of dry products with fat. Dry buttermilk in reconstituted products containing a low fat content may impart a richer flavor.

A dry high acid buttermilk differs from dry sweet cream buttermilk as the name suggests mainly in having a much higher acid content. The acidity is usually ten per cent or above. This type is sold mainly for baked foods.

Bakery

Dry sweet cream buttermilk is used in many baked foods such as bread, doughnuts, cookies, cakes, crackers, and icings. The heat treatment must be sufficient to destroy the loaf depressant factor in buttermilk. The amount of dry buttermilk used in baked products is generally 2 to 15%. Crawford (1947) reported that buttermilk can improve the flavor, texture, appearance, and nutritive value of nearly every product of the baker's oven.

Buttermilk serves as a fermentation stabilizer by providing the dough with more tolerance toward fermentation changes. For example, it will withstand more mixing and handling without deleterious effects. The phospholipids aid in the emulsification of shortening and increase the smoothness and unformity.

Reger *et al.* (1951) in trials with six per cent spray dried buttermilk compared to the same amount of spray dried nonfat dry milk (based on flour) consistently observed larger loaf volume of bread with dry buttermilk. The results between roller dried buttermilk and roller dried nonfat dry milk showed larger loaf volume for the dry buttermilk bread, but less consistently than with the spray dried buttermilk.

A satisfactory topping can be prepared for baked products by recombining spray dried buttermilk, butter, stabilizer, and flavoring with water. This topping has taste appeal, stability in the whipped form and is economical.

Dry high acid buttermilk produced a more distinctive flavor in rye and whole wheat breads as well as white bread. These bread formulas usually limit the amount to three to six per cent of the flour weight. In chocolate and fudge cakes dry high acid buttermilk intensifies the flavoring and color. Other kinds of cakes, biscuits, cookies, and crackers also are improved with high acid dry buttermilk.

Dairy

Ice Cream Products.—Spray dried sweet cream buttermilk has several advantages for dairy product applications. Frozen dessert formulae can include dry buttermilk with distinct improvements. Sommer (1946) stated that buttermilk in concentrated form is especially desirable in ice cream products with a low fat content and in mixes made with butter. Dry buttermilk restores the phospholipid components to mixes made from butter that would normally be present in mixes with cream as the source of fat.

Frandsen and Arbuckle (1961) mentioned buttermilk solids can supply 20% of the milk solids-not-fat with improvement in whipping and richness; Williams *et al.* (1950) manufactured good ice cream with buttermilk solids furnishing up to eight per cent of the milk solids-not-fat. In general, good quality dry buttermilk can be used up to one-fourth of the milk solids-not-fat in frozen desserts with the advantages of faster rate of whipping, drier appearance of the ice cream product from the freezer, and possibly a slightly richer flavor especially if the frozen dessert is low in fat content. This would apply to sherbets, soft serve, and mellorine products.

Cheese and Cheese Products.—Spray dried buttermilk can be used as part of the milk solids in preparing reconstituted milks and creams for Neufchatel and cream cheese. Results of reconstituted milk and cream with nonfat dry milk supplying 100% of the milk solids-not-fat compared with similar products which have 25% dry buttermilk and 75% nonfat dry milk, the latter always had a more stable fat emulsion. Cream and Neufchatel cheese made with some dry buttermilk had less fat loss in the whey and the flavor seemed slightly improved. However, if more than 50% of the solids-not-fat was from dry buttermilk the body and yield were adversely affected. These effects were attributed to the high-heat treatments during cream processing for butter and buttermilk drying.

The dried low-heat buttermilk also can be used to supply a portion of the milk solids for reconstituted cheese milks intended for the ripened cheeses such as Cheddar. These buttermilk solids-not-fat should be limited to not more than approximately 15% of the total solids-not-fat. The fat emulsion in the reconstituted cheese milk is improved or a lower homogenizaion pressure may be used by including buttermilk solids.

Dried buttermilk, either regular or high acid, may be used for the preparation of cheese foods and cheese spreads. The amount varies with the specific product formula, but usually is between 1 and 12%.

Recombined Milks.—Recombined beverage milk and recombined cream have more fat emulsion stability if 10 to 25% of the solids-not-fat is derived from spray dried sweet cream buttermilk. The other ingredients are dry milk fat, nonfat dry milk, and water. The beneficial effect is more apparent if the homogenization is not thorough as occurs in certain types of "mechanical cows." The fat was more effectively emulsified in recombined milk by the lecithin in buttermilk than the equivalent amount of soybean lecithin in nonfat dry milk. Recombined whole milk (3.5% fat) from nonfat dry milk and dry milk fat tends to be rather bland. The substitution of 25% dry buttermilk for nonfat dry milk resulted in the flavor being more comparable to fresh whole milk. Also, Norman (1955) confirmed the fact that dried buttermilk improved the palatability of reconstituted homogenized milk.

Confection Manufacturers

Dry buttermilk sales for candies have been comparatively small. But the favorable characteristics of dry buttermilk increases the potential sales for confections. Since both dry milk and commercial lecithin is used in numerous candies, dry buttermilk should have a natural advantage by supplying a portion of the lecithin along with the milk solids. In general, the flavor is improved, nutritional qualities are enhanced and crystallization may be retarded. Dry buttermilk is being used in the formulae of fudge, caramels, hand roll creams, coatings, and private brand candies.

DRY WHEY

Animal and Poultry Feeds

The demand for dry whey and modified dry whey products is far short of the huge potential supply. The largest outlets for dry whey are in poultry and livestock feeds. The superiority of lactose as a carbohydrate feed for mammals soon after birth is acclaimed by animal nutritionists and livestock experts (Becker *et al.* 1954). The B complex vitamins and other nutritive factors in whey have been recognized for a long time as valuable components in poultry feeds.

Human Food Usage

Since whey is considered a waste product in many cheese plants or is salvaged for animal feeds, its general image as a human food needs to be improved. Consequently, the selection and manufacture of high quality dry whey must be given special attention in order to promote its greater use in food products. Dry whey has advantages for a number of processed food products.

Frozen Desserts.—A review by Webb and Whittier (1948) covered the main utilization possibilities of whey for human consumption. Most of these are applicable to dry whey; otherwise it can be recombined with water for use. Dry whey for ice cream and sherbet mixes is receiving more attention because of the low cost of whey solids. In ice cream mixes as much as 25% of the solids-not-fat can be of sweet, good quality, dry whey. Up to 50% of the milk solids-not-fat in soft serve mixes and up to 100% in sherbets can be supplied by dry whey (Alesch 1958). A limiting factor from an excessive percentage of dry whey is sandiness due to lactose crystallization. There is the possibility of the dry whey also causing an unnatural flavor in mild flavored frozen desserts.

Processed Cheese Products.—Dry whey is a natural product for processed cheese foods, cheese spreads, and dips. The addition of whey solids tends to contribute toward the restoration of some milk components lost during cheese manufacture. A pleasing flavor in these products may be enhanced by dry whey.

Bakery Products.—Interest has been increasing in the use of whey solids in bakery goods. The most common is bread because of the improvement of crust color, crumb structure, and tenderness when dry whey is used at the rate of three per cent of the flour. Other baked products that respond favorably to dry whey in the formula are cookies and crackers, 2 to 6%; cake mixes and cakes, 4 to 15%; pancakes and waffles, 5 to 15%; and pie crust, 2 to 6%.

Miscellaneous Uses.—The confection industry uses a limited amount of whey solids (up to one-third of the sucrose) for alleviating excessive sweetness, prolonging the shelf life or freshness, and improving the nutritive value to some extent. The suggested usage of dry whey in caramels is 16.2%; in fudge, 8.1%; and in fondants, 8.1% based on the weight of sugar (Thiessen 1957).

Starch pudding with 6 to 14% whey solids on a dry basis, has a smooth texture and body. Dry soups and gravies have a longer shelf life with 5 to 25% dry whey if it is added to the mixture before drying. Instant potatoes with two to six per cent dry whey have improved whippability and less crusting of reconstituted product.

Other possibilities for dry whey utilization include: flavor enhancer in canned vegetables, dressings and barbecue sauces, canned fruit, preserves and fruit flavored soft drinks. Dry whey may be a source of whey proteins. Several cheeses are made from the solids of whey (Webb and Whittier 1948).

Whey solids serve as a substrate for the growth of numerous species of microorganisms. Yeasts, molds, and bacteria are grown in whey solutions for the production of alcohol, vinegar, lactic acid, and antibiotics.

Lactose

A large amount of lactose is prepared from whey. The three common grades in decreasing degree of purity are U.S.P., technical, and crude. These lactoses have numerous applications for therapeutic purposes and for special infant foods. Babies fed lactose are less likely to have gastrointestinal ailments or diarrhea, if they are inclined to have these illnesses on a diet of other carbohydrates. Mother's milk is much higher in lactose than milk from a cow. Many pharmaceutical uses are known; one is a diluent for drugs. Lactose may be mixed with the drug in loose or pill form or used as a coating on tablets (Call 1958).

REFERENCES

ALDRICH, P. J., and MILLER, G. A. 1958. A new milky way for your own favorite quantity recipes. Mich. State Univ. Agr. Expt. Sta. Circ. Bull. 225. East Lansing, Mich.

ALESCH, E. A. 1958. Utilization of whey solids in food products. J. Dairy Sci. 41, 699–700.

AMERICAN DRY MILK INSTITUTE. 1951. Nonfat dry milk solids in dairy manufacturing. Bull. 650. Chicago, Ill.

AMERICAN DRY MILK INSTITUTE. 1951–1970. 1950–1969 Census of dry milk distribution and production trends. Bull. 1000. Chicago, Ill.

AMERICAN DRY MILK INSTITUTE. 1953. Meat products improved with nonfat dry milk solids. Bull. No. 804. Chicago, Ill.

AMERICAN DRY MILK INSTITUTE. 1962. Standards for grades for the dry milk industry including methods of analysis. Bull. 916. Chicago, Ill.

AMERICAN DRY MILK INSTITUTE. 1963. 1962 Census of dry milk distribution and production trends. Bull. 1000. Chicago, Ill.

ANON. 1960. How whey solids can lift your profit two ways. Food Eng. 32, No. 9, 92–93.

BECKER, D. E., ULLREY, D. E., and TERRILL, S. W. 1954. A comparison of carbohydrates in synthetic milk diet for the baby pig. Arch. Biochem. Biophys. 48, 178–183.

BLAKELY, L. E., and STINE, C. M. 1964. Foam spray-dried cottage cheese whey as a source of solids in sherbet. Quart. Bull. Mich. Agr. Expt. Sta. 47, 142–148.

BURGWALD, L. H., and STROBEL, D. R. 1959. Recombined condensed milk recombined sterilized milk toned milk. U.S. Dept. Agr. Foreign Agricultural Service FAS-M-66. Washington, D.C.

CALL, A. O. 1958. Utilization of lactose. J. Dairy Sci. 41, 332–334.

COOK, H. L., and DAY, G. H. 1947. The Dry Milk Industry. American Dry Milk Institute, Chicago, Ill.

COOK, H. L., and HALVERSON, H. 1950. Industrial uses and preferences for nonfat dry milk solids. Wis. Agr. Expt. Sta., Univ. Wis. Res. Bull. 169.

CRAWFORD, J. E. 1947. Improving quality through use of buttermilk. Bakers Weekly 135, No. 2, 55–56.

CUSTER, E. W., HERZER, F. H., and CARDWELL, J. J. 1958. The effects of various levels of solids-not-fat on the flavor acceptability of fluid milk. Miss. Agr. Expt. Sta. Bull. 561.

FRANDSEN, J. H., and ARBUCKLE, W. S. 1961. Ice Cream and Related Products. The Avi Publishing Co., Westport, Conn.

GLABAU, C. A. 1964. Re-evaluating the use of nonfat dry milk in bread production. Bakers Weekly 202, No. 49, 33–35.

GOTT, P. P., and VAN HOUTEN, L. F. 1958. All about candy and chocolate. National Confectioners' Association, Chicago, Ill.

GREENE, V. W., and JEZESKI, J. J. 1955. The activity of lactic acid starters in reconstituted nonfat dry milk solids and fluid milks. J. Dairy Sci. 38, 587.

GROVES, W. G., and GRAF, T. F. 1965. An economic analysis of whey utilization and disposal in Wisconsin. Agr. Econ. 44, Dept. Agr. Econ., Univ. Wis., Madison, Wis.

HALL, H. H., and FAHS, F. J. 1949. Milk solids give candies longer-lasting freshness. Food Inds. 21, No. 8, 46–47.

HEDRICK, T. I., ARMITAGE, A. V., and STINE, C. M. 1964. A comparison of high-heat and low-heat nonfat dry milk as the sole source of serum solids in ice cream and ice milk. Quar. Bull. Mich. Agr. Expt. Sta. 47, 153–158.

HOLLANDER, H. A. 1948. Special milk powders for the manufacture of milk chocolate. Ph.D. Thesis, Univ. of Wis., Madison, Wis.

JANZEN, J. J. 1962. What solids should be added to fluid dairy products. J. Dairy Sci. 45, 1562–1565.

MAURON, J., MOTTU, F., BUJARD, E., and EGLI, R. H. 1955. The availability of lysine, methionine and tryptophan in condensed milk and milk powder. In vitro digestion studies. Arch. Biochem. Biophys. 59, 433–451.

MILLER, E. B. 1965. Cooperative marketing of nonfat dry milk to commercial outlets. U.S. Dept. Agr. Farmer Cooperative Service, General Report 129. Washington, D.C.

MULLER, L. L., and KIESKER, F. G. 1965. Studies on recombined sweetened condensed milk. Aust. J. Dairy Tech. 20, 130–135.

NORMAN, G. H. 1955. Dried buttermilk improves palatability of reconstituted milk. Milk Prod. J. 46, No. 1, 38–39.

OKAMOTO, M., THOMAS, J. W., and JOHNSON, T. L. 1959. Utilization of carbohydrates by the young calf. U.S. Dept. Agr. ARS 44-54. Washington, D.C.

PANGBORN, R. M., and DUNKLEY, W. L. 1964. Sensory discrimination of fat and solids-not-fat in milk. J. Dairy Sci. 47, 719–726.

PEARSON, A. M., SPOONER, M. E., HEGARTY, G. R., and BRATZLER, L. J. 1965. The emulsifying capacity and stability of soy sodium proteinate, potassium caseinate and nonfat dry milk. Food Technol. 19, No. 12, 103–107.

PETERS, I. I. 1960. Ripening of Cheddar cheese made from raw, pasteurized homogenized and reconstituted milk. Milk Prod. J. 51, No. 5, 8, 20–21.

PETERS, I., and WILLIAMS, J. 1961. Studies to improve the quality of reconstituted milk cheese. I. Variables: rennet, salt and ripening. Food Technol. 15, No. 11, 486–488.

PONT, E. G. 1960. Observations on the manufacture and properties of recombined sweetened condensed milk. Aust. J. Dairy Technol. 15, No. 1, 17–19.

RANDOLPH, H. E., and KRISTOFFERSEN, T. 1961. Characteristics of commercial nonfat dry milk for cottage cheese. J. Dairy Sci. 44, 833–843.

REGER, J. V., COMBS, W. B., COULTER, S. T., and KOCH, R. B. 1951. A comparison of dry sweet cream buttermilk and non-fat dry milk solids in bread-making. J. Dairy Sci. 34, 136–144.

REID, W. H. E., and MOUGHAN, M. O. 1948. Cottage cheese from nonfat dry milk solids. Nat'l. Butter & Cheese J. 39, No. 6, 39, 64, 66.

RONGEY, E. H. 1961. The effect of various binders and meats on the palatability and processing characteristics of bologna. Ph.D. Thesis. Michigan State University, East Lansing, Mich.

SAFFLE, R. L. 1964. Nonfat dry milk in sausage. Proceedings of 39th Annual Meeting, American Dry Milk Institute, Chicago, Ill., April 17.

SARGENT, J. S. E., BIGGS, D. A., and IRVINE, D. M. 1959. Effect of hard water on the heat stability of skimmilk powder. J. Dairy Sci. 42, 1800–1805.

SOMMER, H. H. 1946. Theory and Practice of Ice Cream Making. Published by the author, Madison, Wis.

STULL, J. W., and HILLMAN, J. S. 1960. Relation between composition and consumer acceptance of milk beverages. J. Dairy Sci. 43, 945–950.

SWICKARD, M. T. 1949. How to use whole and nonfat dry milk. U.S. Department of Agriculture, ARA, Bur. of Human Nutrition and Home Ec. AIS-86. Washington, D.C.

SWORTFIGUER, M. J. 1962. Nonfat dry milk in the continuous mix process. Baker's Digest 36, No. 2, 39–42, 45, 46.

THIESSEN, E. J. 1957. Dried whey in baked products and confections. Wyo. Agr. Expt. Sta. Bull. 347. Laramie, Wyo.

THOMAS, E. L., and COMBS, W. B. 1944. Observations on the use of roller process sweet cream buttermilk powder in ice cream. J. Dairy Sci. 27, 419–42C.

U.S. DEPT. AGR. 1965. Dairy Situation. E.R.S. DS-308, Washington, D.C.

U.S. DEPARTMENT OF AGRICULTURE AND NATIONAL CONFECTIONERS ASSOCIATION. 1948. Progress in Candy Research. Report 14. Chicago, Ill.

VELU, J. G., KENDALL, K. A., and GARDNER, K. E. 1960. Utilization of various sugars by the young dairy calf. J. Dairy Sci. 43, 546–552.

WAHID-UL-HAMID, S. S., and MANUS, L. J. 1960. Effect of changing the fat and nonfat solids of milk. J. Dairy Sci. 43, 1430–1434.

WATT, B. K., and MERRILL, H. L. 1963. Composition of Foods, Agriculture Handbook No. 8 (revised) U.S. Dept. Agr., Washington, D.C.

WEBB, B. H., and WHITTIER, E. O. 1948. The utilization of whey: a review. J. Dairy Sci. 139–164.

WHITAKER, R. 1956. Selection and use of nonfat dry milk solids in the manufacture of cottage cheese. J. Dairy Sci. 39, 231–233.

WILLIAMS, D. H., POTTER, F. E., and HUFNAGEL, C. F. 1950. Concentrated buttermilk in ice cream. J. Dairy Sci. 33, 593–598.

U.S. Patents

Equipment for Drying, Instantizing and Handling of Dry Dairy Products

Name	Subject	U.S. Patent	Year
Brinton, H. F.	Machine for cooking and drying food and grain	230,525	1880
Gere, W. B.	Powder and apparatus for producing the powder	592,906	1897
Passburg, E.	Vacuum-pan	726,742	1903
Ekenberg, M.	Drum for vacuum-pans	727,317	1903
Passburg, E.	Filling or emptying apparatus for vacuum-driers	748,414	1903
MacLachlan, J. C.	Desiccating apparatus	827,172	1906
Just, J. A.	Apparatus for desiccating milk	841,153	1907
Rosenberger, A.	Apparatus for producing Kefir powder	957,104	1910
Holt, H. L.	Apparatus for preparing food products	969,955	1910
Drzymalla, T.	Two-roll drying apparatus for pulp-like paste	1,014,956	1912
MacLachlan, J. C.	Desiccating apparatus	1,038,773	1912
Gray, C. E. and Jensen, A.	Apparatus for desiccating liquids	1,078,848	1913
Merrell, I. S.	Spraying-nozzle	1,183,393	1916
Stutzke, R. W. G.	Process and apparatus for desiccating fluid substances	1,215,889	1917
Rogers, C. E.	Desiccating apparatus	1,226,011	1917
Goodhue, J. G.	Apparatus for treating food or other materials	1,290,734	1919
Campbell, C. H.	Apparatus for producing dried milk	1,292,577	1919
Hill, D.	Drier and evaporator	1,296,519	1919
Collis, N. P.	Apparatus for desiccating liquids	1,317,777	1919
Rea, R.	Fruit, vegetable, and other food dehydrating furnace and its cooperating apparatus	1,328,396	1920
Peebles, D. D.	Drying apparatus and method	1,761,149	1930
Douthitt, F. H.	Apparatus for the desiccation of substances	1,797,055	1931
Peebles, D. D.	Desiccating apparatus and method	1,830,174	1931
Peebles, D. D.	Separating apparatus	1,850,333	1932
Chappell, F. L.	Method of and apparatus for drying casein	1,892,233	1932
Burner, J. A.	Process and apparatus for producing dehydrated products	1,905,263	1933
Peebles, D. D. and Barlow, A. E.	Centrifugal atomizer	1,939,364	1933

Name	Subject	U.S. Patent	Year
Poole, R. T.	Drying apparatus and method	1,968,910	1934
Peebles, D. D.	Apparatus for drying liquid containing materials	2,054,441	1936
Peebles, D. D.	Drying systems and method	2,080,059	1937
Bowen, W. S.	Apparatus for spray drying	2,081,909	1937
Peebles, D. D.	Dehydrating apparatus and method	2,090,984	1937
Gentele, J. G. W.	Method of and apparatus for drying substances which contain liquids	2,132,897	1938
Lavett, C. O.	Drum drier	2,143,019	1939
Spellacy, J. R.	Drying milk whey	2,163,331	1939
Beach, G. W.	Method of and apparatus for dry concentration	2,186,599	1940
Allen, E. F.	Mechanical dehydration apparatus	2,196,650	1940
Allen, E. F.	Mechanical dehydrator and roller mill	2,196,651	1940
Gray, J. E.	Dehydrator	2,199,258	1940
Hagemeyer, H. F.	Drying apparatus and method	2,201,038	1940
Dietrich, J. F.	Drying apparatus	2,211,767	1940
Cady, J. E.	Drying apparatus	2,213,303	1940
Hall, J. M.	Dehydrating system	2,216,815	1940
Hall, J. M.	Concentrating and dehydrating apparatus and method	2,217,547	1940
Pierce, L.	Method of and mechanism for operating a drying apparatus	2,224,608	1940
Henry, G. J.	Dehydrator	2,225,990	1940
Peebles, D. D.	Desiccating apparatus	2,240,854	1941
Spears, J. M.	Rotary drier	2,264,646	1941
Adt, J. B.	Rotary drier	2,267,259	1941
Freund, B. G.	Dehydration apparatus	2,288,616	1942
Freund, B. G.	Dehydrating method and apparatus	2,293,728	1942
Reichel, J.	High vacuum, low temperature drying apparatus	2,302,253	1942
Peebles, D. D.	Drying apparatus and method	2,310,650	1943
Wild, A. F.	Dehydrating compartment	2,311,425	1943
Peebles, D. D.	Milk secondary drying	2,312,474	1943
Peebles, D. D.	Spray drying apparatus and method	2,314,159	1943
Baird, J. B.	Spray head	2,314,754	1943
French, H. C.	Rotary drier	2,319,674	1943
Peebles, D. D.	Powder classifying apparatus	2,330,793	1943
Loomis, E. G.	Dehydrating press	2,331,126	1943
Peebles, D. D. and Manning, P. D. V.	Desiccating apparatus and method	2,333,333	1943
Beardslee, A. C.	Apparatus and method for producing dry products	2,336,461	1943
Newman, I.	Drying machine	2,338,348	1944
Kirchmann, H., Jr.	Rotary drier	2,347,487	1944
Peeps, D. J.	Spray nozzle	2,356,944	1944
Rosecky, J. E.	Machine and a process for making sugared milk powder	2,358,418	1944

Name	Subject	U.S. Patent	Year
Vose, E. W.	Spraying device for liquids	2,399,081	1946
Patrick, W. A., Jr.	Dehydrating method and apparatus	2,399,246	1946
Patrick, W. A., Jr.	Dehydrating apparatus	2,399,247	1946
Patrick, W. A., Jr.	Dehydrating method and apparatus	2,399,504	1946
Hickman, K. C. D.	Method and apparatus for dehydrating in the frozen state	2,402,401	1946
Birdseye, C.	Heated endless conveyor structure for dehydrating foods	2,414,580	1947
Purpura, A. C.	Rotary drum drying machine	2,416,405	1947
Birdseye, C.	Dehydration apparatus having conveyors, agitators, radiant heaters, and gas circulating means	2,419,876	1947
Christensen, N. C.	Liquid rotor spray mechanism	2,448,297	1948
Kloda, S. A.	Sprayer for dehydrating apparatus	2,450,599	1948
Mercier, J.	Accumulator bag	2,465,908	1949
Harris, B. R.	Method and apparatus for cooling hot hygroscopic solids	2,478,889	1949
Hall, J. M.	Method of and apparatus for dehydrating liquid products	2,481,418	1949
Komline, T. R.	Spray drier	2,506,646	1950
Turnbow, G. D. and Osborne, A. V.	Mixing device	2,513,382	1950
Pieper, O. T.	Spray drying device	2,515,665	1950
Erisman, J. L.	Rotary drier or cooler	2,522,025	1950
Nyrop, J. E.	Spray drier	2,559,989	1951
Mojonnier, J. J.	Spray drying apparatus	2,562,473	1951
Stieger, H. J.	Rotary drum drier	2,571,778	1951
Hall, J. M. and Tucker, H. E.	Method and apparatus for spray drying	2,572,857	1951
Peebles, D. D.	Spray drying equipment and method	2,575,119	1951
Horsley, C. B. and Danser, H. W., Jr.	Sonic spray drying	2,576,297	1951
Bill, C. E.	Rotary drier	2,578,166	1951
Nyrop, J. E.	Drying, concentrating by evaporation, or distilling heat-sensitive substances	2,585,825	1952
Vincent, D. B.	Dehydration	2,600,945	1952
Britcher, C. W.	Dry spray equipment	2,614,528	1952
Roell, W. J. and Berreau, T. O.	Spray nozzle with pressure operated clean out	2,614,885	1952
Shaw, G. E.	Forced draft vacuum stack for milk driers	2,644,391	1953
Wenzelberger, E. P.	Dehydration by freezing	2,657,555	1953
Peebles, D. D. and Turner, C. P.	Centrifugal atomizer	2,661,984	1953
Peebles, D. D. and Turner, C. P.	Centrifugal atomizer	2,668,080	1954
Benscheidt, N. H.	Method of and apparatus for dehydrating by freezing	2,685,783	1954

Name	Subject	U.S. Patent	Year
Carter, C. F.	Weighing and filling machine (for powder)	2,687,271	1954
Jehlicka, J.	Apparatus for drying by atomization, particularly of organic substances	2,699,822	1955
Vincent, D. B.	Dehydrating apparatus	2,705,842	1955
Nyrop, J. E.	Method and apparatus for the drying, concentration, or crystallization of liquid materials	2,707,990	1955
Andersen, A. L.	Vacuum drying cabinet	2,710,456	1955
Lauck, F. W.	Method and apparatus for the slow drying of stored material	2,716,289	1955
Irving, H. F.	Drier	2,716,290	1955
Carter, C. F.	Filling machine (for powder)	2,720,375	1955
Bancroft, G. H.	Vacuum dehydration apparatus	2,731,734	1956
Fleming, M. T.	Dryer for powdered material	2,746,171	1956
Moore, D. P.	Apparatus for reducing food liquids to powders	2,750,998	1956
Wenzelberger, E. P.	Freezing apparatus (freeze drying)	2,764,880	1956
Templeton, R. A. S.	Apparatus for prepared dried food and other products	2,788,732	1957
Wenzelberger, E. P.	Apparatus for dehydration of liquids by freezing	2,794,327	1957
Henszey, R. O. and Henszey, R. R.	Horizontal spray drier	2,815,071	1957
Bogaty, S.	Apparatus for drying materials in paste form	2,821,030	1958
Coulter, S. T. and Townley, V. H.	Method and apparatus for spray drying	2,887,390	1959
Griffin, H. L.	Agglomeration process and apparatus	2,893,871	1959
Scott, E. C.	Method and apparatus for producing granulated food products	2,900,256	1959
Huston, C. R. and Hartman, G. H.	Cyclone discharge chamber	2,912,768	1959
Conley, W. E. and Fixari, F.	Method of and apparatus for dehydrating material	2,924,272	1960
Conley, W. E. and Kopp, W. P.	Dehydrating apparatus	2,924,273	1960
Hartman, G. H., Harder, H. F., Rezba, A. J., and Reeve, R. K.	Method and apparatus for producing clusters of lacteal material	2,934,434	1960
Toulmin, H. A., Jr.	Method and apparatus for crystal production	2,957,773	1960
Gidlow, R. G. and Mills, A. D.	Process and apparatus for agglomerating pulverulent materials	2,995,773	1961
Spiess, N. E., Jr. and Sullivan, N. E.	Powder agglomerating method and apparatus	3,042,526	1962

U.S. Patents

Equipment and Accessories for Evaporating Moisture
from Dairy Products

Name	Subject	U.S. Patent	Year
Yaryan, H. T.	Vacuum distillation apparatus	300,185	1884
Ellin, R.	Vacuum-evaporator for milk	347,584	1886
Yaryan, H. T.	Vacuum evaporating and distilling apparatus	355,289	1886
Yaryan, H. T.	Vacuum evaporating and distilling apparatus	355,290	1886
Von Roden, O.	Apparatus for condensing and carbonating milk	376,496	1888
Yaryan, H. T.	Vacuum evaporating apparatus	383,384	1888
Bassler, J. H.	Apparatus for condensing liquids	394,432	1888
Bassler, J. H.	Apparatus for condensing liquids	394,433	1888
Bassler, J. H.	Apparatus for condensing liquids	394,434	1888
Lillie, S. M.	Condenser with multiple effect evaporating apparatus	422,234	1890
Lillie, S. M.	Evaporating apparatus	422,235	1890
Lillie, S. M.	Evaporating apparatus	440,231	1890
Yaryan, H. T.	Vacuum evaporating apparatus	485,315	1892
Lillie, S. M.	Evaporating apparatus	498,938	1893
Lillie, S. M.	Process and apparatus for evaporating liquids in vacuum	521,215	1894
Gere, W. B. and Merrell, I. S.	Evaporator	631,568	1899
Ordway, C.	Vacuum evaporating apparatus	714,513	1902
Lillie, S. M.	Tube for evaporators	740,449	1903
Trump, E. N.	Vacuum pan	743,351	1903
Ekenberg, M.	Apparatus for evaporating liquids	764,995	1904
Just, J. A.	Evaporating apparatus	765,315	1904
Ordway, C.	Vacuum evaporating apparatus	775,577	1904
Lillie, S. M.	Multiple-effect evaporating apparatus	777,114	1904
Lillie, S. M.	Multiple effect evaporating apparatus	789,159	1905
Ordway, C.	Film evaporating heating coil	837,582	1906
Zaremba, E.	Evaporator	882,043	1908
Kestner, P.	Apparatus for concentrating liquids	882,322	1908
Just, J. A.	Evaporating apparatus	888,018	1908
Prache, C. L.	Apparatus for evaporating and concentrating liquids	896,460	1908
Ordway, C.	Recovering crystalline substances from liquor and drying the same	905,568	1908
Faller, O.	Evaporator	907,109	1908

Name	*Subject*	*U.S. Patent*	*Year*
Lillie, S. M.	Evaporating apparatus	939,143	1909
Parker, J.	Evaporating apparatus	940,473	1909
Ordway, C.	Evaporating apparatus	942,407	1909
Lillie, S. M.	Multiple-effect evaporating apparatus	948,376	1910
Zastrow, C. W.	Evaporating apparatus	964,358	1910
Kestner, P.	Evaporator	965,388	1910
Morris, A. S.	Evaporator	965,395	1910
Kestner, P.	Concentrating liquids	965,822	1910
Kestner, P.	Apparatus for concentrating liquids	971,383	1910
Morris, A. S.	Feeding device for evaporators	971,394	1910
Prache, C. L.	Evaporation-boiler with inclined tubes	972,572	1910
Lillie, S. M.	Evaporating apparatus	972,880	1910
Lillie, S. M.	Multiple-effect evaporating apparatus	984,226	1911
Eijdmann, F. H.	Evaporating apparatus	984,754	1911
Lillie, S. M.	Evaporating tube	988,477	1911
Kestner, P.	Evaporator	989,982	1911
Parker, J.	Evaporator	989,996	1911
Kestner, P.	Evaporating apparatus	1,003,912	1911
Scheinemann, F.	Evaporating apparatus	1,004,087	1911
Kestner, P.	Evaporating apparatus	1,005,553	1911
Parker, J.	Evaporator	1,005,571	1911
DeBeers, F. M.	Evaporator	1,006,363	1911
Ordway, C.	Vacuum evaporating apparatus	1,009,782	1911
Kestner, P.	Apparatus for concentrating liquids	1,013,091	1911
Kestner, P.	Evaporating apparatus	1,016,160	1912
Kestner, P.	Process of evaporation and apparatus	1,028,737	1912
Kestner, P.	Evaporating apparatus	1,028,738	1912
Power, J. A.	Evaporating apparatus	1,028,777	1912
Weir, W.	Sea-water evaporator	1,049,014	1912
Mantius, O.	Evaporator	1,054,926	1913
Kestner, P.	Evaporating apparatus	1,060,607	1913
McGregor, J.	Concentration of syrupy liquids	1,068,789	1913
Morris, A. S.	Evaporator	1,069,566	1913
Prache, C. L.	Evaporating apparatus with superposed compartments and progressive level	1,071,341	1913
Kestner, P.	Apparatus for concentrating liquids to high density	1,090,628	1914
Moore, H. K.	Evaporating and concentrating apparatus	1,098,825	1914
Zastrow, C. W.	Evaporating apparatus	1,124,096	1915
Row, R. R.	Evaporator	1,131,738	1915
Stade, G.	Vacuum evaporating apparatus	1,168,758	1916
Naudet, L.	Process and apparatus for evaporation of sugar-juice	1,190,317	1916
Kestner, P.	Evaporator	1,191,108	1916
Wheeler, F. G.	Evaporating process and apparatus	1,222,340	1917
Benjamin, G. H.	Evaporator	1,255,502	1918
Reavell, J. A.	Evaporating apparatus	1,263,467	1918

Name	Subject	U.S. Patent	Year
Benjamin, G. H.	Evaporator	1,280,641	1918
Benjamin, G. H.	Evaporator	1,288,480	1918
Dick, S. M.	Evaporating apparatus	1,298,470	1919
Garrigues, W. E.	Evaporator	1,298,925	1919
Jones, R. C.	Evaporator	1,299,955	1919
Benjamin, G. H.	Evaporator	1,302,625	1919
Davis, H. C. and Row, R. R.	Evaporator feed-water heater and the like	1,304,379	1919
Newhall, E. A.	Evaporating apparatus and operating same	1,318,793	1919
Barbet, E. A.	Evaporator	1,325,461	1919
Braun, C. F.	Evaporator	1,334,014	1920
Benjamin, G. H.	Evaporator and operating same	1,355,935	1920
Witte, F.	Evaporator	1,373,041	1921
Bull, H. J.	Treatment of liquids containing sulfate of lime	1,399,845	1921
Holle, A.	Process of preventing the deposition of scale or sludge from the cooling water in surface steam condensers	1,405,783	1922
Webre, A. L.	Evaporator	1,436,739	1922
Dyson, C. W.	Evaporator	1,440,723	1923
Martin, F. G.	Cleansing of condensers and the like apparatus	1,447,096	1923
Field, C.	Apparatus for transferring heat	1,451,901	1923
Field, C.	Separating a brittle product from a base	1,451,902	1923
Field, C.	Heat transferring apparatus	1,451,903	1923
Field, C.	Heat transferring apparatus	1,451,904	1923
Engel, G.	Concentrating evaporator	1,466,357	1923
Engel, G.	Vacuum pan	1,467,331	1923
Dyson, C. W.	Evaporator	1,477,328	1923
Field, C.	Crystallizing fluids	1,480,382	1924
Fothergill, H.	Evaporator	1,498,440	1924
Brown, S.	Evaporator coil	1,501,646	1924
Hughes, B. S.	Film evaporator	1,506,001	1924
Sebald, L. E.	Flash evaporator	1,516,314	1924
Arato, V.	Fuel heating or vaporizing attachment	2,016,952	1935
Hall, J. M.	Mechanism for removing moisture from liquid products	2,287,795	1942
Meyenberg, J. P.	Making modified evaporated milk	2,288,825	1942
Schoeller, K. F.	Vapor condenser	2,358,940	1944
Cross, J. A.	Evaporation method and apparatus	2,570,210	1951
Cross, J. A.	Falling film evaporator	2,570,211	1951
Cross, J. A.	Milk evaporation process and apparatus	2,570,213	1951
Fenske, M. R.	Process and apparatus for liquid-liquid extraction	2,580,010	1951
Schumacher, F. A.	Freezer evaporator including check valve in header	2,641,113	1953
Cross, J. A.	Milk evaporation apparatus	2,703,610	1955

U.S. Patents

Dry Dairy Products and Related Processes

Name	Subject	U.S. Patent	Year
Birdseye, C. D.	Process of preparing cream	7,644	1850
Dodge, T. H.	Kettle bails	9,744	1853
Borden, G., Jr.	Concentration of milk	15,553	1856
Twing, A. T., Wood, E., and Elderhorst, W.	Vacuum packing (preserved milk)	28,424	1860
Stabler, F.	Process for preserving animal and vegetable substance (reissued 2/62)	50,965	1865
Lamont, C. A.	Preserving (desiccating)	51,263	1865
Percy, S. R.	Process of drying and concentrating liquid substances by atomizing (spray-drying fluid and solid substances)	125,406	1872
Cole, J. R.	Desiccating milk	172,090	1876
Catlin, C. A.	Preserving milk (drying)	214,489	1879
Ellin, R.	Preserving milk	349,574	1886
Von Roden, O.	Condensing and preserving milk	376,495	1888
McIntyre, B. F.	System of drying milk by freezing out the water and centrifuging (condensing and preserving milk)	523,677	1894
Gere, W. B.	Soup powder and making same	625,880	1899
Axtell, L.	Method of and apparatus for treating casein curd	629,644	1899
Stauf, R.	Method of desiccating milk, etc.	666,711	1901
Campbell, J. H.	Desiccated milk and making same	668,159	1901
Campbell, J. H.	Condensed milk and obtaining same	668,161	1901
Campbell, J. H. and Campbell, C. H.	Desiccating milk	668,162	1901
Hall, W. A.	Producing dry condensed milk	694,100	1902
Just, J. A.	Preserving milk in dry form	712,545	1902
Campbell, J. H.	Making desiccated milk	718,191	1903
Dunham, H. V.	Producing milk powder	723,254	1903
Campbell, C. H.	Food product	762,277	1904
Just, J. A.	Dried milk powder	764,294	1904
Just, J. A.	Evaporating liquids	765,343	1904
Campbell, C. H. and Campbell, P. T.	Concentrating and remaking milk	771,609	1904
MacLachlan, J. C.	Desiccating process	806,747	1905
Gathmann, L.	Reducing milk to a dry powder	834,516	1906
Merrell, L. C. and Gere, W. B.	Separating the moisture from the constituent solids of liquids	860,929	1907
Just, J. A.	Obtaining milk sugar	868,443	1907
Just, J. A.	Preparing casein soluble to a neutral solution	868,445	1907

Name	Subject	U.S. Patent	Year
Just, J. A.	Evaporating	868,447	1907
Just, J. A.	Producing dried milk	888,016	1908
Just, J. A.	Condensing milk	888,017	1908
Just, J. A.	Utilizing dried milk	939,138	1909
Just, J. A.	Fibrous desiccated milk	939,139	1909
Govers, F. X.	Desiccating milk	939,495	1909
Osborne, W. S.	Desiccating fluid substances	962,781	1910
Merrell, L. C.	Food obtained from whey (powder)	985,271	1911
Swenarton, W. H.	Desiccating the solid content of milk	995,303	1911
Campbell, C. H.	Desiccating milk	996,832	1911
Bergh, F. P., Loebinger, H. J., and Neuberger, H. C.	Evaporating fluids	997,950	1911
Ellis, C.	Desiccating milk	999,707	1911
Andrews, H. I.	Desiccating milk	1,012,578	1911
Swenarton, W. H.	Desiccated milk and making the same	1,056,719	1913
Nicolai, O.	Drying milk	1,063,581	1913
Davis, J. D.	Manufacture of desiccated milk	1,070,781	1913
Jebsen, G. and Finckenhagen, C.	Separating solid substances from solutions by evaporation	1,074,264	1913
Dunham, A. A.	Producing desiccated milk	1,074,419	1913
McLaughlin, W. B.	Desiccating fluid substances	1,090,740	1914
Gray, C. E.	Process of desiccating	1,107,784	1914
Gray, C. E.	Method of and apparatus for desiccating liquid substances	1,157,935	1915
Gere, W. B.	Treating cream, whole milk, or skimmed milk	1,188,755	1916
Hemig, F. J.	Maltose and making the same	1,214,160	1917
Campbell, C. H.	Dried milk and producing the same	1,233,446	1917
Rogers, C. E.	Means for and method of desiccating fluids	1,243,878	1917
MacLachlan, J. C.	Desiccating liquids	1,258,348	1918
Stutzke, R. W. G.	Desiccating milk and the like	1,350,248	1920
Collis, N. P.	Drying buttermilk	1,356,340	1920
Coulson, A. R.	Milk powder (dried modified cultured)	1,374,138	1921
Dick, S. M.	Dehydrated milk and process of producing same	1,374,555	1921
Weimar, A. C.	Extracting soluble albumin from whey	1,381,605	1921
MacLachlan, J. C.	Granular food product (milk)	1,394,035	1921
Mellott, H. S.	Producing a condensed-milk product	1,423,810	1922
Merrell, I. S.	Desiccated buttermilk	1,430,312	1922
Fest, A. D.	Method for commercially obtaining water-soluble milk albumin and milk sugar	1,444,178	1923
Peebles, D. D.	Drying milk	1,491,166	1924
Christensen, N. H.	Manufacture of powdered milk	1,574,233	1926
McLaughlin, W. B.	Process of producing dry milk	1,616,631	1927

Name	*Subject*	U.S. *Patent*	*Year*
Merrell, I. S.	Food product (baked milk)	1,689,357	1928
McDougall, S. A.	Making food product (malted milk)	1,705,332	1929
Sheffield, W. H.	Manufacture of casein (not dry)	1,716,799	1929
Kronberg, N. M.	Method of producing a pulverized milk serum product	1,721,867	1929
Lamont, D. R.	Controlling characteristics of spray-processed products	1,734,260	1929
Baker, E.	Cold-milk-dehydrating process	1,738,275	1929
O'Connor, T.	Self-preserving milk product	1,749,153	1930
Simmons, N. L.	Making whey products	1,763,633	1930
Nielsen, C.	Food for invalids or infants	1,767,185	1930
Born, P.	Solid milk and cream preparations	1,786,559	1930
Merrell, I. S. and Schibsted, H.	Treatment of milk powder	1,786,858	1930
Börnegg, C. B. von	Method for making a stable powder from whole milk	1,808,730	1931
Sharp, P. F.	Preparing lactose	1,810,682	1931
Horlick, W., Jr.	Food product and manufacture thereof	1,813,574	1931
Wilson, C. P. and Stewart, E. D.	Food product and process of making same	1,814,994	1931
Musher, S.	Food product	1,841,842	1932
Scott, A. A.	Malted milk process	1,851,988	1932
Grindrod, G.	Method of stabilizing food products	1,854,189	1932
Washburn, R. M.	Preparation of dry nonhygroscopic crude lactose	1,870,270	1932
Johnson, L. D. and True, N. F.	Powdered protein milk and process of preparing same	1,882,637	1932
Johnson, L. D. and True, N. F.	Powdered acid milk and process of preparing same	1,882,638	1932
MacLachlan, J. C.	Molasses product and making the same	1,897,729	1933
Monrad, K. J.	Sweetened and flavored dessert made with rennin	1,902,415	1933
Heideman, A. G.	Special product (cheese-like)	1,908,512	1933
Peebles, D. D.	Desiccating method and apparatus	1,914,895	1933
Eldredge, E. E.	Preparing whey concentrate	1,923,427	1933
Finley, J. A.	Food product and making the same	1,925,441	1933
Peebles, D. D. and Manning, P. D. V.	Manufacture of lactose-containing material	1,928,135	1933
Chapin, E. K.	Comminuted shortening	1,928,781	1933
Sanna, A. R.	Producing milk powder	1,929,450	1933
Otting, H. E.	Milk product	1,937,527	1933
Wilson, C. P.	Beverage product and producing the same	1,940,036	1933
Supplee, G. C. and Flanigan, G. E.	Producing beta lactose	1,954,602	1934
Sharp, P. F.	Preparing lactose	1,956,811	1934
Dahlberg, A. C.	Milk products (dry casein) and producing same	1,962,552	1934

Name	Subject	U.S. Patent	Year
Peebles, D. D.	Method and apparatus for treatment of nongaseous materials	1,964,858	1934
Moss, H. V. and Wheelock, T. H.	Free flowing powdered milk	1,966,513	1934
Wilson, C. W.	Drying organic materials	1,975,998	1934
Black, T. and Drew, J.	Manufacture of powdered molasses	1,983,434	1934
Robinson, R. W. and Olsen, A., Jr.	Fudge powder	1,983,568	1934
Peebles, D. D.	Method and apparatus for treatment of nongaseous materials	1,984,381	1934
Ambrose, A. S.	Isolation of milk constituents	1,991,189	1935
Chappell, F. L.	Manufacture of casein	1,992,002	1935
Peebles, D. D.	Dried food products	2,005,238	1935
Supplee, G. C., Flanigan, G. E., and Bender, R. C.	Production of vitamin free casein	2,006,700	1935
Berlatsky, A.	Food product	2,011,558	1935
Chuck, F. Y.	Stabilizing milk powder and similar colloidal products	2,016,592	1935
Axelrod, A.	Milk baking preparation	2,018,394	1935
Šírek, J.	Whey preparation	2,023,359	1935
Kraft, G. H.	Comminuted shortening	2,035,899	1936
Visser, J. M.	Preparing dry powder from emulsions and solutions	2,044,194	1936
Baier, W. E.	Food product and producing the same	2,055,782	1936
Fechner, E. J.	Making fat-containing powder	2,065,675	1936
Fechner, E. J.	Making fat-containing powder	2,065,676	1936
Clickner, F. H.	Lacteal derivative	2,076,400	1937
Peebles, D. D. and Manning, P. D. V.	Manufacture of stable powdered products containing milk sugar	2,088,606	1937
Iverson, C. A.	Making milk confection	2,091,149	1937
Clickner, F. H.	Treating whey	2,091,629	1937
Spellacy, J. R.	Manufacturing casein	2,099,379	1937
Whitaker, R., Weisberg, S. M., and Hilker, L. D.	Food product and process for making same	2,101,633	1937
Coffey, J. R.	Method of producing homogeneous casein	2,112,558	1938
Leviton, A.	Recovery of lactose from whey	2,116,931	1938
Sanna, A. R. and Sanna, F. L.	Milk products	2,117,681	1938
Sanna, F. L.	Chocolate milk product and making the same	2,117,682	1938
Kraft, G. H.	Drying whey	2,118,252	1938
Webb, B. H. and Ramsdell, G. A.	Preparing a useful milk serum or whey product	2,119,614	1938
Riggs, L. K.	Protein mineral complex and making same	2,123,203	1938
Wanshenk, O. B.	Making therapeutic compounds	2,123,218	1938

Name	Subject	U.S. Patent	Year
Peebles, D. D.	Manufacture of stable powdered food products containing milk sugar	2,126,807	1938
Leviton, A.	Recovery of a soluble protein powder from whey	2,129,222	1938
Miner, C. S.	Process of making fondant	2,129,859	1938
Otting, H. E.	Flavoring of alimentary products	2,169,278	1939
Lavett, C. O.	Drying whey	2,172,393	1939
Supplee, G. C.	Treatment of whey	2,173,922	1939
Chuck, F. Y.	Manufacture of lactose containing materials	2,174,734	1939
Salzberg, H. K.	Soluble milk powder	2,181,003	1939
Peebles, D. D. and Manning, P. D. V.	Manufacture of a noncaking dried whey powder	2,181,146	1939
Sharp, P. F. and Hand, D. B.	Making beta lactose	2,182,618	1939
Sharp, P. F. and Hand, D. B.	Preparing beta lactose	2,182,619	1939
Hall, J. M.	Method of desiccating fluid mixtures	2,188,506	1940
Lavett, C. O.	Drying solutions containing lactose	2,188,907	1940
Turnbow, G. D.	Impalpable sugar composition and method of producing the same	2,193,950	1940
Lavett, C. O.	Drying lactose	2,197,804	1940
Musher, S.	Modified antioxygenic milk solids and making and using the same	2,198,198	1940
Nitardy, F. W.	Protection of autooxidizable materials	2,206,113	1940
Dietrich, J. F. and Dietrich, R. H.	Method of treating concentrated milk products	2,209,328	1940
Harford, C. G.	Recovery of solids from buttermilk	2,209,694	1940
Kronberg, N. M.	Preparing dried milk	2,213,283	1940
Powers, R.	Milk product	2,223,269	1940
Offen, B.	Drying method	2,226,319	1940
Chuck, F. Y.	Dehydrating process	2,227,246	1940
Lavett, C. O.	Drying whey	2,232,248	1941
Aeckerle, E.	Improving the stability of hygroscopic substances	2,238,149	1941
Epstein, C. H. and Gotthoffer, N. R.	Gelatin milk dessert composition and method of preparing the same	2,253,614	1941
Pittman, E. E. and Bottoms, R. R.	Treatment of milk waste	2,254,241	1941
Ingle, J. D.	Dried milk product	2,273,469	1942
Jordan, B. A., Clickner, F. H., and Erekson, A. B.	Cheese compositions and preparing the same	2,289,576	1942
Irwin, J. C., Jr.	Dehydrating process	2,292,447	1942
Clark, F. M.	Dehydrating treatment	2,293,453	1942
Vilbrandt, F. C., Sieg, R. D., and Farrar, T. F.	Making confection products from apple products, with milk	2,295,020	1942

Name	Subject	U.S. Patent	Year
Pagenkopf, W. H.	Dehydrating and treating apparatus	2,295,912	1942
Spellacy, J. R.	Casein manufacture	2,304,429	1942
Le Gloahec, V. C. E.	Milk product	2,311,343	1943
Ingle, J. D.	Milk product	2,319,186	1943
Wouters, O. J.	Process for the production of solid milk products	2,319,362	1943
Sharp, P. F.	Stable crystalline anhydrous alpha lactose product and process	2,319,562	1943
Bertram, K.	Drying whey	2,335,380	1943
Peebles, D. D.	Milk treatment process (process of manufacturing a concentrated milk product)	2,336,634	1943
Huber, L. J.	Milk-protein gel dessert and producing the same	2,344,090	1944
Thorneloe, K. C.	Preparing condensed low lactose skimmilk for storage and recovery of lactose therefrom	2,349,227	1944
Kremers, K.	Separating albuminous products from milk	2,349,969	1944
Shipstead, H. and Brant, A. P.	Packaging spray dried milk powders	2,363,445	1944
Peebles, D. D.	Manufacture of preserved food products	2,368,945	1945
Whitatker, R., Myers, R. P., and Homberger, R. E.	Manufacture of evaporated milk	2,372,239	1945
Lemmel, P. T.	Preparation of concentrated fluid milk products	2,381,761	1945
Mook, D. E.	Dry milk product and preparation thereof	2,383,070	1945
North, G. C., and Alton, A. J.	Dairy process	2,392,401	1946
Reeves, E. D.	Controlling flow of powder	2,392,765	1946
North, G. C., Alton, A. J., and Little, L.	Powdered shortening composition (buttermilk used with shortening for improved dry product)	2,392,995	1946
North, G. C. and Alton, A. J.	Dairy product	2,399,565	1946
Buxton, L. O.	Butter product and process for producing the same	2,404,034	1946
Folsom, T. R.	Freezing and drying liquids and semi-solids	2,411,152	1946
Sharp, P. F.	Producing dried milk powder	2,412,635	1946
Peebles, D. D.	Atomizing and desiccating substances and apparatus therefor	2,415,527	1947
Jakobsen, J. L.	Dehydrated butter	2,418,645	1947
Birdseye, C.	Improving and preserving food products	2,419,877	1947
Brandner, J. D. and Goepp, R. M., Jr.	Dehydration of foods by means of hydrophilic liquids	2,420,517	1947

Name	Subject	U.S. Patent	Year
North, G. C., Alton, A. J., and Little, L.	Shortening (dried with milk)	2,431,497	1947
North, G. C., Alton, A. J., and Little, L.	Shortening (dried with milk)	2,431,498	1947
Hipple, I. B. and Sadtler, S. S.	Spray dried ice cream mix	2,433,276	1948
Shipstead, H.	Vacuum treatment of milk powder	2,453,277	1948
Butterfield, G. P. and Thorne, L. A.	Treating lactalbumin	2,457,642	1948
Mead, R. E. and Clary, P. D., Jr.	Making lacteal food products	2,465,907	1949
Beardslee, A. C.	Removing ultimate moisture from powdered products	2,465,963	1949
Howard, L. B., Ramage, W. D., and Rasmussen, C. L.	Preserving foods	2,477,605	1949
Chrysler, L. H. and Almy, E. F.	Preparation of a stabilized cream product	2,503,866	1950
Kaiser, H. S.	Dehydration of liquids	2,525,224	1950
Baur, L. S. and Gerber, M. P.	Noncurding high calcium milk product and producing same	2,541,568	1951
Daniel, F. K.	Modified milk product	2,542,633	1951
Oberg, E. B. and Nelson, C. E.	Preparing a whipping agent from casein	2,547,136	1951
Hansen, F. F.	Milk powder and its preparation	2,553,578	1951
Simmons, N. L.	Package for whey concentrate and other dairy products	2,557,576	1951
Sharp, P. F.	Producing a reducing sugar and milk product	2,558,528	1951
Wenzelberger, E. P.	Freeze dehydration of liquid bearing substances	2,559,204	1951
Wenzelberger, E. P.	Freeze dehydration of liquid bearing substances	2,559,205	1951
Coulter, S. T. and Becker, J. L.	Apparatus for reconstituting dry powders	2,566,555	1951
Cross, J. A.	Milk evaporation process	2,570,212	1951
Peebles, D. D. and Hensley, G. P.	Apparatus for rapid solution and/or suspension of powdered solids	2,580,316	1951
Erisman, J. L.	Rotary drier or cooler	2,581,756	1952
Meade, R. E.	Milk product method of manufacture	2,602,746	1952
Meade, R. E.	Milk product process of manufacture	2,602,747	1952
Long, H. F. and Erickson, J. S.	Cheese-milk fat food product and producing the same	2,617,730	1952
Levin, E.	Drying and defatting tissue	2,619,425	1952
Meade, R. E.	Manufacture of stabilized milk	2,627,463	1953

Name	Subject	U.S. Patent	Year
Maki, C. J.	Process for the dehydration of fluids	2,629,460	1953
Otting, H. E.	Producing concentrated milk products	2,633,424	1953
Kempf, C. A.	Milk product containing milk fat and producing same	2,638,418	1953
Kempf, C. A.	Dried milk product for use in coffee or the like and producing same	2,645,579	1953
Wahl, A. S.	Milk-papaya powders, beverages made therefrom and producing the same	2,650,165	1953
Peebles, D. D. and Girvin, M. D.	Milk product and process	2,650,879	1953
Toulmin, H. A., Jr.	Dried or evaporated food product, especially milk, and making it	2,656,276	1953
Peebles, D. D. and Girvin, M. D.	Reconstituted milk process	2,657,142	1953
Howard, H. W. and Muller, J. F.	Infant food	2,659,676	1953
Meade, R. E.	Drying lacteal fluids	2,661,294	1953
Francis, L. H. and Rodgers, N. E.	Whey products and process of manufacture	2,661,295	1953
Whitaker, R.	Process for concentrating milk and product	2,663,642	1953
Hansen, F. F.	Milk powder and its preparation	2,663,643	1953
Hansen, F. F.	Preparation of milk powder	2,663,644	1953
Stimpson, E. G.	Frozen concentrated milk products	2,668,765	1954
Fear, E. D. and Harris, L. E.	Cultured milk powder and manufacturing same	2,671,729	1954
Turnbow, G. D.	Preserving the fresh natural flavor of butterfat	2,673,155	1954
Stimpson, E. G.	Conversion of lactose to glucose and galactose	2,681,858	1954
Brereton, J. G.	High protein milk product	2,682,467	1954
Stuart, G. H., Howard H., Watson, J. T., Clickner, F. H., and Sommer, W. A.	Food product and method of making same	2,682,469	1954
Traisman, E. and Kurtzhalts, W.	Process of making grated cheese	2,683,665	1954
Silberman, S.	Production of heat-treated cheese products from milk of low fat content	2,701,202	1955
Stimpson, E. G.	High protein deionized milk and process of making the same	2,708,633	1955
Peebles, D. D. and Clary, P. D., Jr.	Milk treatment process	2,710,808	1955
Cranston, H. A.	Baby food	2,717,211	1955
Rollins, J. H. and Turner, H. R.	Dehydrated whole milk product and process of producing the same	2,719,792	1955
Rivoche, E. J.	Dehydration of food products	2,723,202	1955

Name	Subject	U.S. Patent	Year
Lushbough, C. H. and Miller, P. G.	Method for producing malted milk balls and the resulting product	2,726,959	1955
Sharp, P. F.	Process for drying solutions containing crystallizable material, and product produced thereby	2,728,678	1955
De Vries, T. R.	Processes for the preparation of inoculating materials for concentrated milk products and for improving the crystallization of lactose in concentrated milk products	2,730,449	1956
Zepp, A. P., Zepp, C. J., and Zepp, F. P.	Process for making ice cream mixes	2,757,092	1956
Elden, H. S.	Dry soluble casein mix and method of preparing same	2,758,034	1956
Stamberg, O. E.	Gelled condensed whey and process	2,780,548	1957
Reich, I. M. and Johnston, W. R.	Spray drying foamed material	2,788,276	1957
Dorsey, W. R.	Continuous process for dehydrating liquid or semi-liquid concentrates under sub-atmospheric pressure	2,825,653	1958
Boyd, J. M.	Sterile concentrated milk (evaporated)	2,827,381	1958
McDonald, G. W. and Weinstein, B.	Milk powder process and product	2,831,771	1958
Scott, E. C.	Solubilization of milk proteins	2,832,685	1958
Louder, E. A. and Hodson, A. Z.	Instantly soluble milk powder and process for making same	2,832,686	1958
Peebles, D. D.	Dried milk product and method of making same	2,835,586	1958
Barzelay, M. E.	Spray drying process	2,835,597	1958
Notter, G. K.	Packaging of dehydrated foods	2,838,403	1958
Brochner, H. S.	Cream tablet and process of manufacture for it	2,839,407	1958
Langworthy, M. F.	Milk solids compositions and method of making	2,844,481	1958
Turnbow, G. D.	Processes for preserving the fresh natural flavor of butterfat and products produced therefrom	2,847,310	1958
McCarthy, J. L.	Reconstituted dry milk product	2,851,358	1958
Peebles, D. D.	Lactose product and process of manufacture	2,856,318	1958
Krehl, W. A. and Snyder, B.	Food composition	2,860,051	1958
Keville, J. F., Jr.	Process of concentrating milk and milk products	2,860,988	1958
Kempf, N. W.	Heat resistant chocolate incorporates skimmilk solids	2,863,772	1958
Leviton, A.	Process for freeze-drying of milk	2,885,788	1959

Name	Subject	U.S. Patent	Year
Peebles, D. D.	Milk manufacturing method and product	2,911,300	1959
Winder, W. C. and Kielsmeier, E. W.	Process of drying milk	2,911,301	1959
Steigmann, E. A.	Dry gelatin-containing dairy products and method of making same	2,913,341	1959
Heinemann, B.	Egg-milk product	2,920,966	1960
Sharp, P. F. and Kempf, C. A.	Method for the preparation of a powdered milk product	2,921,857	1960
Kennedy, J. G. and Spence, E. R.	Easily reconstituted milk powder	2,928,742	1960
Schram, C. J.	Process for preparing solid fat compositions	2,931,730	1960
Ortman, C. K.	Dry creaming powder	2,933,393	1960
Morgan, A. I., Jr. and Randall, J. M.	Dehydration of lacteal fluids	2,934,441	1960
McIntire, J. M. and Loo, C. C.	Fat containing dried dairy product and method of manufacture	2,941,886	1960
Bissell, R. H.	Process for increasing the solubility of powdered milk	2,949,363	1960
Sanna, F. L.	Method for drying food materials	2,953,457	1960
Sjollema, A.	Process for modifying powdered milk products	2,953,458	1960
Roundy, Z. D. and Ormond, N. R. H.	Cheese products and method for the manufacture thereof	2,956,885	1960
Cuthbertson, W. F. J.	Human dietary preparations	2,961,320	1960
Roundy, Z. D.	Manufacture of cheese	2,963,370	1960
Sinnamon, H. I., Aceto, N. C., and Eskew, R. K.	Dried fat-containing milk products of easy dispersibility	2,964,407	1960
Bauman, H. E., MacMillan, J. L., and Stein, J. A.	Process for rapid manufacture of cheese product	2,965,492	1960
Williams, A. W., Beckman, R. H., and Mook, D. E.	Milk product	2,966,409	1960
Terrett, J. P., Shields, J. B., and Nava, L. J.	Process for producing crystalline spray dried material	2,970,057	1961
Loewenstein, M.	Protein-emulsifier powder and process of producing same	2,970,913	1961
Greenfield, C.	Dehydration of fluid fatty mixtures	2,979,408	1961
Ginnette, L. F., Graham, R. P., and Morgan, A. I., Jr.	Process of dehydrating foams	2,981,629	1961
Kviesitis, B. and Rogerson, W. E.	Method of and means for dehydrating flowable matter	2,991,179	1961
Rice, M. A.	Method of preparing instant dry milk	2,994,612	1961

Name	Subject	U.S. Patent	Year
Cipolla, R. H., Davis, D. W., and Vander Linden, C. R.	Free flowing dried dairy products	2,995,447	1961
Loewenstein, M.	Milk protein-hydrocolloid powder (milk)	3,001,876	1961
Winder, W. C. and Bullock, D. H.	Process of modifying dried milk	3,008,830	1961
Kneeland, J. A.	Process for producing reconstituted milk product like evaporated milk	3,011,893	1961
Carlson, E. E., Plagge, I. F., and Swanson, A. M.	Manufacture of dry chocolate drink product	3,013,881	1961
Peebles, D. D. and Kempf, C. A.	Sweetening product and method of manufacture	3,014,803	1961
Shenkenberg, D. R.	Process for the manufacture of a chocolate flavored powder	3,027,257	1962
Leviton, A. and Pallansch, M. J.	Process for preparing a sterilized, concentrated milk product	3,031,315	1962
Stewart, R. A. and Rock, W. A.	Method of preparing canned infant milk formula	3,052,555	1962
Calbert, H. E., Swanson, A. M., and Giroux, R. N.	Method of preparing concentrated milk and resulting product	3,054,674	1962
Hale, J. F. and Smith, W. B.	Miscible malted milk powder	3,054,675	1962
Rogers, R. H. and Rogers, E. C.	Process of manufacturing dehydrated powdered cheese	3,056,681	1962
Shields, J. B.	Method of manufacturing an instantized product	3,057,727	1962
Forkner, J. H.	Method for dehydrating food products	3,057,739	1962
Lindblad, R. L.	Food supplement composition	3,058,828	1962
Obenauf, C. F. and Tatter, C. W.	Lecithinated product	3,060,030	1962
Wenner, V. and Hirtler, E.	Whole milk powder	3,065,076	1962
Leviton, A. and Pallansch, M. J.	Process for preparing concentrated, sterilized milk products	3,065,086	1962
Perini, L. A. and Leber, H.	Process for making a dairy product	3,066,027	1962
Perini, L. A. and Leber, H.	Process for making a dairy product	3,066,028	1962
Oakes, E. T., Doom, L. G., McElligot, P. A., and Sundheim, P. E.	Preparation for soluble milk powder	3,072,486	1963
Leviton, A. and Pallansch, M. J.	Process for preparing high temperature-short time sterilized concentrate milk products	3,072,491	1963

Name	Subject	U.S. Patent	Year
Peebles, D. D. and Clary, P. D., Jr.	High protein milk product and process	3,074,796	1963
Peebles, D. D., Clary, P. D., Jr., and Kempf, C. A.	Milk treatment process	3,074,797	1963
Rivoche, E. J.	Milk food compositions and method for preparing and using same	3,076,709	1963
Rice, M. A.	Method of preparing instant dry milk containing fat and product	3,078,167	1963
Hodson, A. Z. and Miller, C. B.	Process for making milk powder	3,080,235	1963
Ferguson, E. A., Jr.	Instant yoghurt	3,080,236	1963
Swanson, A. M. and Amundson, C. H.	Agglomeration process	3,083,099	1963
Viall, G. K. and Conley, W. E.	Dehydrating process	3,085,018	1963
Noznick, P. P., Bundus, R. H., and Eggen, I. B.	Method for making dried sour cream	3,090,688	1963
Noznick, P. P. and Bernardoni, E. A.	Cheese product	3,097,950	1963
Noznick, P. P., Tatter, C. W., and Abenauf, C. F.	Process for preparing compressed chocolate chip product using whole milk solids	3,098,746	1963
Noznick, P. P. and Tatter, C. W.	Whipping and powdered shortening compositions	3,098,748	1963
Sanna, C. A.	Process for producing instantly soluble nonfat dry milk	3,102,035	1963
North, G. C., Noznick, P. P., and Bundus, R. H.	Concentrated milk	3,105,763	1963
Leviton, A. and Pallansch, M. J.	Sterile homogenized concentrated milk concentrates	3,119,702	1964
McIntire, J. M. and Loo, C. C.	Fat-containing dried dairy product	3,120,438	1964
Jokay, L.	Cheese-base survival food	3,121,014	1964
Bauer, C. D. and Marks, R. M.	Spray-drying process	3,121,639	1964
Noznick, P. P., Bundus, R. H., and Eggen, I. B.	Sweetened condensed product	3,126,283	1964
Spilman, H. A. and Nava, L. J.	Dry milk process of manufacture	3,126,289	1964
Patton, S.	Flavor stabilization of milk fat	3,127,275	1964
Peebles, D. D., Hutton, J. T., and Clary, P. D., Jr.	Process for agglomerating dry milk	3,151,984	1964
Rubenstein, I.	Dry milk powder mixture and method and apparatus for making the same	3,159,492	1964

Name	Subject	U.S. Patent	Year
Shields, J. B., Nava, L. J., and Kempf, C. A.	Dry milk product and process of manufacture	3,164,473	1965
Morgan, A. I. and Schwimmer, S.	Preparation of dehydrated food products	3,170,803	1965
Hanrahan, F. P., Bell, R. W., and Webb, B. H.	Process for making puff spray dried nonfat dry milk and related products	3,185,580	1965
Morgan, H. A., Jr.	Milk food product and method for making same	3,190,760	1965
Hutton, J. T., Nava, L. J., Shields, J. B. and Kempf, C. A.	Process for producing instantized products	3,231,386	1966
Jertson, E. C. and Glabe, E. F.	Continuous bread mix with NDM	3,234,027	1966
Mykleby, R. W.	Method of producing a low calcium nonfat dried milk product	3,235,386	1966
Damisch, G. A., Jr., and Johnson, R. A.	Process for modifying nonfat dry milk solids	3,238,045	1966
Brochner, H. S.	Process for making a dry milk product	3,241,975	1966
Strashun, S. I. and Talburt, W. F.	Continuous dehydration of edible liquids	3,241,981	1966
Torr, D.	Dried honey-milk product	3,244,528	1966
Hansen, P. and Linton-Smith, L.	Powdered high fat product	3,271,165	1966
Williams, A. W. and Busch, A. A.	Process of making powdered milk containing lecithin	3,278,310	1966
Peebles, C. C.	Process for forming aggregates	3,279,924	1966
Tumerman, L. and Maddock, W.	Dry milk product with improved dispersibility	3,291,614	1967
Nava, L. J., Hutton, J. T., Shields, J. B. and Kempf, C. A.	Method for the manufacture of soluble dry milk products	3,300,315	1967
Loo, C. C.	Wettable dried whole milk	3,301,682	1967
Segal, S.	Method of preparing free-flowing water-insoluble solid particles	3,311,477	1967
Thompson, E. G., Dubbels, E. C. and Ostrom, W.	Agglomerating process for powdered solids	3,313,629	1967
Clausi, A. S., Vollink, W. L. and Michael, E. W.	Breakfast cereal with milk proteins	3,318,705	1967
Clerk, R. E., Spence, E. R. and Stribley, R. C.	Infant's formula made with electrodialyzed milk	3,320,072	1967

Evaporator Comparison Flow Chart Series[1]

"Recapitulation Table—Evaporator Comparison Series" (Form EN-31) and attached **Flow Charts** are a means of comparing a variety of evaporator arrangements with one another to determine relative efficiency in use of steam and cooling water.

The sheet numbers used on the flow charts are an identification method. The first figure tells how many effects are used, "T" stands for Thermocompressor, "V" stands for Vapor Heater, and "B" stands for Bleeder Heater. Thus the chart number tells the type and arrangement of the evaporator components.

All the flow charts are doing the same job, that is evaporating 809 lb. per hr. of water from 1000 lb. of 8.6% solids skimmilk, producing 191 lb. per hr. of 45% solids condensed skimmilk. The steam and water requirements for the two stage ejector are not included.

In all cases:

The skimmilk feed is at 165° for low heat powder.

The steam to the first tube chest is at 175°F.

The temperature of the last effect is 115°F.

The temperature of the cooling water is 60°F.

High-pressure steam from the thermocompressor is 100 p.s.i.g.

Low-pressure steam is at 175°F. saturation temperature.

It is assumed that 1 lb. of vapor will evaporate 1 lb. of water. This is not completely accurate.

It is assumed that 1 lb. of vapor will heat 1000 lb. of milk 1°F. This is not completely accurate.

Note particularly the decrease in steam and water requirements as the evaporator becomes increasingly complex. "Pounds of Evaporation per Pound of Steam" columns give an economy comparison on a unit basis. Observe, however, that as the evaporator becomes more economical of steam usage, the steam required for heating the feed becomes increasingly significant. This becomes so important that in comparing columns "G" and "I," it is found that some double effect evaporators with product heaters are more efficient than plain triple effects, when considered on an over-all steam requirement.

In column "J" the order of economy is based on using the Number 1 as the least efficient evaporator, and Number 16 as the most efficient one. This leaves latitude to add or insert additional evaporator arrangements.

Evaporator efficiency can be further increased by: (1) increasing the design pressure of steam to the thermocompressor; (2) increasing the temperature of the product feed; (3) adding effects; (4) adding heaters; and (5) decreasing temperature differences and increasing heating areas.

---

[1] By A. M. Walker, Marriott Walker Corporation, Birmingham, Michigan.

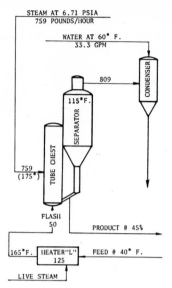

STEAM AT 6.71 PSIA
759 POUNDS/HOUR

WATER AT 60° F.
33.3 GPM

CONDENSER

809

115°F.

SEPARATOR

TUBE CHEST

759
(175°)

FLASH
50

PRODUCT @ 45%

165°F.   HEATER"L"   FEED @ 40° F.
         125

LIVE STEAM

FLOW CHART FOR SINGLE EFFECT EVAPORATOR

CAPACITY

Feed at	8.6% Solids	1000 lb./hr
Product at	45% Solids	191 lb./hr
Total evaporation		809 lb./hr

PERFORMANCE FACTORS

	Evaporator Group Only	Evaporator Plus Live Steam Heater
Total steam used, pounds per hour	759	884
Pounds of evaporation per pound of steam	1.065	0.915
Pounds of steam per 1000 lb. of evaporation	938	1092

Figures above include heating the cold feed through a temperature rise of 125 °F.

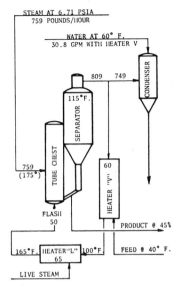

STEAM AT 6.71 PSIA
759 POUNDS/HOUR

WATER AT 60° F.
30.8 GPM WITH HEATER V

CONDENSER

809   749

115°F.

SEPARATOR

TUBE CHEST

759
(175°)

60

HEATER "V"

FLASH
50

PRODUCT @ 45%

165°F. HEATER"L" 100°F.
65

FEED @ 40° F.

LIVE STEAM

FLOW CHART FOR SINGLE EFFECT EVAPORATOR WITH VAPOR
HEATER

CAPACITY

Feed at	8.6% Solids	1000 lb./hr
Product at	45% Solids	191 lb./hr
Total evaporation		809 lb./hr

PERFORMANCE FACTORS

	Evaporator Group Only	Evaporator Plus Live Steam Heater
Total steam used, pounds per hour	759	824
Pounds of evaporation per pound of steam	1.065	0.982
Pounds of steam per 1000 lb. of evaporation	938	1018

Figures above include heating the cold feed through a temperature rise of 125 °F.

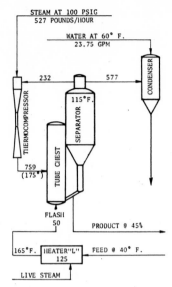

FLOW CHART FOR THERMOCOMPRESSION SINGLE
EFFECT EVAPORATOR

CAPACITY

Feed at	8.6% Solids	1000 lb./hr
Product at	45% Solids	191 lb./hr
Total evaporation		809 lb./hr

PERFORMANCE FACTORS

	Evaporator Group Only	Evaporator Plus Live Steam Heater
Total steam used, pounds per hour	527	652
Pounds of evaporation per pound of steam	1.535	1.240
Pounds of steam per 1000 lb. of evaporation	652	806

Figures above include heating the cold feed through a temperature rise of 125 °F.

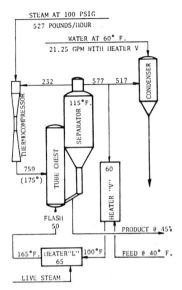

FLOW CHART FOR THERMOCOMPRESSION SINGLE EFFECT
EVAPORATOR WITH VAPOR HEATER

CAPACITY

Feed at	8.6% Solids	1000 lb./hr
Product at	45% Solids	191 lb./hr
Total evaporation		809 lb./hr

PERFORMANCE FACTORS

	Evaporator Group Only	Evaporator Plus Live Steam Heater
Total steam used, pounds per hour	527	592
Pounds of evaporation per pound of steam	1.535	1.366
Pounds of steam per 1000 lb. of evaporation	652	732

Figures above include heating the cold feed through a temperature rise of 125 °F.

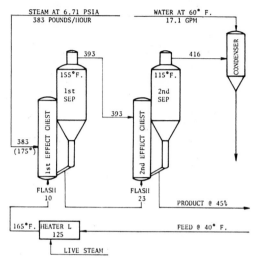

FLOW CHART FOR DOUBLE EFFECT EVAPORATOR

CAPACITY

Feed at	8.6% Solids	1000 lb./hr
Product at	45% Solids	191 lb./hr

Total evaporation	809 lb./hr

PERFORMANCE FACTORS

	Evaporator Group Only	Evaporator Plus Live Steam Heater
Total steam used, pounds per hour	383	508
Pounds of evaporation per pound of steam	2.110	1.593
Pounds of steam per 1000 lb. of evaporation	474	628

Figures above include heating the cold feed through a temperature rise of 125 °F.

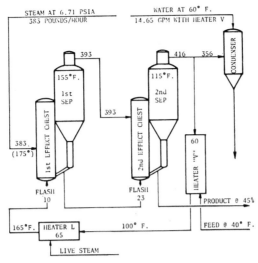

FLOW CHART FOR DOUBLE EFFECT EVAPORATOR
WITH VAPOR HEATER

CAPACITY

Feed at	8.6% Solids	1000 lb./hr
Product at	45% Solids	191 lb./hr
Total evaporation		809 lb./hr

PERFORMANCE FACTORS

	Evaporator Group Only	Evaporator Plus Live Steam Heater
Total steam used, pounds per hour	383	448
Pounds of evaporation per pound of steam	2.11	1.805
Pounds of steam per 1000 lb. of evaporation	474	544

Figures above include heating the cold feed through a temperature rise of 125 °F.

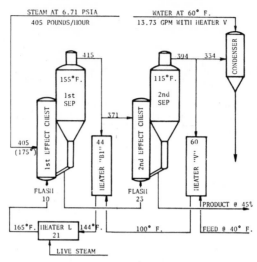

STEAM AT 6.71 PSIA
405 POUNDS/HOUR

WATER AT 60° F.
13.73 GPM WITH HEATER V

CONDENSER

415   394   334

155°F.   115°F.

1st SEP   2nd SEP

371

405 (175°)   44   60

1st EFFECT CHEST   2nd EFFECT CHEST

HEATER "B1"   HEATER "V"

FLASH 10   FLASH 23

PRODUCT @ 45%

165°F. HEATER L 21 144°F.   100° F.   FEED @ 40° F.

LIVE STEAM

FLOW CHART FOR DOUBLE EFFECT EVAPORATOR
WITH VAPOR AND BLEEDER HEATERS

CAPACITY

Feed at	8.6% Solids	1000 lb./hr
Product at	45% Solids	191 lb./hr
Total evaporation		809 lb./hr

PERFORMANCE FACTORS

	Evaporator Group Only	Evaporator Plus Live Steam Heater
Total steam used, pounds per hour	405	426
Pounds of evaporation per pound of steam	2.00	1.90
Pounds of steam per 1000 lb. of evaporation	500	526

Figures above include heating the cold feed through a temperature rise of 125 °F.

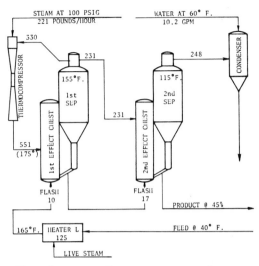

FLOW CHART FOR THERMOCOMPRESSION DOUBLE
EFFECT EVAPORATOR

CAPACITY

Feed at	8.6% Solids	1000 lb./hr
Product at	45% Solids	191 lb./hr
Total evaporation		809 lb./hr

PERFORMANCE FACTORS

	Evaporator Group Only	Evaporator Plus Live Steam Heater
Total steam used, pounds per hour	221	346
Pounds of evaporation per pound of steam	3.680	2.340
Pounds of steam per 1000 lb. of evaporation	272	427

Figures above include heating the cold feed through a temperature rise of 125 °F.

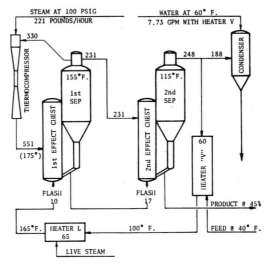

FLOW CHART FOR THERMOCOMPRESSION DOUBLE EFFECT
EVAPORATOR WITH VAPOR HEATER

CAPACITY

Feed at	8.6% Solids	1000 lb./hr
Product at	45% Solids	191 lb./hr
Total evaporation		809 lb./hr

PERFORMANCE FACTORS

	Evaporator Group Only	Evaporator Plus Live Steam Heater
Total steam used, pounds per hour	221	286
Pounds of evaporation per pound of steam	3.68	2.83
Pounds of steam per 1000 lb. of evaporation	272	353

Figures above include heating the cold feed through a temperature rise of 125 °F.

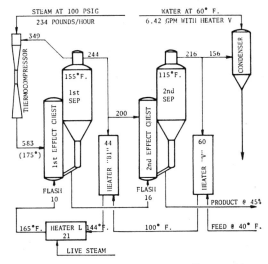

FLOW CHART FOR THERMOCOMPRESSION DOUBLE EFFECT
EVAPORATOR WITH VAPOR AND BLEEDER HEATERS

CAPACITY

Feed at	8.6% Solids	1000 lb./hr
Product at	45% Solids	191 lb./hr

Total evaporation	809 lb./hr

PERFORMANCE FACTORS

	Evaporator Group Only	Evaporator Plus ·Live Steam Heater
Total steam used, pounds per hour	234	255
Pounds of evaporation per pound of steam	3.46	3.17
Pounds of steam per 1000 lb. of evaporation	289	315

Figures above include heating the cold feed through a temperature rise of 125 °F.

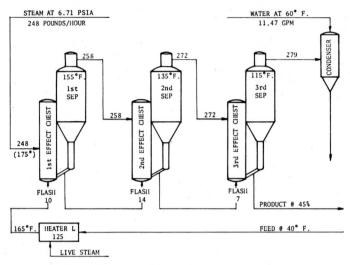

FLOW CHART FOR TRIPLE EFFECT EVAPORATOR

CAPACITY

Feed at	8.6% Solids	1000 lb./hr
Product at	45% Solids	191 lb./hr
Total evaporation		809 lb./hr

PERFORMANCE FACTORS

	Evaporator Group Only	Evaporator Plus Live Steam Heater
Total steam used, pounds per hour	248	373
Pounds of evaporation per pound of steam	3.26	2.17
Pounds of steam per 1000 lb. of evaporation	307	461

Figures above include heating the cold feed through a temperature rise of 125 °F.

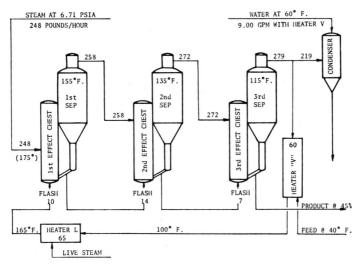

FLOW CHART FOR TRIPLE EFFECT EVAPORATOR
WITH VAPOR HEATER

CAPACITY

Feed at	8.6% Solids	1000 lb./hr
Product at	45% Solids	191 lb./hr
Total evaporation		809 lb./hr

PERFORMANCE FACTORS

	Evaporator Group Only	Evaporator Plus Live Steam Heater
Total steam used, pounds per hour	248	313
Pounds of evaporation per pound of steam	3.26	2.58
Pounds of steam per 1000 lb. of evaporation	307	387

Figures above include heating the cold feed through a temperature rise of 125 °F.

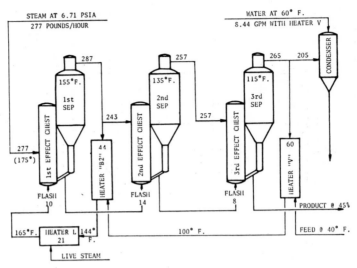

FLOW CHART FOR TRIPLE EFFECT EVAPORATOR
WITH VAPOR AND BLEEDER HEATERS

CAPACITY

Feed at	8.6% Solids	1000 lb./hr
Product at	45% Solids	191 lb./hr
Total evaporation		809 lb./hr

PERFORMANCE FACTORS

	Evaporator Group Only	Evaporator Plus Live Steam Heater
Total steam used, pounds per hour	277	298
Pounds of evaporation per pound of steam	2.92	2.715
Pounds of steam per 1000 lb. of evaporation	342	368

Figures above include heating the cold feed through a temperature rise of 125 °F.

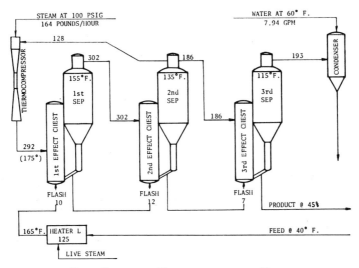

FLOW CHART FOR THERMOCOMPRESSION TRIPLE
EFFECT EVAPORATOR

CAPACITY

Feed at	8.6% Solids	1000 lb./hr
Product at	45% Solids	191 lb./hr
Total evaporation		809 lb./hr

PERFORMANCE FACTORS

	Evaporator Group Only	Evaporator Plus Live Steam Heater
Total steam used, pounds per hour	164	289
Pounds of evaporation per pound of steam	4.93	2.80
Pounds of steam per 1000 lb. of evaporation	203	357

Figures above include heating the cold feed through a temperature rise of 125 °F.

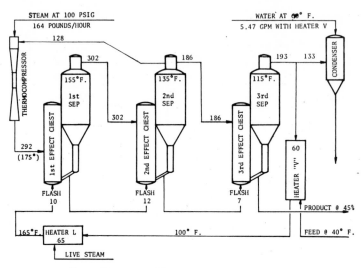

FLOW CHART FOR THERMOCOMPRESSION TRIPLE
EFFECT EVAPORATOR WITH VAPOR HEATER

CAPACITY

Feed at	8.6% Solids	1000 lb./hr
Product at	45% Solids	191 lb./hr
Total evaporation		809 lb./hr

PERFORMANCE FACTORS

	Evaporator Group Only	Evaporator Plus Live Steam Heater
Total steam used, pounds per hour	164	229
Pounds of evaporation per pound of steam	4.93	3.53
Pounds of steam per 1000 lb. of evaporation	203	283

Figures above include heating the cold feed through a temperature rise of 125 °F.

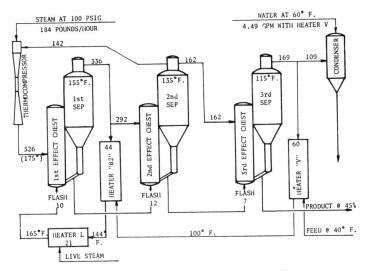

FLOW CHART FOR THERMOCOMPRESSION TRIPLE EFFECT
EVAPORATOR WITH VAPOR AND BLEEDER HEATERS

CAPACITY

Feed at	8.6% Solids	1000 lb./hr
Product at	45% Solids	191 lb./hr
Total evaporation		809 lb./hr

PERFORMANCE FACTORS

	Evaporator Group Only	Evaporator Plus Live Steam Heater
Total steam used, pounds per hour	184	205
Pounds of evaporation per pound of steam	4.40	3.95
Pounds of steam per 1000 lb. of evaporation	227	253

Figures above include heating the cold feed through a temperature rise of 125 °F.

RECAPITULATION TABLE—EVAPORATOR COMPARISON SERIES

A Flow Chart No.	B Evaporator Type	C Thermo-compressor	D Vapor Heater	E Bleeder Heater	F Total Steam	G Lb. Evap. per Lb. Steam	H Total Steam	I Lb. Evap. per Lb. Steam	J Order of Economy	K gpm 60°F. Water[1]
			Arrangement			Evaporator Only		Evaporator Plus Live Steam Heater		Water Cooling
1	Single effect	...	...	Not applicable	759	1.065	884	0.915	1	33.30
1V	" "	...	Yes	"	759	1.065	824	0.982	2	30.80
1T	" "	Yes	...	"	527	1.535	652	1.240	3	23.75
1TV	" "	Yes	Yes	"	527	1.535	592	1.366	4	21.25
2	Double effect	...	...	...	383	2.110	508	1.593	5	17.10
2V	" "	...	Yes	...	383	2.110	448	1.805	6	14.65
2VB	" "	...	Yes	Yes	405	2.000	426	1.900	7	13.73
2T	" "	Yes	...	...	221	3.680	346	2.340	9	10.20
2TV	" "	Yes	Yes	...	221	3.680	286	2.830	13	7.73
2TVB	" "	Yes	Yes	Yes	234	3.460	255	3.170	14	6.42
3	Triple effect	...	...	...	248	3.260	373	2.170	8	11.47
3V	" "	...	Yes	...	248	3.260	313	2.580	10	9.00
3VB	" "	...	Yes	Yes	277	2.920	298	2.710	11	8.44
3T	" "	Yes	...	...	164	4.930	289	2.800	12	7.94
3TV	" "	Yes	Yes	...	164	4.930	229	3.530	15	5.47
3TVB	" "	Yes	Yes	Yes	184	4.400	205	3.950	16	4.49

[1] For cooling water temperatures other than 60° use the following multipliers for water quantity: 50°F., 0.83; 70°F, 1.25; 80°F, 1.67; and 90°F., 2.50.

Courtesy of Marriott Walker Corporation, Birmingham, Michigan.

Baumé Reading and Total Solids of Condensed Skim Milk

Baumé	110°F. Solids %	120°F. Solids %	130°F. Solids %
8.0	17.55	18.04	18.50
8.5	18.50	18.98	19.44
9.0	19.45	19.92	20.38
9.5	20.39	20.87	21.34
10.0	21.34	21.82	22.30
10.5	22.29	22.77	23.25
11.0	23.26	23.73	24.19
11.5	24.19	24.67	25.15
12.0	25.14	25.62	26.10
12.5	26.09	26.57	27.04
13.0	27.03	27.53	27.97
13.5	27.98	28.47	28.94
14.0	28.93	29.41	29.90
14.5	29.88	30.35	30.84
15.0	30.83	31.30	31.78
15.5	31.78	32.24	32.73
16.0	32.77	33.19	33.63
16.5	33.68	34.14	34.62
17.0	34.62	35.09	35.57
17.5	35.57	36.05	36.52
18.0	36.52	37.00	37.47
18.5	37.47	37.95	38.42
19.0	38.42	38.90	39.37
19.5	39.37	39.90	40.32
20.0	40.32	40.79	41.27
20.5	41.27	41.73	42.21
21.0	42.21	42.69	43.16
21.5	43.16	43.64	44.11
22.0	44.11	44.58	45.05
22.5	45.06	45.53	46.01
23.0	46.01	46.48	46.96
23.5	46.96	47.43	47.90
24.0	47.91	48.38	48.85
24.5	48.85	49.33	49.80
25.0	49.80	50.28	50.75

Courtesy of C. E. Rogers Company, Detroit, Michigan.

Baumé Reading and Total Solids of Condensed Whole Milk

Baumé	100°F. Solids %	110°F. Solids %	120°F. Solids %
8	26.6	27.5	28.3
9	29.3	30.0	30.8
10	32.0	32.7	33.4
11	34.5	35.2	35.9
12	37.0	37.8	38.5
13	39.5	40.3	41.0
14	42.0	42.8	43.5
15	44.5	45.3	46.0
16	47.0	47.8	48.5
17	49.5	50.3	51.0
18	52.0	52.8	53.5

Courtesy of C. E. Rogers Company, Detroit, Michigan.

Baumé Reading and Total Solids of Condensed Whey

Total Solids %	Density at 115°F.		Total Solids %	Density at 115°F.	
	°Bé.	Sp. Gr.[1]		°Bé.	Sp. Gr.[1]
30	15.6	1.121	54	28.9	1.249
32	16.7	1.130	56	30.1	1.262
34	17.8	1.140	58	31.2	1.274
36	18.9	1.150	60	32.3	1.287
38	20.1	1.161	62	33.4	1.299
40	21.2	1.171	64	34.5	1.312
42	22.3	1.182	66	35.6	1.325
44	23.4	1.192	68	36.7	1.339
46	24.5	1.203	70	37.8	1.353
48	25.6	1.214	72	38.9	1.367
50	26.7	1.226	74	40.1	1.382
52	27.8	1.237	76	41.2	1.397

Source: Webb, B. H., and C. F. Hufnagel, 1946. Nat'l Butter and Cheese J., *37:* No. 12: 34, 35, 68, 70, 72

Sp. Gr. $= \dfrac{145}{145 - °Bé.}$

Dairy Farm Report

Proper Methods and Clean Equipment Protect Milk Quality

Patron No. _____ Name _____

Address _____ Date _____

	Satisfactory	
	Yes	No

MILK UTENSILS
1. Utensils proper construction _____ good repair _____
2. Utensils clean _____
3. Single service strainer used exclusively_____
4. Brush used for cleaning utensils _____
5. Soapless dairy cleanser used _____
6. Utensils properly stored _____
7. Utensils sanitized before use _____
8. Milking machine clean _____ properly stored _____
   (clean head _____ claw _____ tube _____ inflations _____)

COOLING AND STORAGE OF MILK
9. Adequate facilities _____ proper methods _____
10. Milk house: Construction _____ Repair _____
    Facilities _____ Clean _____
11. Bulk tank: Construction _____ Installation _____
    Facilities _____ Clean _____

MILKING BARN
12. Clean floors and gutters _____ (impervious _____)
13. Lighting _____ ventilation _____ no overcrowding _____
14. Walls and ceilings clean _____
    (painted _____ whitewashed _____)
15. No housing poultry _____ swine _____ sheep _____
    goats _____
16. Bull pens clean _____ calf pens clean _____

COW YARD
17. Clean _____ drained _____ (no swine _____)

COWS AND MILK HANDLING
18. Clean flanks _____ udders _____ teats _____
19. Abnormal milk discarded _____ (strip cup used _____)

PROTECTION OF MILK
20. Reviewed proper use of antibiotics _____
    pesticides _____ sanitizers _____ detergents _____

REMARKS

No. milking cows _____

_____
Field service

Recommended for use in connection with section A-4 of "Recommended Sanitary/
Quality Standards Code for the Dry Milk Industry."
ADMI Bulletin #915 (Revised, 1961).

UNITED STATES DEPARTMENT OF AGRICULTURE
AGRICULTURAL MARKETING SERVICE
DAIRY DIVISION
WASHINGTON 25, D. C.

MINIMUM STANDARDS FOR MILK FOR MANUFACTURING PURPOSES AND ITS PRODUCTION AND PROCESSING RECOMMENDED FOR ADOPTION BY STATE REGULATORY AGENCIES

[Reprinted from Federal Register of June 26, 1963]

DAIRY FARM CERTIFICATION REPORT
(For milk for manufacturing purposes)

Date..

Time..a.m.
....p.m.

Producer...

Address...

Person interviewed..

Name of receiving plant..................................

Farm certification requires that the facilities listed be satisfactory and that the applicable methods being followed receive a total rating of not less than 85 percent for the applicable items in B—Methods with no individual subitems rating less than 75 percent of the maximum score allowed. Subitems may be rated in quarter points.

A—FACILITIES REQUIRED

	Satis-factory	Un-satis-factory*
1. Health of herd:		
(a) Herd appears healthy		
(b) Tuberculin tested, date		
(..................)................		
(c) Brucellosis tested:		
Ring tested, date		
(..................)................		
Blood tested, date		
(..................)................		
2. Barn or milking area:		
(a) Adequate size, construction		
(b) Cow yard graded, well drained		
3. Milkhouse or milkroom:		
(a) Adequate size, location and construction		
(b) Equipped with adequate facilities		
(c) Bulk tank, installation		
4. Combination milking parlor and milkroom, if used:		
(a) Adequate size, construction		
(b) Equipment and arrangement		
5. Utensils and equipment:		
(a) Design, construction		
(b) Cleaning (brushes and cleansers) and storage facilities and supplies available		

	Satis-factory	Un-satis-factory*
6. Water supply:		
(a) Safe, clean		
(b) Supply ample		

B—METHODS

	Maximum score	Score given*
7. Premises:		
(a) Clean, well kept	3	
(b) Fowl, swine and other animals properly confined	1	
(c) Manure properly handled and disposed of	1	
8. Barn and milking area:		
(a) Floors and gutters clean, good repair	5	
(b) Walls and ceilings clean, painted or white-washed	5	
(c) Pens and alleyways clean	2	
(b) Yard or loafing area clean	3	
9. Milking procedures:		
(a) Mastitis program practiced	5	
(b) Cows clean, udders and flanks clipped	5	
(c) Udders and teats washed or wiped before milking	5	
(d) Milk stools and surcingles clean, properly stored	2	
(e) Milker's clothing clean, hands clean and dry	5	
(f) Feed bin kept clean and free from foul odors	1	
10. Milkhouse or milkroom:		
(a) Used for handling milk and utensil care only	2	
(b) Clean, flies and insects controlled to minimum	8	
11. Cooling:		
(a) Facility clean, good operating order	7	
(b) Milk cooled promptly, properly held. (Temperature of milk ......° F.)	8	

	Maximum score	Score given
12. Utensils and equipment: (a) Good condition, clean, properly stored: Cans	5	
Milking machines (head, claw, pulsator, inflations, tubes, air hose, etc.)	7	
Pails, strainers and other utensils	3	
Bulk tank	5	
(b) Bactericidal-treated before use	8	
(c) Vacuum lines clean	2	
(d) Supplies properly stored	2	
	100	

* Indicate under "Remarks" specific deficiency or reason for disrating.

Remarks:

On the basis of this inspection, the farm (is) (is not) eligible for farm certification.

....
(Signature)

....
(Title)

....
(Signature) (Producer)

SEC. 99. *Plant Inspection Report Form.*
The following form shall be used by inspectors in determining eligibility for plant licensing:

PLANT INSPECTION REPORT
Date...

Name of plant...
Owner or manager...
Address...
License No...
Products manufactured...
a.m.
Time of inspection.........................p.m. before, during, after processing.

Plant licensing requires a rating of not less than 85 percent of the maximum score allowed for the total of each applicable numbered group of items. Subitems may be rated in quarter points. In addition, not more than 10 percent of the cans (including lids) shall show open seams, cracks, rust, milkstone, or any unsanitary condition; when pasteurization is intended or required, HTST units shall have a flow-diversion valve and holding tube or its equivalent; and a safe water supply is required, with no cross-connections between safe and unsafe lines.

	Maximum score	Score given**
Premises, buildings, and facilities 1. Premises and surroundings Clean, 0.5 Orderly, 0.5 Properly drained, 0.5 Free from foul odors or smoke, 0.5	2	
2. Buildings Sound construction, 1 Clean, good repair, 1	2	
3. Doors and windows Clean, 0.5 Screened or protected, 1 Outer doors open outward and self-closing, 0.5	2	
4. Conveyor and service-pipe openings covered or protected	1	
5. Floors Smooth and impervious, 1 Good repair, 1 Grains properly trapped, 1 No sewage backflow, 1	4	
6. Wall and ceilings Smooth, 1 Impervious, 1 Washable, 1 Light color, 1	4	
7. Processing rooms Adequate size, 1 Clean, 1 Orderly, 1 No undue condensation or objectionable odors, 1 Ample light, well distributed, 1 Free from unnecessary equipment or utensils, 1	6	
8. Coolers and freezers Adequate size, 1 Clean, 0.5 Dry, 0.5 Orderly, 0.5 Sufficient refrigeration and air circulation, 2 Adequately lighted, 0.5	5	
9. Dry storage space (product) Adequate size, 1 Clean, 0.5 Dry, 0.5 Orderly, 0.5 Adequately lighted and ventilated, 0.5 Free from insects and rodents, 1	4	
10. Supply rooms Clean, 0.5 Dry, 0.5 Orderly, 0.5 Adequately lighted and ventilated, 0.5 Free from insects and rodents, 1	3	
11. Toilet and dressing rooms Properly located and separated, 0.5 Good repair, 0.5 Self-closing doors, 0.5 Clean, 0.5 Orderly, 0.5 Adequately lighted and well ventilated, 0.5	3	
12. Boiler and tool rooms separated from other rooms and adequately lighted	1	
13. Laboratory Sufficient size, 1 Adequately equipped, 1 Adequately staffed, 1 Adequately lighted and ventilated, 1	4	

	Maximum score	Score given**
14. Water supply.............	2	....
Ample hot and cold water, 1		
Conveniently located, 0.5		
Current bacterial tests on file, 0.5 (date tested ....)		
15. Steam............	2	....
Clean and nontoxic, 1		
Adequate supply and pressure, 1		
16. Drinking-water facilities sanitary and convenient....	1	....
17. Hand-washing facilities...........	2	....
Properly equipped and clean, 1		
Convenient, 0.5		
Self-closing waste containers provided, 0.5		
18. Waste disposal............	2	....
Sewer of sufficient capacity, 0.5		
Nonpublic disposal methods approved, 0.5		
Refuse in covered containers, 0.5		
Waste paper properly handled, 0.5		
Equipment and utensils		
19. Construction and maintenance........	7	....
Product contact surfaces of stainless steel or other equally corrosion-resistant material, 3		
Good condition, 2		
Accessible for cleaning, 2		
20. Pasteurizers...........	2	....
Good operating order, 1		
Equipped with thermometers and recorders, 1		
21. Thermometers and recorders........	4	....
Adequate, 2		
Sufficiently accurate, 1		
Recorder charts in order and on file, 1		
22. C–I–P and welded sanitary lines properly engineered and installed...........	2	....
23. Portable equipment and utensils suitably stored.....	1	....
24. Can washers...............	2	....
Operating properly, 1		
Clean, good repair, 1		
25. Stacks, elevators, conveyors in good condition...........	1	....
26. Vacuum cleaner in good condition, used regularly, refuse disposal satisfactory....	1	....
27. Farm trucks....	2	....
Enclosed type, 1		
Clean, 1		
28. Transport tanks..............	4	....
Good condition, interior smooth, enclosed tight-fitting cabinet, 1		
Piping and tubing capped, 0.5		
Washing facilities available, 1		
Tanks clean, 0.5		
Bactericidal-treated before use, 0.5		
Current cleaning and sanitizing tag in place, 0.5		

	Maximum score	Score given*
Plant operations		
29. Cleaning and sanitizing plant equipment and utensils...........	5	....
Equipment not designed for C–I–P disassembled daily and thoroughly cleaned, 2		
C–I–P system operated properly, 1		
Utensils and other equipment and in-place pipelines thoroughly cleaned each day, 1		
All equipment subjected to an effective bactericidal treatment immediately before use, 1		
30. Raw-product storage, proper temperature maintained until start of processing...........	2	....
31. Pasteurization...........	2	....
Milk and cream properly pasteurized, 1		
Recorder corresponds to indicating thermometer, 1		
32. Processed fluid products cooled promptly.............	2	....
33. Laboratory...............	2	....
Tests adequate, 1		
Records available, 1		
34. Containers clean and sound..	2	....
35. Dry storage...............	2	....
Product and supplies placed on dunnage or pallets, 1		
Arranged in aisles, rows, or sections, 1		
36. Refrigerated storage............	2	....
Proper temperature maintained to protect quality, 1		
Products not placed directly on wet floors, 1		
37. Personnel cleanliness............	3	....
Clean outer garments, 1		
Caps or hairnets worn, 0.5		
No smoking, 0.5		
Good hygiene practiced, 1		
38. Personnel health...............	2	....
No communicable disease, 1		
General good health, 0.5		
Current medical records on file, 0.5		

* Indicate under "Remarks" specific deficiency of reason for disrating.

** Indicate under "Remarks" reason for specific disrating.

Remarks:

On the basis of this survey, this plant (is) (is not) eligible for a license.

(Signature) (Inspector)

(Title)

(Signature) (Plant official)

(Title)

Done at Washington, D. C., this 19th day of June 1963.

G. R. GRANGE,
Deputy Administrator,
Marketing Services.

[F.R. Doc. 63–6661; Filed, June 25, 1963; 8:45 a.m.]

American Dry Milk Institute, Inc.
QUALITY DEVELOPMENT PROGRAM
Plant Sanitation Survey

Company _____   copies to _____

Plant _____   _____

Date _____   _____

Product Mfg.: _____   _____

MILK INTAKE:	Cl.	Rpr.	Rpl.
Trucks/Tank Trucks			
Driveway-dock			
Testing equipment			
Can washer			
Conveyors			
Outgoing cans			
Weigh tank			
Drop tank			

MILK STORAGE:	Cl.	Rpr.	Rpl.
Cooler			
Heater			
Lines, valves			
Pumps			
Separators			
Clarifier			
Storage tanks			

PROCESSING:	Cl.	Rpr.	Rpl.
Preheaters			
Hotwell(s)			
Evaporators			
Tank (			
Pumps			
Lines, valves			

SPRAY:	Cl.	Rpr.	Rpl.
Heater			
H-P pump			
Air filters			
Drying chamber			
Conveyors			
Hopper			
Socks/Packaging			
Equip. storage			

ROLLER:	Cl.	Rpr.	Rpl.
Rollers			
Stack (s)			
Hood (s)			
Condensate drains			
End dams			
Blades			
Conveyors			
Mills			
Socks/Packaging			

PREMISES:	Cl.	Rpr.	Rpl.
Floors			
Walls/Ceilings			
Roof			
Windows			
Doors			
Lighting			
Ventilation			
Warehousing			
Surroundings			

GENERAL:	IC	NIC
Water supply		
Dressing rooms		
Lavatory		
Wash tanks		

	IC	NIC
Flies/insects		
Rodents		
Objectionable odors		
Health examinations		

	IC	NIC
Test records		
Waste disposal		
Clothes Storage		
Equip. Storage		

Symbols: (Cl.) Clean; (Rpr.) Repair; (Rpl.) Replace; (IC) In Compliance; (NIC) Not In Compliance.
Note: All recommendations based upon minimum requirements set forth in latest edition of "Sanitary/Quality Standards Code for the Dry Milk Industry" published by American Dry Milk Institute, Inc. Attention invited to following items numbered and encircled in above recommendations:—

_____
Signature

U. S. DEPARTMENT OF AGRICULTURE
AGRICULTURAL MARKETING SERVICE
DAIRY DIVISION

PLANT SURVEY REPORT

APPLICANT (Name and address)	TYPE OF PLANT	DATE
	MANAGER	Fee _____
		Expense _____
PLANT SURVEYED (Name and address)	PURPOSE OF SURVEY	Total _____
		USDA INSPECTOR

ITEM CODE: S – Satisfactory U – Unsatisfactory	CHECK ONE		REMARKS (State unsatisfactory conditions by Item No. Include comments, as necessary, for other items.)
	S	U	
PLANT AND PREMISES			
1. Construction			
2. Location and Plant Surroundings			
RECEIVING			
3. Floor, Walls and Ceiling			
4. Windows and Doors			
5. Lighting and Ventilation			
6. Can Inlet and Outlet			
7. Weigh Tank, Scale & Drop Tank			
8. Can Washer			
9. Disposition of Can Drippings			
10. Condition of Producer Cans			
11. Milk Route Trucks			
12. Bulk Tank Trucks			
13. Facilities for Tank Trucks			
14. Pumps, Pipelines and Valves			
15. Cooler(s)			
16.			
RAW MATERIAL – QUALITY PROGRAM			
17. Milk & Cream Grading Program			
18. Bacterial Testing Program			
19. Sediment Testing Program			
20. Whey Cream, If any			
21. Disposition – Fluid Buttermilk			
22.			
MILK AND CREAM STORAGE			
23. Floors, Walls and Ceiling			
24. Windows and Doors			
25. Lighting and Ventilation			
26. Milk Storage Tanks			
27. Pumps, Pipelines and Valves			
28. CIP Cleaning			
29.			
SEPARATING AND PASTEURIZING			
30. Floors, Walls and Ceiling			
31. Windows and Doors			
32. Lighting and Ventilation			
33. Fore-Warmer, Temperature			
34. Separator(s)			
35. Cooler(s)			
36. Pasteurizer(s), Type, Time, Temp.			
37. Flow Diversion Valve			
38. Indicating & Recording Therm.			
39. Pumps, Pipelines & Valves			
40.			

DRY MILK

ITEM	CODE: S – Satisfactory U – Unsatisfactory	CHECK ONE		REMARKS (State unsatisfactory conditions by Item No. Include comments, as necessary, for other items.)
		S	U	
SKIM MILK STORAGE				
41. Floor, Walls and Ceiling				
42. Lighting and Ventilation				
43. Storage Tanks - Temperature				
44. Pumps, Pipelines, Valves, Etc.				
CONDENSING				
45. 1st Floor, Walls and Ceiling				
46. 2nd Floor, Walls and Ceiling				
47. Windows and Doors				
48. Lighting and Ventilation				
49. Pre-Heater(s) - Temperature				
50. Hot Wells - Time, Temperature				
51. Evaporator(s)				
52. Pumps, Pipelines and Valves				
53. CIP Cleaning				
54. Cooler and Storage Tanks				
55.				
DRYING				
56. Floor, Walls and Ceiling				
57. Windows and Doors				
58. Lighting and Ventilation				
59. Condensed Milk Pre-Heater				
60. Pumps, Pipelines and Valves				
61. Strainers and H.P. Pump				
62. Spray Jets and Lines				
63. Air Inlet Chambers & Filters				
64. Drying Chamber				
65. Product Removal Equipment				
66. Collectors				
67. Conveyors - Air or Mechanical				
68. Storage Bins				
69. Clothing & Boots (for drier work)				
70. Housekeeping & Pest Control				
71.				
PACKAGING				
72. Floors, Walls and Ceiling				
73. Windows and Doors				
74. Lighting				
75. Room Exhaust or Dust Collector				
76. Powder Cooler & Air Filters				
77. Dry Milk Packaging - Temperature				
78. Sifter				
79. Shaker for Barrels or Drums				
80. Scale				
81. Packaging Technique				
82. Waste Powder				
83. Housekeeping & Pest Control				
84.				
STORAGE OF DRY MILK				
85. Floor, Walls and Ceiling				
86. Lighting and Ventilation				
87. Stacking of Containers				
88. Other Products Stored				
89. Housekeeping & Pest Control				
90.				

ITEM	CODE: S — Satisfactory U — Unsatisfactory	CHECK ONE		REMARKS (State unsatisfactory conditions by Item No. Include comments, as necessary, for other items.)
		S	U	
STORAGE OF SUPPLIES				
91. Floors, Walls and Ceiling				
92. Salt, Color, Starter, Rennet, Etc.				
93. Containers, Liners, Wrappers, Etc.				
94. Other Supplies				
95. Housekeeping				
96. Pest Control				
97.				
LOCKER AND REST ROOMS				
98. Floors, Walls and Ceiling				
99. Windows and Doors				
100. Lockers and Benches				
101. Sanitary Conditions				
102. Hand-Washing Facilities & Sign				
103.				
GENERAL				
104. Pest Control Program				
105. Cleaning Supplies & Equipment				
106. Employees' Appearance				
107. Medical Certificates				
108.				

STATUS OF PLANT — PREVIOUS SURVEY, IF ANY   DATE

STATUS OF PLANT — THIS SURVEY (Indicate product(s))

RECOMMENDATIONS:

Index